MORE-THAN-HUMAN DESIGN IN PRACTICE

This book provides an overview of the diverse multidisciplinary field of more-than-human design, offering a philosophical grounding of more-than-human design in posthumanism while putting practical design examples and methods to the forefront.

There is an urgent need to radically re-imagine design, as its current processes are contributing to global warming, pollution, deforestation, ocean acidification, ozone layer depletion, loss of biodiversity and species extinction. Given this need, 'more-than-human design' has emerged as a perspective that widens our thinking beyond solely human-oriented considerations and needs, such as animals, plants and microbes. The book explores the relationship between sustainability and design, touching on topics such as AI, systems thinking, futures studies and pedagogy, and discusses a range of case study projects that are grounded in more-than-human thinking, demonstrating how this can be incorporated into practice.

This easily accessible and theoretically grounded book will provide design researchers and educators an excellent introduction to more-than-human thinking. It will also be of interest to students and scholars studying design more broadly, sustainability, environmental studies and service design, as well as to practicing designers interested in sustainability.

Anton Poikolainen Rosén is a postdoctoral researcher in Human-Computer Interaction at the Department of Design, Aalto University, Helsinki, Finland, and the Department of Computer and Systems Sciences, Stockholm University, Sweden, studying design for sustainable futures and the more-than-human world. His research themes include critical, circular and multisensory approaches to farming and waste management.

Antti Salovaara is a Senior University Lecturer at the Department of Design, Aalto University, Finland. His research develops methods for HCI researchers to anticipate possible futures, particularly with a goal that futuring in HCI would not only consider technology-based factors.

Andrea Botero is an Associate Professor at the School of Arts, Design and Architecture of Aalto University. Her research and practice aim to understand how collectives (broadly speaking) come to understand the design spaces available to them, what counts as design, what other practices for world making are there, and which ones we need to call into being.

Marie Louise Juul Søndergaard is a designer and researcher exploring feminist design of technologies for human and environmental health. She is curious about how the materiality of human bodies relates with ecologies, and uses research-through-design, participation of communities and speculative storytelling to design for social and environmental justice. She has a PhD in Interaction Design from Aarhus University in Denmark and has been a postdoc at KTH Royal Institute of Technology in Sweden and the Oslo School of Architecture and Design in Norway, where she researched somatic approaches in design and more-than-human design.

MORE-THAN-HUMAN DESIGN IN PRACTICE

Edited by
Anton Poikolainen Rosén, Antti Salovaara, Andrea Botero and Marie Louise Juul Søndergaard

Routledge
Taylor & Francis Group
LONDON AND NEW YORK

earthscan
from Routledge

Designed cover image: Bladderwrack embracing Julia Lohmann. Photographed by Anna-Marie van der Lei.

First published 2025
by Routledge
4 Park Square, Milton Park, Abingdon, Oxon OX14 4RN

and by Routledge
605 Third Avenue, New York, NY 10158

Routledge is an imprint of the Taylor & Francis Group, an informa business

The Introduction, Chapter 10 and Chapter 15 was supported by the Research Council of Finland (Future Methods, 330124).

Chapter 6 was supported by NODUS, *Gaziulusoy.*

Chapter 7 was supported by Design Research Works under UK Research and Innovation grant reference MR/T019220/1.

British Library Cataloguing-in-Publication Data
A catalogue record for this book is available from the British Library

Library of Congress Cataloging-in-Publication Data
Names: Poikolainen Rosén, Anton, 1992- editor. | Salovaara, Antti, editor. | Botero, Andrea, editor. | Juul Søndergaard, Marie Louise, editor. Title: More-than-human design in practice / edited by Anton Poikolainen Rosén, Antti Salovaara, Andrea Botero and Marie Louise Juul Søndergaard. Description: Abingdon, Oxon ; New York, NY : Routledge, 2025. | Includes bibliographical references and index. Identifiers: LCCN 2024032397 (print) | LCCN 2024032398 (ebook) | ISBN 9781032741192 (hardback) | ISBN 9781032741208 (paperback) | ISBN 9781003467731 (ebook)
Subjects: LCSH: Design—Philosophy. Classification: LCC NK1505 .M665 2025 (print) | LCC NK1505 (ebook) | DDC 744.01—dc23/eng/20240918
LC record available at https://lccn.loc.gov/2024032397
LC ebook record available at https://lccn.loc.gov/2024032398

ISBN: 9781032741192 (hbk)
ISBN: 9781032741208 (pbk)
ISBN: 9781003467731 (ebk)

DOI: 10.4324/9781003467731

Typeset in Times New Roman
by codeMantra

CONTENTS

FIGURES

CONTRIBUTORS

Anton Poikolainen Rosén is a postdoctoral researcher in Human-Computer Interaction at the Department of Design, Aalto University, Helsinki, Finland, studying design for sustainable futures and the more-than-human world. His research themes include critical, circular and multisensory approaches to farming and waste management.

Antti Salovaara is a Senior University Lecturer at the Department of Design, Aalto University, Finland. His research develops methods for HCI researchers to anticipate possible futures, particularly with a goal that futuring in HCI would not only consider technology-based factors.

Andrea Botero is an Associate Professor at the School of Arts, Design and Architecture of Aalto University. Her research and practice aim to understand how collectives (broadly speaking) come to understand the design spaces available to them, what counts as design, what other practices for world making are there, and which ones we need to call into being.

Marie Louise Juul Søndergaard is a designer and researcher exploring feminist design of technologies for human and environmental health. She is curious about how the materiality of human bodies relate with ecologies, and uses research-through-design, participation of communities and speculative storytelling to design for social and environmental justice. She has a PhD in Interaction Design from Aarhus University in Denmark and has been a postdoc at KTH Royal Institute of Technology in Sweden and the Oslo School of Architecture and Design in Norway, where she researched somatic approaches in design and more-than-human design.

Michelle Westerlaken is a postdoctoral researcher at the University of Cambridge with a PhD in Interaction Design from Malmö University. Her research and design work investigates possibilities for more relational – multispecies – ways of living on this planet. She collaborated with humans, cats, dogs, ants, penguins, forests and various interactive technologies.

Erik Sandelin makes prototypes that activate and investigate gracious modes of life. In his recent PhD dissertation in Art, Technology and Design at Konstfack and KTH in Stockholm,

Sandelin develops the concept of grace – actively not doing what you are able to do – to decouple action from force and passivity from resignation in design. By reconfiguring situations where people consume other animals, Sandelin experiments with ahuman design through carefully crafted disentanglements.

Katerina Cerna is an assistant professor at Halmstad University, Sweden and uses design ethnographic and participatory approaches to understand learning and design to enable people to take the actions they need in the more-than-human world.

Yuxi Chen is a designer and researcher with a background in product design, interactive design, sustainable design and social innovation. Her practice considers design as a method for taking care of the relations between humans and non-humans.

Oscar Tomico is an Associate Professor at the Department of Industrial Design at Eindhoven University of Technology. Building on his research on 1st person research methodologies he explores living-with and designing-with through cohabitation. He is currently involved in projects ranging from greening interior spaces through multi-species design, to envisioning regenerative ecologies of design and production.

Dawn Sanders is an associate professor at Gothenburg University, Sweden, working across the art/science interface and how such ways of working might allow us to create new attentional frames in teaching and learning contexts towards plants and notions of "plantness".

Nadia Campo Woytuk focuses on critical and intersectional feminist design of technologies for menstrual health and intimate care. She has led and contributed to projects involving new media art, textiles, software art and postcolonial computing. Her PhD work explores touch and sensing of the intimate body and the social and environmental ecologies it entangles.

Danielle Wilde is a full professor in Design for Sustainability at Umeå Institute of Design, Umeå University, Northern Sweden; and in Sustainability Transitions in the Department of Business and Sustainability at The University of Southern Denmark (SDU) Kolding, in the MERE research group, where she undertakes food system transformation research. She, seeks to understand how critical, collaborative, embodied engagement with the challenges of sustainability transition might assist us in finding new ways of living that are intimately situated, as well as personally, locally, regionally and globally sustainable and sustaining.

Tau Lenskjold is an associate professor at the University of Southern Denmark. His work centres on design cultures in green transition and redirective design practices for more-than-human futures.

Thomas Laurien is a designer, curator and senior lecturer/associate professor at HDK-Valand – Academy of Art and Design at the University of Gothenburg, Sweden. He is immersed in several water-related projects, such as "Whose Body of Water?", "Shimmer and Entanglements in the River Basin of Viskan" and "Confluence of European Water Bodies". Laurien is also co-initiator and co-coordinator of the network "Design + Posthumanism".

Anna Schröder is curious about oceanic relationalities and regenerative ways of living with the seas. In her current PhD project at Malmö University, she focuses on affect and ocean resonance in participatory design practices that work with ocean culture around the Baltic Sea.

Inna Zrajaeva is a Participatory Designer focusing on Ecology and Care. She is a master's student in Interaction Design at the Umeå Institute of Design, Sweden.

Christoph Matt is an Austrian eco-social designer based between Reykjavik and Vienna. He founded Studio Matt, an award-winning bureau for environmentally and socially conscious design, working thoughtfully for planet Earth, the human and more-than-human. He works as an interdisciplinary facilitator with a diverse background in design, sustainability, visual arts and posthumanism. As a design researcher, he focuses on how to build relations between nature and humans with design methods and artistic practice.

Delphine Rumo is a doctoral researcher at NODUS Sustainable Design Research Group, Aalto University, Finland. She is a trained designer and a sustainability researcher exploring the materiality of diverse economies and post-anthropocentric practices. Fascinated by the timing of human-geology economies, she writes about the life of soils and rocks in construction and her encounters with spirited geologies.

Gloria Lauterbach is a postdoctoral researcher at NODUS Sustainable Design Research Group, Aalto University, Finland. As an artistic researcher, Gloria is passionate about dialogical post/humanism, post-anthropocentric theories, materials in cooperation, the more-than-human in human habitation and darkness as an urban restoration force.

Marcos Chilet is a Professor at the School of Design, Pontifical Catholic University of Chile (undergraduate and postgraduate courses). He is co-editor of *Design for More-than-human Futures: Towards Post-Anthropocentric Worlding* (Routledge, 2023) and co-curator and gold medallist with the Chilean Pavilion at the London Design Biennale 2021. He is also Adjunct Researcher of the Millennium Nucleus Futures of Artificial Intelligence Research (FAIR).

Martín Tironi is an Associate Professor and Director of the School of Design at the Pontifical Catholic University of Chile. Tironi is the director of the Millennium Nucleus Futures of Artificial Intelligence Research (FAIR) He is a sociologist, at Pontificia Universidad Católica de Chile. Master in Sociology, Sorbonne; PhD, Center de Sociologie de l'Innovation, École des Mines de Paris. Visiting Fellow, Center for Invention and Social Process, Goldsmiths, University of London, 2018. He is co-editor of the book *Design for More-Than-Human Futures: Towards Post-Anthropocentric Worlding* (Routledge, 2023).

Iohanna Nicenboim is a postdoctoral researcher at Delft University of Technology. Her PhD research, funded by a Microsoft Research scholarship, focused on More-than-Human Design in practice within the context of AI. She has published several articles on these topics in leading HCI and design venues, edited a special issue on MTHD for the *HCI* Journal, and chaired the 'MTHD in Practice' track at DRS2024.

Joseph Lindley runs Design Research Works, a UKRI Future Leaders Fellowship. His research is about the value of Design Research, and how that can be applied to the challenges associated with emerging technologies, rapid societal change and living sustainably. He is chairing the DRS 2024 track "More-than-human Design in Practice".

Petra Jääskeläinen is a critical design researcher focusing on sustainability and ethics of Creative AI/AI arts, with a background in Human-Computer Interaction and Design for Change. Her/their current research interest lies in critically examining the ethical underpinnings of AI art technologies and diversifying the values and imaginaries present in constructing their technological-material-cultural futures, with a particular focus on more-than-human design and feminist environmental posthumanities. She is currently (2021–2026) a PhD candidate at KTH Royal Institute of Technology (Sweden) at the Media and Interaction Design department.

Heidi R. Biggs is an Assistant Professor at Georgia Institute of Technology in the School of Literature, Media, and Communication's Digital Media Department. They are a writer and art and design-based human-computer interaction design (HCI/d) researcher. Their research explores how the design of information technologies builds and mediates relationships to the natural environment, how those relationships are more or less sustainable (both environmentally and socioeconomically), and what other types of relationships we might design that imagine new ways of being in the world with more sustainable outcomes. Biggs' work is featured in ACM SIGCHI conference proceedings and HCI journals as well as electronic textile exhibitions and performance arts venues.

Emīlija Veselova is a researcher in multispecies design at Aalto University, Finland. Her research focuses on developing collaborative design theories and practices that contribute to the sustainability of socio-ecological systems.

David Sánchez Ruano is an assistant research professor at the School of Architecture, Art and Design at Tec de Monterrey. Mexico. He currently leads the Advanced Design Processes Research Group (ADP) and organizes workshops, congresses and publications on topics such as biomimicry, biophilia, regeneration and multispecies design.

Suzanna Törnroth is an Associated Senior Lecturer in Design at Luleå University of Technology, focusing on more-than-human design and feminist technoscience queries in renewable technologies.

İdil Gaziulusoy is a sustainability scientist and a design researcher. She is Associate Professor of Sustainable Design and leader of NODUS Research Group at Aalto University. Her research focuses on long-term, structural, systemic, societal transformations for sustainable, resilient and just futures.

Tatu Marttila is a design researcher working as a Senior University Lecturer in Design for Sustainability at Aalto University, Finland. With several years of experience in strategic co-design action in the context of various sustainability transitions, his current research is focusing on services for climate action and carbon impact monitoring.

Svenja Keune, PhD, is a researcher at the Swedish School of Textiles, University of Borås where she focuses on multispecies perspectives to design by employing multispecies autoethnography and practice based artistic research. As a passionate weaver Svenja co-founded and facilitates the I.N.S.E.C.T. Summer camps and initiated the VIBRA Research Network.

Colleen Ludwig is an Associate Professor of Art at Oakland University in Michigan. Her current research centers on co-creation with living materials within communities, as a way to address biodiversity loss and environmental justice.

Anneke ter Schure is a biologist and artist driven by the question: How can we explore, understand, communicate and question our inherent entanglement with the living and nonliving that surround us and are part of us?

Julia Tabet is an architect and urban ecologies researcher at Gehl. Their work explores methods of co-creation of the public space, developing a multispecies approach to urban design.

Julia Lohmann investigates and critiques the ethical and material value systems underpinning our relationship with flora and fauna. She is Professor of Contemporary Design at Aalto University, Finland, and directs her eponymous Helsinki-based design practice.

Gizem Oktay is a PhD candidate at the Industrial Design department of Eindhoven University of Technology in the Netherlands. Her research deals with the entanglement of time and care in more-than-human relations. She holds a Master of Arts from the Maryland Institute College of Art in Baltimore, Maryland, USA.

Minha Lee is an Assistant Professor at the Department of Industrial Design, Eindhoven University of Technology. She researches morally relevant interactions with CUIs, particularly on the promotion of well-being through positive moral emotions like compassion.

Bahareh Barati is an Assistant Professor at the Department of Industrial Design, Eindhoven University of Technology. Her work explores the intersection of biological and digital capabilities and their mutually informed integrations in designing for sustainable and equitable futures.

Ron Wakkary is a Professor of Design at the School of Interactive Arts and Technology at Simon Fraser University in Canada. In addition, he is a professor and Chair of Design for More Than Human-Centered Worlds in Eindhoven University of Technology in the Netherlands. Wakkary's research investigates the changing nature of design and human-computer interaction in response to new understandings of human-technology relations and posthumanism.

Lotte Nystrup Lund, PhD, is an action researcher exploring historical and contemporary societal ideas and practices targeting biodiversity and regenerative transformation. Her work spans from urban environments to rural landscapes. She is affiliated with the Institute of Architecture, Urbanism and Landscape at the Royal Danish Academy in Copenhagen, Denmark.

Shams Hazim is a creative technologist with a passion for sustainability. He graduated from the Eindhoven University of Technology, the Netherlands, where his work focused on designing between the borders of human, nature and technology. He continues his work at Studio Babel, creating parametric climate-adaptive insect shelters.

Camilo Sanchez is a doctoral researcher at Aalto University's Department of Design. His research applies the concept of anticipation to question design in the face of uncertainty and possible futures.

Felix A. Epp carries out postdoctoral research with the Department of Design at Aalto University and the Social Computing Group at the University of Helsinki, where his work focuses on interactive technologies' future consequences – social, political and beyond.

Tim Moesgen is a doctoral researcher in the Department of Information and Communications Engineering at Aalto University researching the design of novel multisensory interactive experiences with a focus on haptic and thermal sensations. Moreover, he is interested in exploring how research methods of Human-Computer-Interaction can become more future-aware.

Eva Eriksson is associate professor at Department of Digital Design and Information Studies, Aarhus university, Denmark. Her research involves value sensitive design in higher education.

Elisabet M. Nilsson is an Associate Professor in Interaction Design at the School of Arts and Communication (K3), Malmö University in Sweden, exploring conditions for teaching and practicing human and more-than-human inclusion and participation in technology design.

Anne-Marie Hansen is an Assistant professor in Interaction Design at the School of Arts and Communication (K3), Malmö University in Sweden, exploring how we might live in communities and collectives with more-than-humans through traditional ecological knowledge and land connectedness.

Daisy Yoo is an Assistant Professor at the Eindhoven University of Technology, focusing on designing for socio-ecological justice by embracing more-than-human perspectives in value sensitive design.

Tilde Bekker is Professor of Digital Technologies for Playfulness and Motivation. She is interested in designing playful interactions between multiple people and multiple objects in contexts of play, health and learning. She examines how theories on for example, play and learning, from diverse disciplines, can contribute to designing motivational playful tools and technologies for learning and development. The outcome of her design research includes a combination of design exemplars/prototypes, design relevant knowledge and design methods and tools for designing playful technologies.

Daniel Metcalfe is a Multispecies Designer and Visiting Assistant Professor of Design at the Technion Institute of Technology. He works on large-scale projects that integrate animal perspectives into the planning process and aim to create mutually beneficial human-nature interactions.

His research and teaching interests include human-animal interactions, somatic design, digital craft and more-than-human design.

Ann Light is an interaction theorist and professor of design working at University of Sussex, UK, and Malmo University, Sweden. Regarding the social and ecological as inextricably linked and believing creative remaking of relations is needed, she is co-creator of the CreaTures Framework CreaTures and is about to embark with ReWorlding.

INTRODUCTION

Editors' Introduction

The topic of sustainability needs no motivation. It is widely recognised that several human activities are causing global warming, pollution, deforestation, ocean acidification, ozone layer depletion, loss of biodiversity and species extinction. In many cases, the way we have designed everyday objects, products, services, systems and technologies is strongly contributing to this unsustainability. There is thus a need to radically re-imagine how we design. Given this need, an increased interest has emerged across several disciplines to widen the perspective of Western material-oriented thinking beyond solely human-oriented considerations and needs (Akama et al., 2020; Bastian, 2017; Clarke et al., 2018; Coulton & Lindley, 2019; Foth & Caldwell, 2018; Giaccardi & Redström, 2020; Heitlinger et al., 2018; Laurien et al., 2020; Poikolainen Rosén et al., 2022; Veselova, 2023; Wakkary, 2021). This design approach has become referred to as "more-than-human design", and it entails designing for the interdependent relationships between humans, technologies and other organisms (such as animals, plants and microbes). This focus on interdependence forefronts that many organisms, including humans, benefit from considering design spaces as holistic and relational and implies that designers need to expand who or what counts as a user or stakeholder, who is included in design processes, and what is considered as design or designable. This perspective is both radical and generative and requires that design practice, methods and theories are augmented, hybridised and remade.

In this book, we – design researchers, interaction designers, architects, textile designers, interior designers, artists and educators – address more-than-human design as a response to this need to design otherwise.

Several aspects of more-than-human design are more or less established: the central worldview grounded in posthuman ethics; the framing of all organisms in the environment as stakeholders in design, and an awareness of the agency that exists in human relationships with organic matter and technologies. However, it is less established how to practise more-than-human design (method) and what concrete implications this approach has for the form and function of design (Nicenboim et al., 2024). Given this gap, we want to give the reader analytical and methodological tools that may help them approach design spaces and research problems from more-than-human perspectives.

DOI: 10.4324/9781003467731-1

Beyond these conceptual research gaps, this book has a more situated and practical origin. It started as a conversation between Anton Poikolainen Rosén who has written a thesis on more-than-human design in urban farming communities and Antti Salovaara, a senior Human-Computer Interaction and design researcher. Antti was curious about more-than-human design, saw its need, and had many students interested in working with these issues. However, he lacked resources that provided a comprehensive overview of the emergent, messy and growing field of more-than-human design in practice. We hope this book can be one such resource while we also recognise that a book on more-than-human design in practice can never be complete or fully comprehensive as the field must continue to evolve. We also hope to emphasise a critical dialogue between more-than-human design and more established design research approaches. We thus invited Marie Louise Juul Søndergaard and Andrea Botero as co-editors to diversify the perspectives of the editorial team. Marie Louise has a background in feminist approaches and theory in design. Andrea has a background in participatory design and feminist technoscience coupled with a curious attitude towards understanding what – if anything – might come out of stringing together the more-than-human with design. All four of us work in Nordic universities, and this reflects the final configuration of themes and authors that ended up being part of the collection, as our invitation and awareness of the topics reached nearby regions, despite efforts to enlarge it. Most chapter authors work in European universities, although some work in North American universities and places like China, Chile, Mexico and Israel feature in some of the examples and author backgrounds.

Notes on Theory, Critical Perspectives and Format

It is not easy to write an introduction to an emergent, contradictory, diverse and multiple field like more-than-human design that resists definition and categorisation – when these attempts risk being reductionist and overtly universalising. Yet, several recurring concepts are being used and mobilised by those engaging in more-than-human design practices that the reader may benefit from being familiar with before diving into the chapters of this book. To point at these concepts openly, while being relevant to practice, we have selected two formats for the introduction. These include a lexicon for more-than-human in practice – inspired by other such attempts in more-than-human design research (Lindley et al., 2023) – and a manifesto for more-than-human design – inspired by other similar manifestos (Haraway, 1984; Höök, 2018; Rams, 2014). We hope the lexicon can concretise key ideas and concerns of more-than-human and point to further critical readings, rather than providing overtly fixed definitions. The manifesto complements the lexicon, by framing in a more open-ended yet assertive way what more-than-human designing could be.

The experimentation on formats for expressing more-than-human knowledge continues throughout this book's chapters. As established by Abram (1997) – one of the first scholars who popularised the term more-than-human – our language often distances us from what we colloquially signify as "nature". To tackle such issues, there is a need to maintain a stance of critical anthropomorphism (Morton et al., 1990) where we recognise that we can only understand the world from our human position, yet this human position implies that we can imagine things otherwise, that we can, for example, recognise shared sameness with other entities in the world or imagine what it is like to be another being – even if we can never fully know. This implies that we must work actively to include the perspectives and presence of other species, organisms and entities in the book. We do this, for example, by including images to make the presence of the more-than-human more pronounced. Some chapters even experiment with the format of the text itself, exploring more poetic ways of layouts of a text for example to emphasise and express the liquid flow of water.

Another trouble that this book needs to stay with continuously is the diverse meanings of "more-than-human". Some strands of more-than-human design have focused on autonomous, dynamic and evolving computational things (Coulton & Lindley, 2019; Giaccardi & Redström, 2020; Wakkary, 2021), while others have focused on design for and with the interdependencies of organisms (such as animals, plants and microbes) (Akama et al., 2020; Clarke et al., 2018; Nijs et al., 2020). As Noorani and Brigstocke (Noorani & Brigstocke, 2018) argue, more-than-human design is:

> a research paradigm that is in principle applicable to almost anything. This is because it insists that human social worlds are always 'more-than-human' social worlds, in the sense that they are composed of relations between humans, non-human life, and lively materials.

We find this a very apt definition, while also recognizing that a focus on "almost anything" easily becomes unclear or shallow – it is thus important to maintain specificity, situatedness and criticality while recognizing the world as entangled and relational – this is something we hope to forefront through the selection and juxtaposition of the chapters in the book.

The Structure of the Book: Introducing the Chapters

This book showcases practical design outcomes of more-than-human thinking, offers methods for designing with the more-than-human world, and highlights the diversity of more-than-human design practice. We believe that it is important to showcase these examples since more-than-human design is still an emerging way of practising design no matter if you are a seasoned design researcher or a first-year student. We present this work in two parts: "Focus Areas" and "Methods and Pedagogy".

Although more-than-human design recognises the world as an interdependent system, it is important to acknowledge the specific character of entities, species, individuals etc. within this system. The first part thus introduces *focus areas* in more-than-human design: animals, plants, the human body as more-than-human, microbes, winds, soils, rocks, water and artificial intelligence. These examples are not all-encompassing – nevertheless, they have been carefully selected to show the breadth of more-than-human design in practice. They illustrate shared ontologies, positions and methods – as well as differences and conflicting perspectives.

The second part – "Methods and Pedagogy" – discusses a variety of *methods and approaches*, firstly for *understanding design spaces as more-than-human*, secondly for *creating more-than-human design and imagining alternative futures* and thirdly *pedagogical examples*.

The first two chapters in this part discuss methods, frames, thought figures and sensibilities that can be used to understand and empathise with the more-than-human world from a situated first-person perspective, i.e., how you as a designer can understand and relate to the more-than-human world. This includes more empirically oriented methods of perceiving, attuning, observing, measuring and *being with* the more-than-human world. It also includes methods that aim to emphasise and imagine how it might be to be more-than-human. The next chapter discusses methods for compiling the insights derived from such methods to more systemic understandings of design challenges and design spaces. Here the authors outline various approaches to systems thinking from a more-than-human perspective and give examples of methods for systems analysis that more-than-human designers can use.

After discussing methods for understanding design spaces as more-than-human, the next set of chapters focus on creating more-than-human designs and imagining alternative futures. These chapters exemplify the complexities and challenges of design intervention while offering methods that facilitate this. Here three chapters address the temporality of more-than-human design, including visions of a more sustainable future, reparations of wrongdoings in the past, and the presence of multiple temporalties in the now. These chapters offer complementing perspectives, involving for example anticipating the systemic consequences of a design intervention over time or imagining the future as radically different.

The book as a whole provides pedagogical material and teaching features. However, the final two chapters explicitly discuss pedagogical lessons learned from including more-than-human perspectives and exercises in university courses. The first of these chapters focuses on how to teach more-than-human values by for example exploring what must be unlearned to change our perspectives and attitudes toward more-than-human worlds. The second chapter focuses on how to step out of the classroom and into the worlds of other species by discussing several shifts that design pedagogy can make in practice.

In sum, this book provides a collection of concrete examples and actionable methods that are underpinned by emerging and conflicting philosophies and worldviews of more-than-human design.

A Lexicon for More-Than-Human Design in Practice

This lexicon intends to support the reader in understanding the chapters of this book – not to be all-encompassing for what more-than-human design is or could become. We thus aim for a schematic introduction, laying grounds for the following chapters that provide a more in-depth discussion and problematisation of the concepts.

Posthumanism

Much – but not all – work within more-than-human design is rooted in posthumanism (Abram, 1997; Barad, 2003, 2007; Bogost, 2012; Braidotti, 2019; Forlano, 2017; Haraway, 2016; Hayles, 1999). This field is critical of the Western enlightenment ideal of the distinct, rational and dominant human individual, and seeks to understand the human subject and its relationship with the world in a new, non-anthropocentric light. In posthuman epistemologies, knowledge is situated, embodied and partial, meaning that it is pluralistic rather than universal. This implies that there are many differences within posthumanism. However, posthuman thinkers have in common that they undermine traditional boundaries and dualities such as nature/culture, mind/body, or human/technology – and recognize the significance of the non-human contribution to our lifeworld.

More-Than-Human

In design research, the term more-than-human has come to signify the broad orientation of designing with the relationships humans have to the rest of the world including other beings and lively matter and technologies. The term "more-than-human" was popularised by philosopher David Abram (1997) in his book "The Spell of the Sensuous: Perception and Language in a More-than-Human World". This book bridges the philosophical tradition of phenomenology with environmental and ecological issues. Although the idea of the more-than-human was theorised and popularised by Abram, this kind of thinking originates in Indigenous ontologies that view humans and 'nature' as an inseparable whole (Escobar, 2018). The idea of the more-than-human is thus far from new but is marginalised in – and marginalised by – contemporary Western society. Nevertheless, scholars are beginning to acknowledge, recognise and take seriously the diversity of more-than-human Indigenous ontologies (Akama, 2015) and their sustainability benefits (Brant, 2021; Latulippe & Klenk, 2020; Vásquez-Fernández & Ahenakew pii tai poo taa, 2020).

There are many ways of using the term more-than-human. Its current popularity stems from a need to replace the word "nature" with a term that explicitly recognises that humans are not separate from nature. In this book, we use the word "more-than-human" to signify a shared relationality as made sense of by humans. In other words, "more-than-human" is not a substantive that can be used interchangeably with words such as "species", "organism", "technology" or "non-human". You cannot be "a more-than-human". Instead, the term makes the most sense in relationships with other words, to signify a momentary focus on phenomena, while recognising its interconnectedness to all other phenomena that exist in the world. We may for example say "more-than-human world" to signify what in other contexts is called "nature" or "environment"; or "more-than-human entities" to signify specific phenomena within this world. One should thus take caution to not use the term more-than-human to obscure phenomena. We should always aim to be explicit and precise with what we mean. To paraphrase a common expression, we may call a spade a more-than-human spade, to explicitly recognize that we are mutually affecting each other.

Non-human and Other-than-human

Everything that is not human is non-human. However, it is problematic to define something as what it is not, as discussed in feminist theory regarding terms such as "non-male" and "non-white", which in their effort to highlight discrimination instead risk rendering invisible or "making other" in a nonconstructive way (Haraway, 2016). Nevertheless, there are cases where a term such as "non-human" may help us in a particular kind of analysis where a momentary separation and recognition of difference supports sensemaking – and although there are big differences between species, families and orders, it is often more practical to refer to the diversity of life with a more general term. For this, some prefer the term "other-than-human", since other organisms are truly "other" in terms of cognitive capacities, senses, shapes, DNA, needs, driving forces, etc. Not less, not more – other. It is only as "the other" that we may approach other beings and relate to them on their terms. As Haraway (2016) suggests, we might still make kin with other beings. Similar to the term "more-than-human" the terms "non-human" and "other-than-human" make most sense when used in combination with other words. For example, when using the expression "non-human animals" to remind us that humans are also animals.

Relationality and Entanglement

More-than-human design frames *systemic relations* as the focus of design. This is often phrased as an entanglement of a vast set of relationships that are interacting, often in ways that are hard to overview. In such entanglements human-made systems are seen as extensions of ecological systems (Metcalfe, 2015). We are "of nature", and so are the things we make. This inseparability is sometimes referred to as *natureculture* (Haraway, 2016). A relational perspective further forefronts interdependence & reciprocity, recognizing how interacting entities in a system affect each other mutually (but not always equally). Philosophically, these ideas of relationality and interdependence are often grounded in *Object Oriented Ontology* (Harman, 2018) which rejects the privileging of human existence over the existence of nonhuman objects, and *Actor-Network-Theory* (Latour, 1996) which emphasises how everything in the social and natural worlds exists in constantly shifting networks of relationships. In design research, concepts such as *entanglement design* (Frauenberger, 2019) have been proposed as explicit design approaches recognising systemic relationality.

Care Ethics

Many more-than-human design approaches go further than merely recognizing interdependent relationality. They see this interdependence as a foundation for building an "ethics of care" (Puig de la Bellacasa, 2011). This ethics, characterized as virtue or practice rather than a distinct theory, posits that the fundamental aspects of human relationships and dependencies to other humans and the rest of the more-than-human world hold moral significance. Care ethics aims to nurture and sustain relationships by prioritizing the well-being of both caregivers and care recipients within a complex web of social connections, including those with the more-than-human world. Care is grounded in the motivation to support those who are reliant and vulnerable, drawing inspiration from personal experiences of receiving care as well as aspirational self-concepts.

Decentering

The idea of decentering is prominent in several more-than-human design approaches where explicit efforts are made to not put the human on the centre stage (Forlano, 2016; Nicenboim et al., 2023). This idea is an aspect of *non-anthropocentrism* or the *"non-human turn"* (DiSalvo & Lukens, 2011). Similar terms are *life-centred design* (Borthwick et al., 2022) and *multispecies design* (Metcalfe, 2015) which are used to signify a focus on the well-being of all life on the planet. In many ways, these terms can be used interchangeably with the term more-than-human design, as they all signify a shift from solely focusing on human needs, to a more holistic understanding of ecologies.

Situated, Embodied and Partial Knowledge

More-than-human design recognizes knowledge as situated, embodied and partial, meaning that it is pluralistic rather than universal. Such situated knowledge is embedded in, and thus affected by, the concrete historical, cultural, linguistic and value context of the knowing person (Haraway, 1988). On an epistemological level, the notion of situated knowledge is an effort to think outside the duality of objectivity-relativism that may be both ineffective and harmful for more-than-human purposes, since it risks enforcing for example overtly dualistic thinking, and perspectives of the already established and powerful. The production of situated knowledge strives for a more adequate, richer, better account of a world to navigate it effectively and engage critically with both our own and others' tendencies towards domination and the inherent disparities in privilege and oppression that characterize all social positions.

Pluriversal Design

Pluriversal design is a reaction to the idea of universal design (Noel, 2020). The critique of universal design is that it is often suited to a particular being (a white, Western, heterosexual cisgender man) while other ways of being are marginalised by this "universal" position. The idea of pluriversal design is to instead recognise and nurture differences by allowing for parallel existing worlds capable of moving along each other, including the worlds of other species, indeed sometimes causing friction, but not the total extinction that a universalising position risk causing.

Agency and Intra-action

There are diverse ideas of what agency is. In most dictionaries, agency is described both as *action or intervention producing a particular effect* and *a thing or person that acts to produce a particular result* (Stevenson, 2010). A common colloquial formulation is that phenomena "have" agency. However, you can only have agency in relation to something to which you are producing a particular "effect" or "result". This means that agency is a relational and multidirectional phenomenon where entities affect each other mutually, but not always to an equal extent. Analysing power and domination involves identifying where this relationship is unbalanced, and figuring out how the agency of the disadvantaged party can be strengthened. More-than-human design explores how such ideas of balancing agency through social justice can be expanded beyond what has historically been included in "social" (i.e., relationships between humans). Relatedly, *intra-action* is a term coined by Barad (2007) used to replace 'interaction' to understand bodies that participate in action with each other. Intra-action understands agency as not an inherent property of an individual or human to be exercised, but as a dynamism of forces (Barad, 2007) – both social and physical.

Given this understanding of agency as a dynamism of forces, categorising notions of agency is a risky endeavour. Nevertheless, for more-than-human design in practice, it can make sense to discuss how agency is characterised in certain situations.

Agency and Materials

Some perspectives on more-than-human design are beginning to recognise matter as lively and vibrant (Bennett, 2010). This implies that matter is not seen as fully "passive" but as being able to "talk back". A ceramist cannot command clay to behave in just any way but must understand, and work with its material properties. From this perspective, craft and design is a skill of sharing control with materials, of creating scaffolds in which they are shaped in desired, but not fully controllable ways. Other examples of material agency are weather phenomena that tangibly shape our everyday experiences and more global climate conditions (see for example the chapters "5 Designing with Bodies of Water in the Hydrocene" and "6 Weathering with Storms and Grounds as a More-than-Human Design Practice- Encountering Winds, Soils and Rocks").

Agency and Technologies

Technologies are a special form of matter that has been arranged in such a way that their leverage on the world is strengthened. More-than-human design thus recognises the active role technologies take in shaping the world and our experiences of it. This includes the increased autonomy of technological networks, for example through Artificial Intelligence (AI). As noted by Giaccardi and Redström (2020) "We still experience technologies as things with a clear presence and tangibility – yet, they are just one element within a system of decentralised interactions that makes them very different from what things used to be like". In more-than-human design, we thus also need to recognise the network of algorithms, protocols, databases, etc. that affect the world (see for example the chapters "7 Designing with Planetary Artificial Intelligence" "8 Creative AI as More-than-Human – Design Practices, Aesthetics and Cultural Imaginaries").

Agency and Living Beings

Dictionaries often define life as *the condition that distinguishes animals, plants, fungi, bacteria etc. from inorganic matter, including the capacity for growth, reproduction, functional activity and continual change preceding death* (Stevenson, 2010). Here, the exemplary boundary case is viruses that display several of these characteristics, but for example, cannot reproduce without using the cells of other organisms – with a prominent example of COVID-19 showing their potential for extreme systemic leverage. The agential effects of some living beings tend to be more familiar to us humans, such as animal locomotion, while others are more unfamiliar, such as the seemingly imperceptible chemical communication of plants (see for example chapters "1 Design By-For-With-About-Without Animals: Tactics for Animal Liberation" "2 Being with Plants through Collective Fabulation, Critical Companionship and Cohabitation" "3 Biomenstrual: Designing with the More-than-Human Body" "4 Trying Out Shit!: Experimental Approaches for Relating with Microbes" "9 Multispecies Ethnography in Design Research and Practice").

Sentience

Sentience is the state of having an experience of the world. It is a conscious awareness of stimuli without requiring association or interpretation. Most more-than-human design posits that we humans have a special moral obligation towards sentient beings, as the process of sentience implies the capacity for both suffering and enjoyment (Humphreys, 2020). However, what these moral obligations are is more debated. Scientific knowledge is further continuously developed on which organisms have sentience.

Life-world,* Umwelt *and Worlds

More-than-human design often talks of multiple "worlds". This should not be understood as a rejection of a shared universe – the term is rather used to emphasise the experiential perspective of *life-worlds*, such as how all sentient beings are in the centre of their own experiences, their world. Sometimes *umwelt* (De Roo & Ganzevles, 2023), a German word for environment, is used to denote an organism's unique sensory world. It is an epistemological challenge to access these "worlds" – nevertheless, this does not mean that we should not aim to consider them when designing.

Critical Anthropocentrism

Our human position of perceiving, understanding, knowing and relating to the world is inescapable. Yet, more-than-human design aims to understand how other beings experience the world. This paradox requires a stance of critical anthropocentrism, where we recognise that our capacities as humans are limited, but also that we can use our imagination in combination with our knowledge to approach what it might be like to be another being (Affifi, 2020).

Representatives: Spokesperson, Ombudsmen and Diplomats

An omnipresent issue to continuously address in more-than-human design is how to represent other beings and things in design processes, while we are inherently trapped in our humanness, never fully able to grasp what it is like to be anything other than human. Design research has experimented with several tactics for this. For example, the idea of being a *spokesperson* (Wakkary, 2021) for non-human others by advocating for their position in society; the idea of being an *ombudsman* (Tompuri, 2015) aims for the same while de-emphasizing focus on human language and the Idea of being a *diplomat* (Peña-Guzmán, 2022) emphasises conflict resolution and also shifts perspective to ask questions such as: how can we as humans represent ourselves in the best way possible towards a species or ecology in focus?

*Co-creation, 'Designing **with**', Emergence and Constituency*

The more-than-human worldview recognizes more–than-human co-creative capacities we can *design with*. Consequently, the role of the designer is reimagined. A designer is not necessarily one person or even a design team, but a coming together of the creative capacities of all kinds of people, organisms, materials and technologies, in what Wakkary (2021) calls *constituencies*.

A More-Than-Human Design Manifesto

Above, we used the format of a lexicon to make the assumptions of more-than-human design more explicit. Another common way in design and the humanities to make a specific worldview clearer is through a manifesto (Haraway, 1984, p. 198; Höök, 2018; Rams, 2014). We offer the following collaboratively written more-than-human design manifesto as an explanation and reminder of underlying principles that guide our diverse work in more-than-human design. The manifesto serves as the tentative and iteratively changing lens through which we can understand more-than-human design. The ambiguous format of a manifesto has been chosen intentionally since it reflects the inherent transitional and open nature of more-than-human design. It is concrete enough to be generative, yet vague enough to invite further explorations. It is open to being wrong and changing accordingly. It is pluralistic rather than dogmatic. The manifesto was initially written by Anton Poikolainen Rosén as part of his thesis work. It was then shared with all the authors of the book and the design+posthumanism network (designandposthumanism.org), who had an opportunity to comment on and edit the manifesto – and sign it if they wanted to stand behind it.

A More-Than-Human Design Manifesto

We cannot separate human needs from those of other organisms in our environment. Many organisms, including humans, benefit from considering design spaces as holistic and relational. More-than-human design focuses on this mutual interdependence.

More-than-human design resists binaries and blurs notions such as nature/culture, self/environment, digital/physical, mind/body, human/technology, and human/non-human.

More-than-human design highlights the sentience, intelligence, and agency of other organisms. This implies that we need to seriously consider the needs of all organisms, their sensory capacities, and their capability for interaction.

We are humble in recognising that we will never fully understand others. More-than-human design recognises that everything is understood from a perspective. We explore how our limited human perspectives can be enriched, strengthened, and augmented in ways that increase our understanding of the more-than-human world.

There are many ways to sense and make sense. Technology can be helpful when we reach the limits of our senses. It can amplify, augment and make perceptible the seemingly imperceptible. Simultaneously, we acknowledge that technology has detrimental impacts on more-than-human worlds. It is important to be aware of the resource use and environmental harm of technologies – and their potential benefits.

We who practice more-than-human design are not a monolithic category but bring to our practice diverse personal and disciplinary backgrounds. Our intersectional positionalities, including privileges, oppressions and lived experiences of race, class, gender, sexuality, and ability shape how we think and practice design.

More-than-human design aims to widen the scope of action through the creative development of methods for noticing, knowing and making the more-than-human world.

More-than-human design works with the co-creative capacities of living entities, lively matters and machines.

We design (into) relational systems, rather than (contributing) single artefacts. We design in ways recognising that we ourselves are relational systems, our bodies being biotopes for more-than-human micro-communities in constant interaction.

More-than-human design recognises the direct and indirect effects of design, over time and across space, for a multitude of species. It is both microscopic and planetary, with impacts beyond.

More-than-human design recognises diverse temporalities: the millennial lifecycle of a boulder, the centennial lifecycle of an oak and the brief lifecycle of a dayfly or a microbe. We work with regenerative systems sustaining themselves but also recognise the linear reality of natural history. Nature is in constant becoming, nothing stays the same. Everything in nature is slowly changing at the pace of biological evolution – but we want to avoid abrupt harmful disturbances and mass extinction. As humans, we must tread lightly to contribute with generative suggestions rather than disturbances. We recognise multiple and complex experiences of time including linear, non-linear and more.

We resist cries for perfection and imaginaries of perfect futures. We consider diverse and conflicting needs and focus on how we make systems resilient for ourselves and other organisms. We "stay with the trouble". We live with tricky questions about justice, acknowledging that the inquiry into more-than-human design is never settled. We are in a constant process of critical reflection and becoming, not always finding answers or solutions.

We recognise that the oppression of human hierarchies and extractive abuses makes all forms of life suffer and that the living world's struggles with human patterns of domination are aligned in politics, if not the impacts experienced. This brings the suffering and abuse of humans into the frame of more-than-human design.

This is only the first part of the manifesto. The second part is inscribed in leaves and rocks, and all the more-than-human world. We need to go out and notice these inscriptions. What can we see, hear, smell, feel and taste? What might be beyond our senses? By being spokespersons, ombudsmen and diplomats, we advocate for what we notice and more – for all that is still, and might forever be, beyond our human ability to notice, understand and grasp.

Endorsed by:

Anna Schröder, Ann Light, Antti Salovaara, Anton Poikolainen Rosén, Camilo Sanchez, Cecilia Åsberg, Christoph Matt, Daniel Metcalfe, Danielle Wilde, Delphine Rumo, Felix Anand Epp, Gloria Lauterbach, Henrik Lübker, Iohanna Nicenboim, İdil Gaziulusoy, Lígia Oliveira, Lotte Nystrup Lund, Marcos Fernando Chilet Bustamant, Marie Louise Juul Søndergaard, Marietta Radomska, Matthew Dalziel, Michelle Westerlaken, Moniek Driesse, Nadia Campo Woytuk, Oscar Tomico, Petra Jääskeläinen, Svenja Keune, Tatu Marttila, Tau Lenskjold, Thomas Laurien, Tim Moesgen, Yuta Ikeya, Yuxi Chen, Åsa Ståhl, Daisy Yoo, Anne-Marie Hansen.

References

Abram, D. (1997). *The Spell of the Sensuous: Perception and Language in a More-Than-Human World* (1st Vintage Books edition). Vintage.

Affifi, R. (2020). Anthropocentrism's fluid binary. *Environmental Education Research, 26*(9–10), 1435–1452. https://doi.org/10.1080/13504622.2019.1707484

Akama, Y. (2015). Being awake to Ma: Designing in between-ness as a way of becoming with. *CoDesign, 11*(3–4), 262–274. https://doi.org/10.1080/15710882.2015.1081243

Akama, Y., Light, A., & Kamihira, T. (2020). Expanding Participation to Design with More-Than-Human Concerns. *Proceedings of the 16th Participatory Design Conference 2020 – Participation(s) Otherwise – Volume 1*, 1–11. https://doi.org/10.1145/3385010.3385016

Barad, K. (2003). Posthumanist performativity: Toward an understanding of how matter comes to matter. *Signs, 28*(3), 801–831. https://doi.org/10.1086/345321

Barad, K. (2007). *Meeting the Universe Halfway*. Duke University Press. www.Dukeupress.Edu/Meeting-The-Universe-Halfway

Bastian, M. (2017). Towards a more-than-human participatory research. Routledge, London.

Bennett, J. (2010). *Vibrant Matter: A Political Ecology of Things*. Duke University Press. https://doi.org/10.2307/j.ctv111jh6w

Bogost, I. (2012). *Alien Phenomenology, or What It's Like to Be a Thing*. University of Minnesota Press. https://doi.org/10.5749/minnesota/9780816678976.001.0001

Borthwick, M., Tomitsch, M., & Gaughwin, M. (2022). From human-centred to life-centred design: Considering environmental and ethical concerns in the design of interactive products. *Journal of Responsible Technology, 10*, 100032. https://doi.org/10.1016/j.jrt.2022.100032

Braidotti, R. (2019). A theoretical framework for the critical posthumanities. *Theory, Culture & Society, 36*(6), 31–61. https://doi.org/10.1177/0263276418771486

Brant, J. (2021). Sand talk: How Indigenous thinking can save the world. *International Journal of Lifelong Education, 40*(4), 427–430. https://doi.org/10.1080/02601370.2021.1954393

Clarke, R., Heitlinger, S., Foth, M., DiSalvo, C., Light, A., & Forlano, L. (2018). More-than-human urban futures: Speculative participatory design to avoid ecocidal smart cities. *Proceedings of the 15th Participatory Design Conference: Short Papers, Situated Actions, Workshops and Tutorial – Volume 2*, 34:1–34:4. https://doi.org/10.1145/3210604.3210641

Coulton, P., & Lindley, J. G. (2019). More-than-human centred design: Considering other things. *The Design Journal, 22*(4), 463–481. https://doi.org/10.1080/14606925.2019.1614320

De Roo, B., & Ganzevles, G. A. (2023). The umwelt-sketch as more-than-human design methodology: Decentering the design process from human-centered towards more-than-human-centered. *Companion Publication of the 2023 ACM Designing Interactive Systems Conference*, 203–206. https://doi.org/10.1145/3563703.3596628

DiSalvo, C., & Lukens, J. (2011). *Nonanthropocentrism and the Nonhuman in Design: Possibilities for Designing New Forms of Engagement with and through Technology*. https://doi.org/10.7551/mitpress/8744.003.0034

Escobar, A. (2018). *Designs for the Pluriverse: Radical Interdependence, Autonomy, and the Making of Worlds*. Duke University Press.

Forlano, L. (2016). Decentering the human in the design of collaborative cities. *Design Issues, 32*(3), 42–54. https://doi.org/10.1162/DESI_a_00398

Forlano, L. (2017). Posthumanism and design. *She Ji: The Journal of Design, Economics, and Innovation, 3*(1), 16–29. https://doi.org/10.1016/j.sheji.2017.08.001

Foth, M., & Caldwell, G. A. (2018). More-than-human media architecture. Proceedings of the 4th Media Architecture Biennale Conference (MAB18), Beijing, China – November 13–16, 2018, 66–75. https://dl.acm.org/authorize?N673075

Frauenberger, C. (2019). Entanglement HCI: The next wave? *ACM Transactions on Computer-Human Interaction, 27*(1), 2:1–2:27. https://doi.org/10.1145/3364998

Giaccardi, E., & Redström, J. (2020). *Technology and More-Than-Human Design* (Vol. 36). Design Issues Autumn 2020.

Haraway, D. (1984). A cyborg manifesto. *Socialist Review (US)*.

Haraway, D. (1988). Situated knowledges: The science question in feminism and the privilege of partial perspective. *Feminist Studies*, *14*(3), 575–599. https://doi.org/10.2307/3178066

Haraway, D. (2016). *Staying With The Trouble*. Duke University Press.

Harman, G. (2018). *Object-Oriented Ontology: A New Theory of Everything*. Penguin UK.

Hayles, N. K. (1999). *How We Became Posthuman: Virtual Bodies in Cybernetics, Literature, and Informatics*. University of Chicago Press. https://press.uchicago.edu/ucp/books/book/chicago/H/bo3769963.html

Heitlinger, S., Foth, M., Clarke, R., DiSalvo, C., Light, A., & Forlano, L. (2018). Avoiding ecocidal smart cities: Participatory design for more-than-human futures. *Proceedings of the 15th Participatory Design Conference: Short Papers, Situated Actions, Workshops and Tutorial – Volume 2*, 51:1–51:3. https://doi.org/10.1145/3210604.3210619

Höök, K. (2018). *Designing with the Body: Somaesthetic Interaction Design*. MIT Press.

Humphreys, R. (2020). *Suffering, Sentientism, and Sustainability: An Analysis of a Non-Anthropocentric Moral Framework for Climate Ethics*. Routledge, Taylor & Francis Group. https://repository.uwtsd.ac.uk/id/eprint/1345/

Latour, B. (1996). On actor-network theory: A few clarifications. *Soziale Welt*, *47*(4), 369–381.

Latulippe, N., & Klenk, N. (2020). Making room and moving over: Knowledge co-production, Indigenous knowledge sovereignty and the politics of global environmental change decision-making. *Current Opinion in Environmental Sustainability*, *42*, 7–14. https://doi.org/10.1016/j.cosust.2019.10.010

Laurien, T., Jönsson, L., Lilja, P., Lindström, K., Sandelin, E., & Ståhl, Å. (2020). An emerging posthumanist design landscape. In S. Herbrechter, I. Callus, M. Rossini, M. Grech, M. de Bruin-Molé, & C. John Müller (Eds), *Palgrave Handbook of Critical Posthumanism* (pp. 1–23). Springer International Publishing. https://doi.org/10.1007/978-3-030-42681-1_42-1

Lindley, J., Benjamin, J. J., Green, D. P., McGarry, G., Pilling, F., Dudek, L., Crabtree, A., & Coulton, P. (2023). Productive oscillation as a strategy for doing more-than-human design research. *Human–Computer Interaction*, *0*(0), 1–26. https://doi.org/10.1080/07370024.2023.2276393

Metcalfe, D. (2015). *Principles of Multispecies Design* [Falmouth University].

Morton, D. B., Burghardt, G. M., & Smith, J. A. (1990). Critical anthropomorphism, animal suffering, and the ecological context. *The Hastings Center Report*, *20*(3), S13–S13.

Nicenboim, I., Oogjes, D., Biggs, H., & Nam, S. (2023). Decentering through design: Bridging posthuman theory with more-than-human design practices. *Human–Computer Interaction*, *0*(0), 1–26. https://doi.org/10.1080/07370024.2023.2283535

Nicenboim, I., Lindley, J., Zaga, C., Berger, A., Forlano, L., & Giaccardi, E. (2024); Editorial: More-than-human design in practice, DRS 2024, in *Proceedings of the Design Research Society* (DRS'24), Boston, https://doi.org/10.21606/drs.2024.114

Nijs, G., Laki, G., Houlstan, R., Slizewicz, G., & Laureyssens, T. (2020). Fostering more-than-human imaginaries: Introducing DIY speculative fabulation in civic HCI. *Proceedings of the 11th Nordic Conference on Human-Computer Interaction: Shaping Experiences, Shaping Society*, 1–12. https://doi.org/10.1145/3419249.3420147

Noel, L.-A. (2020). Envisioning a pluriversal design education. *Pluriversal Design Conference Series*. https://dl.designresearchsociety.org/pluriversaldesign/pivot2020/researchpapers/6

Noorani, T., & Brigstocke, J. (2018). More-than-human participatory research. In Facer, K. and Dunleavy, K. (Eds), Connected Communities Foundation Series. Bristol: University of Bristol/AHRC Connected Communities Programme.

Peña-Guzmán, D. M. (2022). *When Animals Dream: The Hidden World Of Animal Consciousness*. Princeton University Press.

Poikolainen Rosén, A., Normark, M., & Wiberg, M. (2022). Towards more-than-human-centred design: Learning from gardening. *International Journal of Design*, *16*(3), 21–36.

Puig de la Bellacasa, M. (2011). Matters of care in technoscience: Assembling neglected things. *Social Studies of Science*, *41*(1), 85–106.

Rams, D. (2014). *Less, but Better*. Gestalten.
Stevenson, A. (2010). *Oxford Dictionary of English*. OUP Oxford.
Tompuri, T. T. (2015). Metabolic equivalents of task are confounded by adiposity, which disturbs objective measurement of physical activity. *Frontiers in Physiology*, *6*, 226.
Vásquez-Fernández, A. M., & Ahenakew pii tai poo taa, C. (2020). Resurgence of relationality: Reflections on decolonizing and indigenizing 'sustainable development.' *Current Opinion in Environmental Sustainability*, *43*, 65–70. https://doi.org/10.1016/j.cosust.2020.03.005
Veselova, E. (2023). *Designing with Nature for Sustainability: Towards a critical approach for including natural nonhuman stakeholders in collaborative and participatory design when designing for sustainability*. https://research.aalto.fi/en/publications/designing-with-nature-for-sustainability-towards-a-critical-appro
Wakkary, R. (2021). *Things We Could Design: For More Than Human-Centered Worlds*. MIT Press.

PART 1

Focus Areas

1

DESIGN BY/FOR/WITH/ABOUT/ WITHOUT ANIMALS

Tactics for Animal Liberation

Michelle Westerlaken and Erik Sandelin

Introduction

> KARIN: I was very surprised that I was the only one in my class working with animals that we eat, and with the whole ethical dilemma of eating animals and consuming animal products.... It was just surprising that it wasn't a bigger topic. When I read the project description I was very much like: "Ok, sounds like a vegan project to me".
>
> ERIK: What made you think that?
>
> KARIN: It was the "careful entanglement with companion species". That was the main course topic. I was, like, slaughter isn't very careful, right?

Over the last decades, academic scholarship and design research have become engaged with a move towards establishing less anthropocentric relations with the planet and all its living inhabitants. The field of more-than-human design emerged from this orientation and asks designers and researchers to reconsider the importance of other entities, environmental systems, and their relations in design processes and outcomes. Notions of "multispecies design", "more-than-human design", "biocentric design", or "non-anthropocentric design" are prominent in contemporary design research, practice and education (DiSalvo & Lukens, 2011; Clarke et al., 2019; Giaccardi & Redström, 2020; Wakkary, 2021). Various "post-anthropocentric" approaches – often influenced by strands of actor-network theory, posthumanism and new materialist thinking – are increasingly used as theoretical grounding for design work (Forlano, 2017). While such orientations help to shape more-than-human design, we argue that existing work often steers clear of the most uncomfortable aspects of human-animal relationships, namely those relations characterised by explicit human violence, dominance and oppression of other animals. "Karin" (not their real name) was a student in a course on more-than-human design at a Scandinavian design school. Her confusion is a symptom of how a critical position towards human exploitation of other animals can be eclipsed by an, often uncritical, celebration of entanglement, hybridity, care and companionship. In this chapter we will use quotes from an interview with Karin to highlight the frictions and fault lines between animal oppression and more-than-human design.

DOI: 10.4324/9781003467731-3

Our everyday encounters with animals are largely shaped by oppressive interactions by consuming food, wearing clothes, taking medication, consuming animal representations in media, and the effects of these practices in environmental change. The design industry is heavily involved in shaping such practices, but design scholarship on these issues is still limited. This tendency to avoid explicit engagement with design and animal oppression risks overlooking and even normalising a large aspect of our everyday interactions with other living beings. It appears to be more popular in more-than-human research to bring in relational thinking with jellyfish, domestic dogs, forests and mushrooms, as researchers like Donna Haraway and Anna Tsing have done in their important contributions to this field (Haraway, 2016; Tsing, 2015), than to engage with farm animals or other captive or industrial animal usage. Such works add important concepts and methods that help to imagine less anthropocentric futures, but they are far from the only type of relations from which less anthropocentric design work can emerge. In this chapter we will show how the combination of Critical Animal Studies and the field of design can contribute vital understandings and ideas for rethinking and remaking human relations with other living entities. The interdisciplinary field of Critical Animal Studies emerged in the early 00s and is grounded in subjects of environmental and social justice, intersectionality and critical analysis (Sanbonmatsu, 2011; Nocella et al., 2014; Taylor & Twine, 2015).

ERIK: What species did you classmates work with?
KARIN: Bees, different insects, rats.... There were a lot of bacteria, fungi, but also the wind, air, trees. That was about it.

When we focus on animals as units of analysis in this context, this is not intended as a strict ontological distinction between animals, other living beings, and their relations. As outlined in existing feminist, new materialist and biology scholarship, animals – including humans – are in constant relation within complex ecosystems, they even consist of living entities like viruses and bacteria themselves, and these relations cannot be ontologically separated as they sustain each other's lives on this planet (Gilbert, 2017; Haraway, 2016). Instead, we focus on animals as a temporary category to better analyse and understand certain relationships and systems better, in order to give form to design research and practices. We use the notion of 'other animals', or simply 'animals', as temporary distinction from human animals, instead of more general terms like 'non-humans' or 'more-than-humans', to recentre animals and their agencies within this analysis. This focus on animals helps us to analyse human-animal relations in design that are characterised by consumption and exploitation, the relations that were marginalised in Karin's course.

The significant role of design in the lives of animals, especially for production and captive animals, cannot be ignored if more-than-human design aims to tackle broader issues of inequality, unfairness and violence. Production animals are actually involved in nearly all aspects of design practice, even those that seemingly have nothing to do with animals. They are involved, sometimes nearly invisibly, through design approaches such as biomimicry, as design materials in the form of paintbrushes or inks, on people's plates at conference lunches, and many other aspects of design and design research. The field of Critical Animal Studies helps to address this gap as it focuses on understanding the real-world conditions that animals face within human-designed environments. Unlike mainstream animal studies, Critical Animal Studies is not content with analysing animals as symbols or metaphors. Instead, as a strand of critical theory, it is committed to creating real change towards animal liberation, it includes a strong theory-praxis connection that is inextricably imbricated with other modes of oppression such as racism, classism and sexism, and it involves a "firm unwavering commitment" (Pedersen & Stanescu, 2014, p. 263) to ending the exploitation of animals. This focus on activism and real-world impact makes Critical Animal Studies an essential tool for designers engaged in practices and themes of social and environmental justice. For example, design-oriented Critical Animal Studies questions that emerge could include: What does this [design situation] do for the animals that are impacted directly or indirectly? Or how does this [design act] potentially reinforce animal oppression through the use of materials or the normalisation of harmful species hierarchies? Or how can we design alternatives to animal oppression with the aim towards constructing multispecies futures?

By examining various existing design projects in relation to animals and design, we employ Critical Animal Studies to ask critical questions, to engage with the complexity of human-animal relations, and to help generate new design openings. The ability of designers to propose alternatives, imagine different futures, and try out new ideas aligns well with the field of Critical Animal Studies as both can seek to challenge and change the status quo. To this end, we summarise related projects according to five prepositions through which design has already contributed to this goal: 'Design by/for/with/about/without Animals'. It is important to note that many other prepositions in relation to design and animals exist as well but these are more embedded in oppressive understandings. For example, some designs are 'of' animals, like breeding programs. Others are 'on' animals, such as animal testing for cosmetics. And some are 'out of' animals, like the industrial processing of animal products. Our focus, however, will be on those designs that have the potential to work towards animal liberation. Through this text, we argue that developing a Critical Animal Studies literacy in more-than-human design must underpin any orientation that seeks to unmake anthropocentrism, animal oppression, and move towards more sustainable and less violent design practices. This analysis can be of practical use for designer scholar-activists and help not just to criticise current systems, but to imagine and build better ones.

Design by/for/with/about/without Animals

Each of the five sections below articulates a preposition in the design space of human-animal relations. By detailing and exemplifying some of the projects that have emerged within these design trajectories and analysing them through critical questions and generative contributions from the field of Critical Animal Studies, we aim to emphasise the importance of Critical Animal Studies scholarship for the field of more-than-human design.

Design by *Animals*

After a longstanding tradition in which humans were regarded as the sole entities on this planet with the capability to design their environments, less anthropocentric reflections nowadays also articulate how animals can be acknowledged as creative agents in design processes. This creativity becomes evident when analysing animal architectures, such as intricate nests, dens and burrows. These structures are not merely shelters but also embody functional and aesthetic considerations, reflecting the unique needs and preferences of their creators (Dobraszczyk, 2023). Furthermore, animals demonstrate ingenuity in tool building. From the skilful crafting of implements for obtaining food to the use of objects as tools, animals exhibit a capacity for design thinking that extends beyond mere instinctual behaviour (Gigliotti, 2022). This challenges the traditional boundaries of design and inspires humans to consider the diverse ways in which design can be expressed. Also within mating rituals and other social relations, animals' creative expressions, playful explorations, and aesthetic preferences are examples of skilful proposals that can be read in the context of animals as designers.

Growing design scholarship on this topic reveals the fascination of human designers with these works. Besides establishing a less anthropocentric understanding of who constitutes design spaces, this attention towards design created by animals also inspires new design works by humans (Benyus, 2002; DiSalvo & Lukens, 2011). The recognition of animals as designers encourages a shift away from a solely human-centred perspective of design. This shift challenges the assumption that humans are the sole active subjects in design processes. Instead, it prompts us to acknowledge more-than-human creators in design spaces. For example, "Design by Animals" by design group Front (2003) is a series of domestic objects – lamps, vases, tables, wallpaper patterns – whose creation is attributed to animals. An unnamed snake squeezes the soft clay of a mould that is then used (by the human designers) to cast a clothes hanger. Rats chew on wallpaper to create intricate patterns. The designers frame their project as asking for help and receiving an enthusiastic response: "We asked animals to help us. Sure, we'll help you out! they answered. Make something nice, we told them. And so they did" (Front, 2003).

Although this work seems to underscore the agency of animals in creatively shaping their surroundings, Critical Animal Studies would ask: What's in it for the animal? Front's artefacts are not biomimetic solutions inspired by nature. The animals are instrumentalised to provide unexpected input to the surface design of typical design artefacts, supposedly as a critique of the (human) genius of signature design. The animals are used as random seeds or creative resistance for generating ideas and patterns that the humans then refine and finalise. Front surrenders some control to the nonhuman actors but are not really risking much as they are still the stewards of the situation and its ultimate outcome.

Yet, Critical Animal Studies provides a vital framework for understanding the implications of recognizing animals as creators of design work beyond admiration and inspiration. Critical Animal Studies literature prompts us to ask essential questions regarding agency, consent, disturbance, and ownership. Critical Animal Studies asks us not to ventriloquise, not to represent, animals like Front, above, but to focus on actual lived experience. Can animals be acknowledged as creators of their work, and does this require a form of informed consent that may be impossible to obtain (MacCormack, 2014)? Furthermore, how can designers respect the implications of ownership and engage in carefully analysing and documenting animal creations without disturbing animal lives and destroying their environment?

Finally, Critical Animal Studies encourages designers to transcend the mere analysis of animals' design functions and aesthetics. It calls upon human designers not only to engage in the more-than-human 'art of noticing' but also to embark on a deeper exploration of the life worlds of the animals involved and the socio-political effects of their designs. For instance, in urban spaces where wild animals are often regarded as pests, examples of the ways birds build their nests around anti-bird spikes show how their designs demonstrate ways in which urban animals are trying to survive despite their oppression by humans (Hiemstra et al., 2023). Animal adaptations of these environments are responses to human urban designs that involve animal proposals for changing their surroundings. A Critical Animal Studies-involved analysis connects these socio-political insights to design processes to advance the creation of future (urban) spaces in which animals can live alongside humans in healthy and mutually respectful relations.

Design for *Animals*

Over the last decades, fields such as animal science and ethology have produced an increasing body of knowledge about animals' sensory perceptions and capacities, their diverse modes of expression, and their behaviour. These epistemological transformations emerge from an increased consideration of animal phenomenology, their welfare, and their life worlds and predominantly focus on those animal species that show similarities with humans, such as primates, domestic animals and mammals more generally. This growing body of knowledge is also rich with design openings to create artefacts that are specifically catered to specific animals' needs and experiences. As a design field expanding notions of human-centred design, the inclusion of animals as users has created notions of animal-centred, or biocentric, design in which other living entities are specifically considered as users.

The increasing awareness of the conditions of animals in captivity has spurred the design of tools and environments for animal enrichment. In this field, designers may aim to propose new modes of interaction for animals through artefacts like digital toys, foraging technologies, or spatial design in order to improve the welfare of these animals. An example of this is the speculative 'Playing with Pigs' project, in which a videogame for pigs was prototyped as a potential solution for pigs' boredom and their lack of playful activities on farms (Driessen et al., 2014). Other examples include interactive technologies for animals in zoos and shelters (Hirskyj-Douglas et al., 2017; Webber et al., 2017; Wirman, 2014), or tablet games for cats in animal shelters (Westerlaken & Gualeni, 2014). Also within the field of architecture, animals are considered as users of human-built environments. Architectural office Studio Gang, for instance, translated insights from birds' sensory perspective to construct a more bird-friendly skyscraper, called Aqua Tower, featuring a facade designed to minimise the risk of birds accidentally flying into the building (Studio Gang, 2010). Another, particularly insightful, example includes the work of Temple Grandin, who spurred widespread redesigns of slaughterhouses in the US with the goal to lower stress and anxiety for these animals to make the slaughter processes more efficient (Grandin & Johnson, 2006). All of these examples show ways in which animals use artefacts and spaces specifically designed for them.

However, as "animal-centred" as these designs might seem, all these examples also bring up crucial questions about the wider speciesist structures in which design is employed and risk to further normalise animal exploitation. Critical Animal Studies literature helps to better understand how these examples are situated within normalised structures of keeping animals in captivity as well as capitalist or human-centric motivations for the design of these artefacts. For instance, while designers like Temple Grandin might employ a user-centred design rhetoric, aiming to empathise with animals in slaughterhouses and make their final hours less distressing, such designs ultimately serve the animal slaughtering industry and effectively increase its productivity. Even with the design of playful enrichment for captive animals, while such designs may moderately improve the welfare of the animals involved, they may ultimately normalise exploitative animal industries, for example by presenting a false narrative that promotes zoos as enriching environments where visitors are willing to pay to see captive animals play video games (Westerlaken, 2020). Critical Animal Studies scholar Anna Williams critically notes that recognising animal subjectivity can actually serve capitalist modes of industrialised breeding and killing (Williams, 2004). In essence, even the most well-intentioned designs, rooted in a deep understanding of animal behaviours, can bolster systems that intensify animal oppression.

This realisation prompts design introspection and critical questioning. How can idealistic design briefs be turned into projects that can actually challenge the status quo? How does systemic change emerge from understanding the impact of the enrichment design industry? Whose interests truly lie at the heart of these designs? Do they genuinely serve the animals' interests, or are they merely tools that further human-centric agendas? Which types of animals are typically included as users, and who tend to remain excluded? And what are the responsibilities of designers and design researchers in these projects? These types of questions may help direct designers who create artefacts, systems and spaces 'for' animals to reconsider the deeper meaning and effects of user centred design that involve other living beings.

Design with *Animals*

Further expanding the focus of animals as users of design artefacts, the field of participatory design (PD) has also moved towards the inclusion of animals as active participants already earlier on in design processes. The subfield of Animal-Computer Interaction (Mancini, 2011), branching from Human-Computer Interaction, along with other more-than-human participatory design domains (Clarke et al., 2019), have proposed tools, methods and frameworks to facilitate this wider understanding of participation in design. This orientation is fundamentally different from designs 'by' or 'for' animals because this trajectory is more explicitly focused on the types of methods that are incorporated into design processes and are usually facilitated by human designers to specifically create spaces for animals to join design activities. Questions that guide this process include: How do animals express their design proposals, and how can they be understood and transformed into design outcomes created by humans?

Some approaches in PD with animals translate tools from PD with humans towards their use for projects with animals or iterative prototyping and testing methodologies (Mancini & Lehtonen, 2018). Other approaches to participation involving animals explore less prescriptive notions of becoming-with (Haraway, 2008; Westerlaken & Gualeni, 2016), co-evolving (Mancini et al., 2012), interspecies democracy (Meijer, 2019), or focus on design transformations in situated contexts (Hakio et al., 2022). Recurring questions for each of these emerging modes of design revolve around the meaning of participation in these contexts: Do animals need specific capabilities or modes of articulating themselves that are understandable for human designers? Can animals be categorised at a species level in which their preferences can be generalised, or should design take place at the level of individual animals? Can participatory methods be used for all different species, or are there limitations that prevent meaningful participation by certain types of animals? How can human designers become better at understanding animals and what design activities can help with this? The participation of animals in design processes risks creating an expectation for animals to articulate and express their needs and preferences in ways that are understandable for human designers. It is not surprising that participatory design projects tend to focus on including those animals that we are already in communicative relations with, such as pets and larger mammals.

Critical Animal Studies further complicates these questions in important ways by drawing parallels with critiques on participatory design from other disciplines of critical theory. Decolonial and feminist scholarship on participatory design with humans have urged designers to challenge and disrupt the exclusions that often precede design situations (Tlostanova, 2017). Designers are critiqued for their approaches to PD when participation remains focused on dominant or privileged user groups at the expense of others. More democratic engagement in design requires more than a couple of community workshops that ultimately fail at addressing larger socio-political issues and systemic injustices. Also when it comes to animals, Critical Animal Studies researchers have critiqued approaches to participation for offering romanticised and depoliticised understandings of human-animal relations. Haraway's notion of 'becoming-with', for instance, is used to describe situations of mutual reciprocity and response between humans and domestic dogs but has also been critiqued for its failure to address the largely unequal relationship between humans as pet owners or breeders (Weisberg, 2009). In another example, Critical Animal Studies scholar Eva Giraud shows how the researchers' "careful", attentive, and affective interactions with the animals in a dog experimentation laboratory actually mask the inherent power imbalance in human-dog laboratory dynamics. Such participatory scenarios risk overshadowing the pre-existing exclusions that have already dictated the protocols in which humans perform harmful experiments on laboratory dogs (Giraud, 2019).

Critical Animal Studies commitments to designing with animals urge designers to reconsider the circumstances in which participation takes place. A perhaps unresolved question here is to what extent oppressed animals themselves should be included as participants in design contexts. For example, how can farm or zoo animals be enabled to propose multispecies futures without risking to further reinforce the oppressive situations in which they currently live? In a project with a colony of Magellanic penguins who live in a zoo enclosure, Westerlaken writes how, on the one hand, the participation of these specific penguins has helped her articulate important new knowledge about penguins' modes of expression, their playful preferences, and the ways they are shaping design processes (Westerlaken, 2020). On the other hand, this project also revealed how these penguins risk becoming tokens for their species groups, how they are exploited as entertainment devices in commercial settings under the banner of education and conservation. Participants can develop a close and unequal bond with designers during these projects. Designers tend to get attached to their prototypes, and their close relations with participants influence subsequent decision-making (Westerlaken, 2020). A Critical Animal Studies approach can help to notice how these relations are a part of wider oppressive structures and show the non-innocent positions of designers. However, there is a real risk that such attempts at collaborative future-making through design fail to create actual change for oppressed animals.

It is important to note that while Critical Animal Studies scholarship is firm in its rejection of speciesism, there is no uniform agreement on how this should be undertaken. Although vegan approaches are seen as an intrinsic practice of Critical Animal Studies, like many fields of critical theory, it is a heterogeneous and interdisciplinary field. Similarly, some designers may decide to approach animal participation by refusing to engage with animals who live in captivity, or only focus on animals living in sanctuaries or shelters. Other designers may focus on the transformation of systemic issues of captive animals by developing closer relationships with them through design. In both cases, Critical Animal Studies helps designers to critically reflect on complex issues, systemic power-asymmetries, and larger oppressive structures. It can inspire designers to come up with ways in which participatory design can also help to unmake harmful situations, or to determine when 'opting out' is a preferable design decision.

Design about *Animals*

Design that involves animals is also used to raise animal related issues. Here, design helps people to better understand the lives of animals. In relation to animal oppression, these types of projects often invoke ideas of critical design (Malpass, 2017). Such artefacts provoke people to think about human-animal relations and generate societal debate. Design techniques are employed with the aim to generate emotions such as shock, anger, or sadness and often include an agenda towards animal liberation.

Wearing a VR headset, the user of Animal Equality's iAnimal is invited to 360° tours of the everyday lives of farmed animals. Immersed in the situation, but without ability to intervene, the spectator watches as a male calf, the unusable "by-product" of the dairy industry, is crammed into a narrow concrete chute and killed by a plant worker, putting a bolt gun to the calf's forehead and firing with a loud pop. The calf plunges to the ground, and the camera follows the corpse, now suspended on hooks, being bled and flayed by a plant worker. The project aims to make the suffering of animals "less abstract" by "creating a feeling of presence" (Animal Equality, 2016). In another example, Christien Meindertsma's book *PIG 05049* shows the extent of human consumption of animals. It maps the human use of an individual pig by listing 185 products made from different parts of this particular animal: skin (gloves, injectable collagen, wine gums), bones (paints, shoes, x-ray film), meat (mortadella, hamburger, weisswurst), internal organs (pâté, insulin, tambourines), blood (dog treats, cigarettes, blood sausage), fat (lubricant, washing powder, crayons), and miscellaneous (renewable energy, bread, fertiliser). Notably, the book lists several items used by designers and artists, such as pig-bristle paint brushes, gelatin-covered printing papers, and pigments made from pig bone (Meindertsma, 2008).

Hartmut Kiewert's Animal Utopia paintings are instances of another strand of design about animals. Instead of disclosing the current, horrific, conditions of production animals, Kiewert depicts a world after animal liberation (Kiewert, 2015). Animals, cows and sheep, that used to be called "livestock" now graze the grass growing in the concrete cracks in the ruins of derelict meat processing plants. The animals share the urban space with convivial picnickers. While Kiewert offers visual snapshots from a less speciesist world, Terike Haapoja and Laura Gustavsson's interactive performance The Trial offers a site where participants get to engage with and actively rehearse such futures (Gustafsson & Haapoja, 2014). In a fictional court, two hunters are charged with killing a wolf. After the prosecutor and lawyer argues their case, the audience is invited to pass a sentence, through filling out forms. If the hunters are guilty, are they guilty of slaughter or murder?

Such projects show how animals are involved in global industries, or help to understand them as "absent referents" and make their suffering more visible (Adams, 2010). However, projects such as iAnimal also risk maintaining a dividing-line between animals that "are like us" and therefore are worthy of our compassion, while other animals are still outside ethical consideration. Designers, such as Meindertsma, can put their skills of visualising complex and large-scale phenomena to use in showing how the (ab)use of other animals as resources in many industries is ubiquitous in contemporary societies (Twine, 2012). Designers and artists, such as Kiewert and Haapoja & Gustavsson, can put their skills to use in envisioning, trying out and activating less anthropocentric worlds. In a section on "vegan painting technique" on his website, Kiewert (2018) shares his findings on alternatives to animal-based painting material and showcases the Critical Animal Studies designer-activist position. As designers or design researchers, even as we are crafting a prototype or attending a conference lunch, there is no stepping outside of the often-exploitative human-animal relations we are part of. These projects show that when it comes to animals, design inevitably moves beyond representation both through its material engagements as well as its efforts in making animal conditions present through experience.

Design without *Animals*

Design values curiosity and creativity. This is often expressed by the designer moving *into* a situation, *towards* the other. Karin is rebuked by her supervisor for sacrificing designerly curiosity on the altar of her "firm and unwavering" vegan commitment:

> KARIN: My tutors believed I was too subjective, and too opinionated, in a way. They wanted me to empathise with the meat-eaters, even though I was doing interviews with both vegans and meat-eaters.... They really wanted me to go out and buy leather and, almost like, go out and be a meat-eater so that I wasn't too opinionated in one direction.

As seen above, more-than-human design approaches often acknowledge the presence and importance of animals through emphasising embodied entanglement, touch, hybridity and care. Many Critical Animal Studies-influenced projects, such as iAnimal, work through human identification with other animals. In both these cases there is a focus on moving *towards* the animal, physically and/or discursively. Still, even if well-intended, bringing animals into human realms, physically or discursively, is often a kiss of death for the animal (MacCormack, 2014), such as in the example above concerning Temple Grandin's slaughterhouse designs. This proposition, of designing by and for carefully leaving the animal be, challenges this approach towards animals and asks if the animal has perhaps been studied enough. If humans, not animals, are at the root of problems, perhaps "the question of the animal" (Wolfe, 2003) is no longer the most pressing to respond to by more-than-human design. Perhaps we should instead approach humans with a Critical Animal Studies-oriented design lens.

In this vein, recent calls within the field of Critical Animal Studies voices the need for the, creative and careful, *leaving be* of animals. Such "post-animal studies" (Pedersen & Stanescu, 2014; Stanescu & Twine, 2012) suggest two trajectories:

1) "removing the animal from the meat" and
2) turning the gaze back to the human and the human desires that prop up the subjugation of animals.

These approaches open up for more-than human design as moves of "replacing animals" and "undesigning humans".

The strand of replacing animals is concerned with removing animals from destructive relations and replacing them with designed stand-ins or surrogates, that cater to many of the qualities of the original activities, while being less harmful for the animals. Crafting surrogates for animal products is no longer a fringe phenomenon. The market for plant-based meat-imitations is growing and such products can serve an important social function in transitioning to a plant-based diet. Cary Wolfe (Wolfe, 2013) asks: "Does sacrifice make meat taste better? Indeed, is it what makes meat 'meat'?" Experimental mock-meat initiatives, such as Vegan Ortolan (The Center for Genomic Gastronomy, 2012) and Eating E.T. (Sandelin, 2014), try to also cater to some of the darker "complicated pleasures", of the domination involved in eating animals. Eating E.T is the festive barbecuing and eating of a full-scale gluten replica of the famous Hollywood extra-terrestrial. This mock-alien asks what it feels like to eat a creature, albeit fictional, that you may have had a long-standing emotional relation to when younger? The project asks if it is possible to replace that elusive aftertaste of domination that accompanies eating animals? Beyond eating, Sandelin has explored an alternative recreational fishing practice, Fishwatching, where the angler uses a hookless lure and an underwater camera to capture fish on film instead of piercing them on a hook (Sandelin, 2024). In one sense, Fishwatching retains the essence of angling, that of "outwitting nature" (Gilbey, 2008), and caters to many of the qualities of traditional angling, spending time outdoors, tinkering with the fishing gear, and the convivial chatting with fellow anglers on the pier. On the other hand, Fishwatching differs by challenging the hooking and "playing" of the fish, suggesting a slightly looser connection with the animal.

Although such practices can be helpful as tools for humans to transition towards less harmful relations with animals (Nath & Prideaux, 2011), Critical Animal Studies scholarship reveals how such practices still risk supporting animal exploitation industries. It helps to show the risk of moving too smoothly and therefore risk of maintaining or supporting problematic ways of relating to animals. It helps to zoom out from the specific interaction to the situation of animals in society. For example, as Vegan Ortolan and Eating E.T. highlight, a Critical Animal Studies-informed analysis helps to show how some of these 'replacement' techniques become desperate attempts at maintaining oppressive practices by merely replacing animal bodies. Other areas of critical theory have articulated how these forms of replacements risk doing little to change violent practices by the perpetrator: the trajectory of replacing animals may have value in transitioning to less harmful human-animal interactions, but, like the rush towards "green technologies", it risks perpetuating the myth that our host, the world, is infinitely exploitable if we just invent the right tools (Fry, 2008; Serres, 1995).

The second trajectory of post-animal design, "undesigning humans" or ahuman design (MacCormack, 2014), in contrast, turns the designerly gaze back towards the human, and works more directly with human desires, towards more "gracious" (MacCormack, 2012; Sandelin, 2024; Serres, 1995) human relations with animals, characterised by careful and creative withdrawals, and leaving be of the animals. Grace entails actively not doing what you are able to do, not using all the force that is at your disposal. Designers can experiment with deploying carefully crafted *no*s and *not*s into situations of human consumption of animals (Sandelin, 2024). Examples of this practice include Sandelin's Meat Addiction Hypnotherapy (Laurien et al., 2020). He invited people who wanted to stop eating animals but found it hard to quit, to participate in hypnosis sessions crafted to undesign or at least introduce a certain hesitation in situations of consuming animals, at the restaurant or in the supermarket. Participants took part in bespoke hypnosis sessions, based on their aims and ambitions, where they rehearsed saying 'no' to animal products. This "mental prototyping" was anchored in simple gestures like putting an index-finger to your thumb, so called posthypnotic suggestions, intended to affect change after the end of the session. This Critical Animal Studies-oriented project shows how the notion of grace is a vital counterpoint to common more-than-human and posthuman calls for embodied entanglement, of moving closer to the animals. A register of grace can help designers in considering and arguing for creative and careful 'nos' and 'nots' when considering the involvement of other animals in design projects, to learn to oscillate between standing in solidarity with the animal and leaving the animal be.

Conclusion

> KARIN: The other supervisor … encouraged me to make a gratefulness campaign. So that I would make stickers that said 'thank you' and put them on milk and on meat packaging … to

express gratitude. I found that very strange. It's not like the cows chose to give us their meat, or give us a gift in that sense.... It might help the humans to feel better about themselves, but it doesn't help the animal and their suffering.

This chapter further articulated the intersection of more-than-human design and animals, emphasising the need for a sustained engagement with Critical Animal Studies in design research and practice. Without this engagement, the field of more-than-human design is at risk of reinforcing animal oppression and exploitation. Within a more general move towards less anthropocentric design, we argue that an understanding of Critical Animal Studies commitments is not optional, but manifests a core component of decentring humans. This is because oppressed and exploited animals are encountered by most humans on a daily basis, and the design industry is heavily involved with creating and sustaining these practices. Animals are affected by, or involved in, design practices through the materials that designers use, the food that designers consume, or the design outcomes that are created. Together with critical engagements of other forms of oppression including racism, classism, sexism, or otherwise, more-than-human design must widely orient itself towards more sustainable and more justice focused practices that involve animals.

The five prepositions in this chapter show different ways in which animals can be involved in more-than-human design. Design by/for/with/about/without animals each attend to different aspects of human-animal relations in design. By analysing different examples of design within each area, and relating these to Critical Animal Studies insights, each of these sections proposes questions and potential areas to explore in design research and practices. By incorporating Critical Animal Studies scholarship in more-than-human design, designers can challenge the status quo and envision more sustainable and less violent design proposals. Even though these propositions are here presented as separate sections to inspire different relational perspectives, it must be stated that, in practice, designers – and animals – move between complex human-animal relations that will inevitably involve different types of encounters.

While each section proposes different questions that can guide designers forward, living and designing towards animal liberation cannot be reduced to simple checklists or a set of questions, as it engages with far more complex, less resolved, and perhaps infinite multiplicities in which human-animal relations can be sustained in societies across the world. Ultimately, there is no fixed agreement on what exactly constitutes such liberation. Nonetheless, it is important to note that merely acknowledging the existence of Critical Animal Studies-related understandings in design is not the same as actively working towards animal liberation. Thus, more-than-human design projects with animals cannot be reduced to a set of methods, approaches, questions, or tools that can be employed. Such attempts risk becoming a naïve interpretation of how human-animal relations are shaped. Instead, this chapter argued how part of being a designer requires us to become aware of the dilemmas, of problematic norms, of uncomfortable spaces, and commit to challenging the status quo in informed and creative ways.

References

Adams, C. J. (2010). *The Sexual Politics of Meat: A Feminist-Vegetarian Critical Theory* (Twentieth Anniversary). Continuum.

Animal Equality (2016). *iAnimal—A virtual Reality Experience into the Lives of Farmed Animals*. https://ianimal360.com/

Benyus, J. M. (2002). *Biomimicry: Innovation Inspired by Nature*. Perennial.

Clarke, R., Heitlinger, S., Light, A., Forlano, L., Foth, M., & DiSalvo, C. (2019). More-than-human participation: Design for sustainable smart city futures. *Interactions*, *26*(3), 60–63. https://doi.org/10.1145/3319075

DiSalvo, C., & Lukens, J. (2011). Nonanthropocentrism and the nonhuman in design: Possibilities for designing new forms of engagement with and through technology. In *From Social Butterfly to Engaged Citizen*. MIT Press.

Dobraszczyk, P. (2023). *Animal Architecture: Beasts, Buildings and Us*. Reaktion Books. https://reaktionbooks.co.uk/work/animal-architecture

Driessen, C., Alfrink, K., Copier, M., Lagerweij, H., Peer, & Irene. (2014). What could playing with pigs do to us? *Antennae: The Journal of Nature in Visual Culture*, *30*, 79–102.

Forlano, L. (2017). Posthumanism and design. *She Ji: The Journal of Design, Economics, and Innovation*, *3*(1), 16–29. https://doi.org/10.1016/j.sheji.2017.08.001

Front (2003). *Design by Animals*. Front. http://www.frontdesign.se/design-by-animals-project

Fry, T. (2008). *Design Futuring*. London: Berg.

Giaccardi, E., & Redström, J. (2020). Technology and more-than-human design. *Design Issues*, *36*(4), 33–44. https://doi.org/10.1162/desi_a_00612

Gigliotti, C. (2022). *The Creative Lives of Animals*. NYU Press.

Gilbert, S. F. (2017). Holobiont by birth: Multilineage individuals as the concretion of cooperative processes. In A. L. Tsing, N. Bubant, E. Gan, H. Swanson, & A. (Eds), *Arts of Living on a Damaged Planet: Stories from the Anthropocene* (pp. 73–89). Island Press.

Gilbey, H. (2008). *Fishing*. DK Publishing.

Giraud, E. H. (2019). *What Comes After Entanglement?: Activism, Anthropocentrism, and an Ethics of Exclusion*. Duke University Press.

Grandin, T., & Johnson, C. (2006). *Animals in Translation: The Woman Who Thinks Like a Cow* (New edition). Bloomsbury Publishing PLC.

Gustafsson, L., & Haapoja, T. (2014). *The Trial*. www.gustafssonhaapoja.org/the-trial/

Hakio, K., Dolejšová, M., Mattelmäki, T., Choi, J. H., & Ampatzidou, C. (2022). Following Seals and Dogs: Experimenting with Personal Dimensions of Transformative Design. *Proceedings of the Participatory Design Conference 2022 – Volume 2*, *2*, 167–172. https://doi.org/10.1145/3537797.3537869

Haraway, D. (2008). *When Species Meet*. University of Minnesota Press.

Haraway, D. (2016). *Staying with the Trouble: Making Kin in the Chthulucene*. Duke University Press.

Hiemstra, A.-F., Moeliker, C. W., Gravendeel, B., & Schilthuizen, M. (2023). Bird nests made from anti-bird spikes. *Deinsea*, *21*, 17–25.

Hirskyj-Douglas, I., Read, J. C., & Cassidy, B. (2017). A dog centred approach to the analysis of dogs' interactions with media on TV screens. *International Journal of Human-Computer Studies*, *98*, 208–220. https://doi.org/10.1016/j.ijhcs.2016.05.007

Kiewert, H. (2015). *Animal Utopia. Oil painting.*

Kiewert, H. (2018). *About. Hartmut Kiewert*. https://en.hartmutkiewert.de/about

Laurien, T., Jönsson, L., Lilja, P., Lindström, K., Sandelin, E., & Ståhl, Å. (2020). An emerging posthumanist design landscape. In S. Herbrechter, I. Callus, M. Rossini, M. Grech, M. Bruin-Molé, & C. John Müller (Eds), *Palgrave Handbook of Critical Posthumanism* (pp. 1–23). Palgrave Macmillan.

MacCormack, P. (2012). *Posthuman Ethics: Embodiment and Cultural Theory*. Ashgate.

MacCormack, P., Ed. (2014). *The Animal Catalyst: Towards Ahuman Theory*. Bloomsbury Academic.

Malpass, M. (2017). *Critical Design in Context: History, Theory, and Practice*. Bloomsbury.

Mancini, C. (2011). Animal computer interaction: A manifesto. *Interactions*, *18*(4), 69–73.

Mancini, C., & Lehtonen, J. (2018). The emerging nature of participation in multispecies interaction design. *Proceedings of the Designing Interactive Systems Conference (DIS '18*, 907–918.

Mancini, C., van der Linden, J., Bryan, J., & Stuart, A. (2012). Exploring interspecies sensemaking: Dog tracking semiotics and multispecies ethnography. *Proceedings of the 2012 ACM Conference on Ubiquitous Computing*, 143–152. https://doi.org/10.1145/2370216.2370239

Meijer, E. (2019). *When Animals Speak: Toward an Interspecies Democracy*. NYU Press.

Meindertsma, C. (2008). *Pig 05049*. Flocks.
Nath, J., & Prideaux, D. (2011). The civilised burger: Meat alternatives as a conversion aid and social instrument for Australian vegetarians and vegans. *Australian Humanities Review*, *51*, 135–151.
Nocella II, A. J., Sorenson, J., Socha, K., & Matsuoka, A. (2014). Defining critical animal studies: An Intersectional social justice approach for liberation. New York: Peter Lang.
Pedersen, H., & Stanescu, V. (2014). Conclusion: Future directions for critical animal studies. In R. Twine & N. Taylor (Eds), *The Rise of Critical Animal Studies: From the Margins to the Centre* (pp. 262–275). Routledge.
Sanbonmatsu, J., Ed. (2011). *Critical Theory and Animal Liberation*. Lanham: Rowman & Littlefield.
Sandelin, E. (2014). Eating E.T. In E. A. Cederholm, A. Björck, K. Jennbert, & A.-S. Lönngren (Eds), *Exploring the Animal Turn: Human-animal Relations in Science, Society and Culture* (pp. 47–56). The Pufendorf Institute for Advanced Studies.
Sandelin, E. (2024). *Design and Grace: An Ahuman Odyssey* [PhD Thesis.]. KTH Royal Institute of Technology & Konstfack University of Arts, Crafts and Design.
Serres, M. (1995). *Genesis*. The University of Michigan Press.
Stanescu, V., & Twine, R. (2012). Post-animal studies: The future(s) of critical animal studies. *Journal for Critical Animal Studies*, *10*(4), 4–19.
Studio Gang (2010). *Aqua Tower*. https://studiogang.com/project/aqua-tower
Taylor, N., & Twine, R., Eds, (2015). *The Rise of Critical Animal Studies*. London: Routledge.
The Center for Genomic Gastronomy (2012). *Vegan Ortolan*. https://genomicgastronomy.com/work/2012-2/vegan-ortolan/
Tlostanova, M. (2017). On decolonizing design. *Design Philosophy Papers*, *15*(1), 51–61.
Tsing, A. L. (2015). *The Mushroom at the End of the World: On the Possibility of Life in Capitalist Ruins*. Princeton University Press.
Twine, R. (2012). Revealing the 'animal-industrial complex': A concept & method for critical animal studies? *Journal for Critical Animal Studies*, *10*(1), 12–39.
Wakkary, R. (2021). *Things We Could Design: For More Than Human-Centered Worlds*. MIT Press.
Webber, S., Sarah, C., Marcus, S., Sally, S., Wally, J., Zaher, V., & Vetere, F. (2017). Kinecting with orangutans: Zoo visitors. *Empathetic Responses to Animals' Use of Interactive Technology, ACM SIGCHI Conference on Human Factors in Computing Systems*, 6075–6088.
Weisberg, Z. (2009). The broken promises of monsters: Haraway, animals and the humanist legacy. *Journal for Critical Animal Studies*, *7*(2), 22–62.
Westerlaken, M. (2020). *Imagining Multispecies Worlds* [Doctoral Dissertation, Malmö University]. http://urn.kb.se/resolve?urn=urn:nbn:se:mau:diva-17424
Westerlaken, M., & Gualeni, S. (2014). Felino: The philosophical practice of making an interspecies videogame. *Proceedings of the Philosophy of Computer Games Conference*, 1–12. http://gamephilosophy2014.org/wp-content/uploads/2014/11/Westerlaken_Gualeni-2014.-Felino_The-Philosophical-Practice-of-Making-an-Interspecies-Videogame.-PCG2014.pdf
Westerlaken, M., & Gualeni, S. (2016). Becoming with: Towards the inclusion of animals as participants in design processes. *Proceedings of the Third International Conference on Animal-Computer Interaction*, 1–10. https://doi.org/10/f94dsg
Williams, A. (2004). Disciplining animals: Sentience, production, and critique. *The International Journal of Sociology and Social Policy*, *24*(9), 45–57. https://doi.org/10.1108/01443330410790768
Wirman, H. (2014). *Games For/With Strangers: Captive Orangutan (Pongo Pygmaeus) Touch Screen Play*. http://ira.lib.polyu.edu.hk/handle/10397/7611
Wolfe, C. (2003). *Zoontologies: The Question of the Animal*. University of Minnesota Press.
Wolfe, C. (2013). *Before the Law: Humans and Other Animals in a Biopolitical Frame*. University of Chicago Press.

2

BEING WITH PLANTS THROUGH COLLECTIVE FABULATION, CRITICAL COMPANIONSHIP AND COHABITATION

Katerina Cerna, Anton Poikolainen Rosén, Yuxi Chen, Oscar Tomico and Dawn Sanders

Introduction

Plants shape the landscape, provide us with oxygen and are cornerstone species in the global food chain. However, we need to develop new modes of *being with* plants that allow us to coexist with them in a mutually beneficial relationship (Loh et al., 2024). Today, we mainly interact with and exploit plants by using them as food, material or a design medium. Plants are used as raw materials in architecture, fabric, craftsmanship and industrial production – and there is a range of design projects using living plants as a material feature of the design (Fastnacht et al., 2016; Steer et al., 2015). However, in line with more-than-human design, plants should be treated like more than this – they are not only a resource simply made for us humans; we are *dependent* on them as living organisms that share the world with us. It is thus essential to include plants as a deeper concern for design if we are to build more-than-human worlds of mutual interdependence and care.

The idea of "designing *with* plants" opens an exciting opportunity to explore the boundaries of what more-than-human design can be and push them further. When studying animals, it is possible to approach them fairly similarly to humans – who are also animals. Humans can observe how most animals move, communicate with gestures and sound, and affect objects around them. Some animals even interact with interfaces (Jackson et al., 2018; Kleinberger et al., 2023). By contrast, focusing on plants forces us to reimagine frameworks and methodologies, since the plant way of being in the world can be perceived as quite different to humans.

This chapter aims to provide an overview of perspectives, methods and tools for including plants in design by discussing three modes of plant engagement that help us to *be with* plants in an alternative way, seeing them as more than mere design material.

Therefore, we first synthesise the frontiers of plant science, as this detailed knowledge of plant characteristics facilitates design *with* pants. We then explore why it is challenging to consider plants as stakeholders in design by addressing plant awareness disparity (Parsley, 2020). We then discuss the three modes: *Collective fabulating*; *Critical companionship*; and *Cohabitation*. The articulation of these modes emerged from reflexive discussions on our examples – while they also draw on established concepts in more-than-human literature.

DOI: 10.4324/9781003467731-4

We exemplify these modes based on our concrete activities as design researchers who aim to *design with* and *be with* plants – sometimes also including others in these experimental activities through workshops and other participatory engagements. The activities highlight the tension of using plants as an "eco-friendly" material, while at the same time approaching the plants with respect and as having integrity in and of themselves. Aiming to treat plants with such grace (Sandelin, 2024), we use the term *designing with* to signify that plants can take the role of material in design, but that we also need to *be with* them in a more emphatic sense in line with more-than-human philosophy on *being with* (Armstrong et al., 2021) in more respectful and attentive ways. We finish the chapter with implications of the modes for design, where we contrast the different modes and reflect on the need for design and designers to consider plants as actors in the design processes.

For clarity, we discuss plants as a cohesive phenomenon, while in reality many of the capacities we discuss are particular to certain species. It is essential to note that all species of plants do not have all these capacities. It is a broad generalisation to discuss "plants" in a sweeping manner as if they had cohesive properties. As such, actual design projects should benefit from recognising and researching the specific characteristics of each individual organism and species they work with.

Knowing Plants as Active Stakeholders

Historically, in the Western world, plants have been considered semi-alive but passive, "beings without a soul", which only react to the impulses from the outside without any active part on their side (Bos, 2010). More-than-human design resists such hierarchical thinking, posing that species are not more or less valuable than other species, only closer or farther away from your ethical and moral obligations for care and concern. We include this fairly detailed background on the capacities of plants to show their active role in ecosystems, as a basis for knowing plants, and in extension designing for and with them.

Today, a growing body of research points to the complex capacities of plants that have an active role in ecologies because of their diverse characteristics. Plants live their life by converting sunlight into energy. They might seem stationary at first sight, as they are rooted to one position. However, many of their parts are mobile: for example, leaves move with sunlight, respond to touch and curl up to avoid being attacked by herbivores. Utricularia traps (underwater plant carnivores) move faster than we can click our fingers to trap tiny water insects and squirting cucumbers using a form of jet propulsion to disperse their seeds (Sanders, 2019). Most plants have more DNA than humans, which hints at complex biochemical capacities (Messing, 2001). Here, a growing body of literature acknowledges the active capacities of plants (Kohn, 2013; Sheikh et al., 2021). According to this literature, plants have between fifteen and twenty distinct senses, including analogues of our five senses: smell and taste (plants sense and respond to chemicals in soil and air), sight (plants react differently to various wavelengths of light, as well as to shadow), touch (a vine or a root can identify when it encounters a solid object) and sound (some plant roots can grow towards the sound of water or their parent plant). The capacity of plants to sense and react to their environment is called tropism. However, plants also display more complex behaviour than mere reactions to environmental stimuli. Plants can remember previous states in the environment via a special cell in the root tips (Thellier & Lüttge, 2013). They can recognise kin by sharing nutrients through the roots with offspring from the same mother plant, thus outcompeting other plants from the same species (Karban et al., 2013). Plants can further communicate through chemical signalling, for example warning nearby plants of mechanical trauma (e.g., an animal eating the plant) (Moreira et al., 2016).

In sum, plants have unique, diverse and specific capacities that can be designed for and with and we can take them seriously as subjects of our design. They have the same needs as most living organisms (nutrition, water, sunlight, oxygen/carbon dioxide etc.) and even driving forces such as spreading their seeds. By pointing at some of the broad range of capabilities which plants have, we illustrated their:

1) importance for design,
2) capacity for a deeper engagement with humans and
3) capabilities which might be relevant when designing with plants.

Plant Awareness Disparity has Consequences for Design

A central challenge for knowing plants is Plant Awareness Disparity (PAD), also known as "plant blindness" – an inevitable consequence of a human-centred worldview. The concept of "plant blindness" coined by Wandersee and Schussler (2001) suggests that there are three representations of plant blindness: (1) the inability to be aware of the critical role plants play in the planet and humans' everyday life (2) the failure to appreciate the beauty of plants as a life form (3) the placing of plants into an anthropocentric ranking that defines plants as inferior to animals and therefore not worth considering (Allen, 2003). Building on this work, Parsley (2020) articulated the more academic concept of Plant Awareness Disparity noting that it consists of four components:

- *Attention* (not noticing plants),
- *Attitude* (not liking plants),
- *Knowledge* (not understanding why plants are important),
- *Relative interest* (being less interested in plants than in animals).

If these are the key constituents of Plant Awareness Disparity, design research should examine the implication of all four facets in detail. We aim to do this in what follows.

Including Plants in Design Practice

If we want our methods to counter Plant Awareness Disparity and go beyond the anthropocentric point of view on plants in design, we must create designs that are better aligned with our ecologies. Therefore, we provide three modes of engagement with plants that include plants in empirical activities in modes of more-than-human design. Before discussing the modes, we want to clarify our approach to *Critical anthropocentrism* (Affifi, 2020). This approach recognises that finding "human" aspects in plants, such as sleeping or intelligence risks manifesting yet another form of anthropocentrism. At the same time, such language is an efficient way of communicating capacities that we humans understand to be somewhat like "sleeping" or "intelligence". In practice, we need familiar language to discuss the complex capacities of plants and take them seriously in design. We have no other choice than to understand the world from the position of being humans within it. As our modes will show, there are insights to be made by naming plants, bringing them to human contexts or ghostwriting as a plant on social media, i.e., temporary and consciously anthropomorphising plants to make certain discursive points.

Mode 1: Collective Fabulation with Plants

One mode of being with plants we have worked with is *Collective fabulation.* Fabulation is a fairly established approach in design (Nijs et al., 2020; Rosner, 2020; Søndergaard & Hansen, 2018) that utilises the creative capacities of storytelling, imagination and speculation to draw attention to phenomena of interest and triggers reflection and conversation about them emphasising radically different futures or ways of being (Haraway, 2016). Examples of design research that has focused on fabulating around plants include a robotic pot that seeks beneficial environmental conditions for a plant (Smith & Qaurooni, 2020) and three suggestions of four plausible worlds visualised with exemplary scenes where plants play a central role in design (Budinger & Heidmann, 2019). Formulating this mode, we add "collective" to emphasise the discursive benefits of fabulating together.

We illustrate the mode of collective fabulation through Chens' workshops series that involved drawing and writing exercises to know plants. The workshops explored if multimedia fabulation could trigger conversation and foster opportunities to explore human-plant relationships. Aiming to mitigate plant awareness disparity (Pany et al., 2022) the workshops propose various methods to re-see, re-connect, re-imagine and re-construct human and plant relations by integrating visual information and bodily experiences to form an immersive context for understanding. Chen reflects on the experience of conducting the first instance of the workshop:

> This workshop used drawing to uncover the hidden stories of everyday plants, Divided into four stages – preparation, greeting, drawing, and reflection – it took place in Miro, enabling global participation from Sweden, Germany, China, Japan, and the U.S. Participants brought photos of nearby plants and drawing tools. After sharing plant photos and introductions, they drew their plant's hidden story, considering its intrinsic values and roles in daily life. The drawings illustrated unique human-plant relationships, discussed collectively in a reflective session. The fabulating with plants through drawing and direct observation helped expand understanding and raise awareness of plants previously perceived in a more ambient position. Drawing enabled the articulation of unspoken narratives and emotions associated with the human-plant relationship. Starting the workshop with a photo of plants provided an intuitionistic impression for the participants to uncover sensory experiences, memories, and connections that might be challenging to express solely through written language. The participants often found it easier to convey subtle nuances, sentiments, and personal connections visually, transcending the limitations of verbal expression. However, there are limitations to only using drawing as storytelling material. Therefore, the final step in the workshop, the reflective session, provided a chance to communicate the fabulated story in text. This illustrates the benefit of including different media in the workshops to reveal the latent human-plant relationships.

The second part of the workshop series focused more on writing exercises. Chen reflects on this experience:

> I created a de-anthropocentric fabulation using informal text, freewriting, and epistolary novel techniques. Through the Instagram account @pfor_potato, posing as a potato influencer, I shared potato-centric views, self-portraits, manifestos, and aphorisms, fostering dialogue via Instagram stories. I then hosted a fictional storytelling workshop on Miro, focused on pushing boundaries on human-plant cohabitation. Half of the participants came from the

FIGURE 2.1 Participant reflections. Left image: "This is a small plant 'Elefantöra' (Elephant Ears) which I have by the windowsill. I bought this plant at a gardening store and sometimes I wonder what kinds of conversations, sounds or noise this plant heard or experienced in their 'past life'." Right image "This Peperomia plant is always helping me to remember when it's time to give my plants water. When this plant is 'sad' I've learned that all my plants are 'thirsty'. How does the water travel in this plant? It's a hidden story to the eye. When drawing plants I almost always use the colour green. That's partly why I chose this plant. Why is it red, or rather, why do I have such a predetermined mindset that plants are green?"

previous workshop, while others were new, with no pre-introduction to more-than-human concepts. This arrangement was intended to study how experience with more-than-human activities affects the participant's fabulation process. Thorough instructions on Miro facilitated the independent participation of the participants. This included three sets of words that provide time, place, and character, which could help the participants set a tone and background for their stories. Writers were prompted to select words from the list and add new words to inspire others. This virtual space provides an intriguing foundation for short fictional stories shared on @pfor_potato. The virtual space aimed to inspire imaginative short stories, resulting in works emphasising sensory engagement.

The following are some fabulations created in the virtual space:

> Story A: Inspired by the selected words: 'Museum, Potato, Famine'. As humans continue to encroach on agricultural land and forests for industrial development, wildfires like the one in California in 2020 occurred many times over the next century. Dust swept the world, plants were unable to survive. The last potato was frozen and shown in the museum. It tells the story of the planet's past to the new human who can survive on a charge.
>
> Story B: Inspired by the selected words: 'Sand, Museum, Potato'. It was named the Potato Festival, but what it meant was that the leadership tried to silence the museums. Each year, from April to October, sand was poured out on the museum floor. Not only in one museum but in all museums. Then potatoes were grown all over the place. Talks walks, and other activities were arranged, everything centering around the importance of potatoes. It was of course, interesting but it also meant that most of the year the museum staff had to work with potato-related culture, and nothing else.

This example shows how fabulation can be used to construct possible human-plant futures from both the potatoes' presumed view and the humans' view. Here, the perspective-shifting of narratives helps to reconstruct, recount, recall and resume the potential human-plant relationship. *Reconstruct* indicates existing social norms on how humans perceive the plants' existence, and 're' signifies the active process of rebuilding or reshaping the existing perceptions and understanding of the human-plant relationship. It implies a deliberate effort to revisit and transform the established societal norms and viewpoints regarding plants. *Recount* refers to the act of narrating or multimedia storytelling. When designing with plants, it suggests retelling stories, experiences, or histories related to plants, providing a fresh perspective that might challenge or complement conventional narratives. *Recall* implies a summoning process of memories or knowledge that possibly draws attention to forgotten or taken-for-granted aspects, fostering deeper connections. *Resume* refers to the act of continuing or picking up again. It suggests a continuation of the narrative, emphasising an ongoing and evolving relationship between humans and plants, rather than a static or fixed understanding.

In sum, the mode of collective fabulation particularly mitigates plant awareness disparity by working with people's *attitudes* (not liking plants) by for example asking the participants to notice plants and understand their full instrumental and intrinsic value during the reflective drawing process; *Relative interest* (being less interested in plants than in animals) by for example emphasising that a nearby plant could take centre stage in a fabulated story; and *Knowledge* (not understanding why plants are important), by actively researching biological, historical and sociological facts about their story's main character (potato, apple, willow, etc.) to enrich the storyline.

Mode 2: Critical Companionship with Plants

Our second mode of being with plants is connected to experiencing plants as our companions in human activities through *Critical companionship*. We often encounter plants in homes, flower beds or forests – but what happens if we bring a plant to a conference? By enabling the participation of plants in a novel context, we can learn more about the nature of participation, and our own and community's relationship to plants, and highlight the topic of more-than-human design. By participation we mean that the plant is physically moved into a novel context and takes part in it in its own ways (such as consuming carbon dioxide and producing oxygen); however, it does not take part in the context intentionally, as a consequence of its own decision.

Inspired by Donna Haraway's (2003) take on companions, and in line with an embodied view of knowledge, Cerna visited two conferences focused on human-computer interaction and brought with her a small plant, Vratislav (NordiCHI, 2022) and Stanislav (CHI, 2023). In the following excerpt, Cerna reflects on an experience that became the starting point for a more intentional exploration of companionship with plants at conferences – the moment that she received a plant as a gift and decided to name the plan:

> When I visited one of my closest friends, just before I left her apartment, she gifted me a tiny money plant. I was not surprised, as exchanging plants is a common practice among my friends. I asked jokingly "Is there a name?" and she said "Stanislav".
>
> The moment when I received the plant/Stanislav, though seemingly so small, is of high significance to those who took part in it. My friend knows I love plants and this was an expression of her knowing me; by giving me this particular gift she knew it would make me happy. Thus, sharing resources is an expression of our small community of friends, who care for sustainability and who got used to sharing their belongings (instead of throwing them away). Finally, and here quite crucially, the name "Stanislav" can be found both in Polish (the native language of my friend) and Czech (my native language). It refers to our common cultural heritage, something we often jokingly refer to that has had an actual impact on us: being part of a Slavic nation and the same political unit (Soviet Union). Vratislav (my second plant) is also a money plant and I chose a Czech name with the same root, suggesting their relatedness. So even though I did not receive Vratislav as a gift, he is connected to Stanislav through their common story of naming them. The names additionally came with a gender. Money plants reproduce asexually, i.e., the plants do not have a particular sex. In the Slavic languages, every word is gendered; Stanislav is a man's name. I choose to keep the name related to gender. Though I (correctly) suspected that the majority of the conference participants speak English I will not have an intuitive understanding of this context, I decided to keep the male gender for the following reasons: when talking about them, it emphasises the agency that resides in the plant side off the relationship (in contrast to calling him "it" as an object), this aligns with more-than-human approaches of bringing marginalised agency to the front. In addition, other conference participants would often inquire about the reasons for his gender and it would allow us to dive into a concrete discussion about more-than-human design.

Cerna reflects on the practice of carrying around a companion plant at a conference:

> Carrying a plant around is an atypical activity at conferences. I often held Stanislav or Vratislav in my hand during coffee breaks or when moving between rooms, conference dinners or getting a beer in the evening. Their presence prompted others to approach me and inquire about "my plant". I often set the scene by stating "This is Stanislav/Vratislav, my participant" and continued explaining how they connect to more-than-human design and my design work. These conversations sometimes led to deeper discussions connected to plants in design, such

> as if they are participants, living beings or materials in design; and the meaning of plants in pots we keep at home. As the conference progressed, caring for Stanislav/Vratislav became a collaborative effort. Many started to recognise him and inquired about how he was or how he enjoyed the conference, either to me or directly to him (though I always provided the answer). Eventually, if I did not have Stanislav in my hand, others would inquire about where he was. As it was difficult to constantly manipulate both him and other objects relevant to conference attendance (my phone, coffee mug, laptop, bag etc.), I often left him standing on a table and asked the closest person standing by to keep an eye on him or hold him. I could also observe changes in myself – during the five days, holding the pot with Stanislav in my hand became a habit, and I would sometimes tap one of the leaves as a type of calming behaviour.

The accounts above show how Stanislav and Vratislav gained a new symbolic meaning – they represent a nexus of different contexts coming together, opening a space for novel insights. By engaging with plants through a critical companionship, we make the first step to learning to pay attention to them more substantially, as we need to choose which plant to bring with us in the first place – and how. By bringing plants into a broader social context, curiosity is awakened in other people. However, bringing a plant into a novel social context does not automatically mean we will positively change our attitude towards plants. This form of engagement might involve a range of experiences, such as frustration, boredom or tiredness but also joyful moments and interesting conversation. However small, it will create a new way to relate to plants. We call this form of relating "critical companionship" (Ziewitz & Singh, 2021), which originally comes from nursing research and builds on mutually constituting learning trajectories. The plant did not have to learn to be with her companion at the conference. Instead, the human companion had to learn how to be with Stanislav and Vratislav in much more active ways than if they were "just" standing on her window. A new way of being with them has led Cerna to have a stronger relationship with her plants, not only as a consequence of gaining new biological knowledge but also because she experienced a special situation with the plants.

To sum up, the mode of critical companionship particularly mitigates plant awareness disparity by working with peoples' *attitude* (not liking plants) by, for example, showing that plants have a role as a companion; *relative interest* (being less interested in plants than in animals) by for example highlighting the advantages of having a plant as a companion to a community event; and *knowledge* (not understanding why plants are important) by for example having the plant as a conversation starter for conversations around the plant and more-than-human design importance.

Mode 3: Cohabitation with Plants

The third mode we discuss focuses on living and designing with plants through *Cohabitation*. Cohabitation with plants is beginning to be established as a mode for design in design research (Bardzell et al., 2021; Keune, 2021; Tomico et al., 2023) and requires sharing space and resources in a mutually beneficial manner and creates opportunities for mutualism, symbiosis and companionship. Cohabitation with plants approaches worldmaking – collectively designing emerging habitats – as collaborative survival (Liu et al., 2018), recognising the mutual interdependence of humans and plants. The purpose of cohabiting is to unpack social-ecological-technological systems from a critical and ethical perspective while looking for alternative ways of designing – rather than testing or validating hypotheses. Such alternatives include life-affirming ecologies of practice that have the potential to support biodiversity, maintain caring, and develop new more-than-human sensitivities and repertoires (Oogjes & Wakkary, 2022) for designing with plants.

Living and designing with plants through cohabitation is further a form of first-person design research (Desjardins et al., 2021) that explicitly supports non-representational research (Ulmer, 2017) i.e., research that instead of studying and representing social relationships, focuses on practices – how human and nonhuman formations are enacted or performed through events, relations, affect, practices and backgrounds (Vannini, 2015). This allows us to experience the hybridity (Smith et al., 2017) existing in our natureculture (Haraway, 2003) through a myriad of embodied & situated ways of acting, looking at things, and being involved. For example, through embodied design ideation, multispecies autoethnography or autobiographical design (see also Figure 2.3).

As most of us have plants in our homes, living and designing with plants through cohabitation is at reach for almost everyone. When the number of plants reaches a critical mass, we could consider the space that we are living in a cohabitation lab. A cohabitation lab is both a design outcome and an emergent process. In the case described below, Tomico documented frictions, surprises, ideas and activities relevant to him while setting up a cohabitation lab (Tomico et al., 2023). This documentation practice helped him to support his design goal of greening interior spaces through multispecies design. But more importantly, it was used as a material to think with, to continuously question the notion of a home and how he wanted to live in it, in line with the sensitising concept of "wanting to live here" (Bardzell et al., 2021). In practical terms, the cohabitation lab was a flat in Barcelona with more than 250 plants, IoT infrastructure (an assemblage of smart lighting, humidity and watering systems), furniture and decorative art pieces (see Figure 2.2).

The following excerpts comprise the last five reflection notes taken before August 2022, when the research was paused because of moving to another country. These reflection notes are

FIGURE 2.2 Tomico's home design studio – or cohabitation lab. Photos show the terrace, living room and studio where creating environments for the plants to thrive reshapes the space as they grow, reproduce and decay.

selected because they focused on foreseeing the implications of his actions and noticing the life cycles he was part of concerning growth, reproduction, decay and death.

> Reflection 42: *More circular material flows emerge as you reach a critical mass of plants.* Repotting becomes less resource intensive as pots, soil and other support structures can be reused from plants that have died, overgrown their pots or from unused stored things. Repotting in spring means reorganising and replacing. Chains of actions are created. Multiple pots are exchanged in sequence. Plants are combined. Cuttings are planted. The soil used is a mix of exceeding old soil from existing plants and new ones.
>
> Reflection 43: *Unforeseen double uses.* Watching my phone in bed at night attracted small insects to my screen. I thus realised that there was an increase of soil bugs in the plants around. A mobile phone screen indirectly helped me to realise that I'm overwatering the plants and that root rot is attracting soil bugs.
>
> Reflection 44: *Propagation and pruning, beauty and growth in circularity.* Timing pruning in early spring with propagation and repotting in late summer allowed the transformation of overgrown branches into cuttings that can be repotted as new plants with the mother plant. Finding a space for propagation was difficult, especially with small branches that require a lot of humidity. Long-pruned branches are easy to propagate if used as a decorative element in flower vases during the transition months.
>
> Reflection 45: *When do we know a plant is thriving?* After two years and multiple relocations, a hibiscus is flowering for the first time. How important is flowering as a variable related to flourishing? Is it more important than the size of the leaves or the time leaves stay on the plant? What other ways of noticing am I learning while sharing a space with plants in some kind of symbiotic relation?
>
> Reflection 46: *Home chores and plant cycles.* There are cycles for pruning, watering, feeding and repotting. These daily, monthly, or yearly cycles vary with the seasons, weather, and stage of development. I must continuously iterate on habits and patterns that are part of my everyday life to be coherent for the plants. The biweekly watering done by hand in summer to complement the automated watering systems is combined with adding fertiliser to the soil and cleaning the floor around the pots. In winter, complimentary watering is done once a month, which means that fertiliser and cleaning happen every two months. A more than appreciated slowing down.

As these quotes illustrate, sharing space and resources with more than 250 plants in Tomico's home, allowed him to pay attention to circular material flows and create chains of action to reuse and reduce resources while repotting. It helped him to explore unexpected multispecies relations like combining watering plants and cleaning the floor, or using propagating cuttings as decorative bouquets. His attitude towards designing with plants changed as he became more attuned with growth and decay patterns. He began noticing changes in how plants flowered and leaves developed and became more alert to water dampening and root rot. *Designing became a never-ending continuous process* and the action of *prototyping became distributed and iterative*, tuned into the plant's life cycles. Understanding the interrelatedness of more-than-human care practices was forged by experiencing how taking care of plants helps to take care of yourself. The relative interest in plants of other people around him increased with the sharing of cuttings and the collaborative diagnosis of why plants display symptoms of not being well in 'plant cliniques'.

In sum, cohabitation with plants supports a more-than-human design approach by decentering the human, highlighting what we are living in and around – and designing as part of relational infrastructural arrangements (Puig de la Bellacasa, 2016) based on interdependence and cooperation. The multispecies collaborations emerging from cohabiting with plants in an ecosystem are based on continuous engagement and attunement to (mundane) everyday actions. They can be considered a series of symbiotic encounters (Liu et al., 2019) through which more-than-human inter- and intra-actions are constantly renegotiated, foregrounding coadaptation and codependency (Ávila, 2022). For designers, cohabitation mitigates plant awareness disparity by increasing *knowledge* and generating a complex systemic understanding of what is needed for an ecosystem to thrive. This involves working with your *attention* by living in the ecosystem you are designing with, allowing for experiencing what it means to design for a shared aliveness. This frames designing *with* as a form of giving back rather than taking from. Cohabitation labs can increase biodiversity by greening the spaces we inhabit, breaking once and forever the imaginary plant and human divide. Human dwellings can become open and welcoming habitats, transforming cities into biodiversity sanctuaries.

Conclusion

We initiated the reflections of this chapter in a growing body of research that points to the complex capacities of plants. Plants might seem "passive" because they do not manifest all the same features as animals (locomotion, etc.). However, as cornerstones of our ecosystems, they are active in global entanglement. This implies that we should consider plants as relevant actors in our design process, who have unique, diverse and specific capacities that we can design for and with – that we need to learn to engage with more in-depth.

We discussed the meaning and consequences of plant awareness disparity in design processes. We argued for the need to foster more-than-human design by developing new ways of *being with* plants. We suggested three modes of alternative plant engagement that go beyond framing plants as only material or food. As illustrated in Figure 2.3, each mode reflects an increased closeness to the plant, *collective fabulating* being the most distant or speculative, not necessarily even requiring the presence of a plant, *critical companionship* introducing actual individual plants in novel social settings, thus including them in a closer social sphere – while *cohabiting* involves closely integrating the plants into one's lifeworld. The modes may fit different goals. Collective fabulating and critical companionship imagine new opportunities and aim to transform the thinking of people (attention, attitude and relative interest) while cohabiting more closely focuses on observing plant needs (knowledge) – while the performative aspects of cohabiting are also a way of structuring life in a partly non-normative way. Figure 2.3 additionally illustrates how the cohabitation mode utilised a variety of sensors to generate knowledge about the plants cohabited with. We believe that this combination of modes and more concrete tools and methods can inspire designers to work creatively with how they approach plants in the design process.

In sum, we have argued for the need to go beyond the traditional Western use of plants as only "material" and wonder more about a novel way of *being with plants*. Such approaches are critical if we are to mitigate extractivism and its catastrophic consequences for the environment. We hope that the reader may further reflect on their being with plants by considering the following questions:

- How can we foster deeper relationships with plants as part of collaborative design?
- What hinders or enables the consideration of plants in design processes?

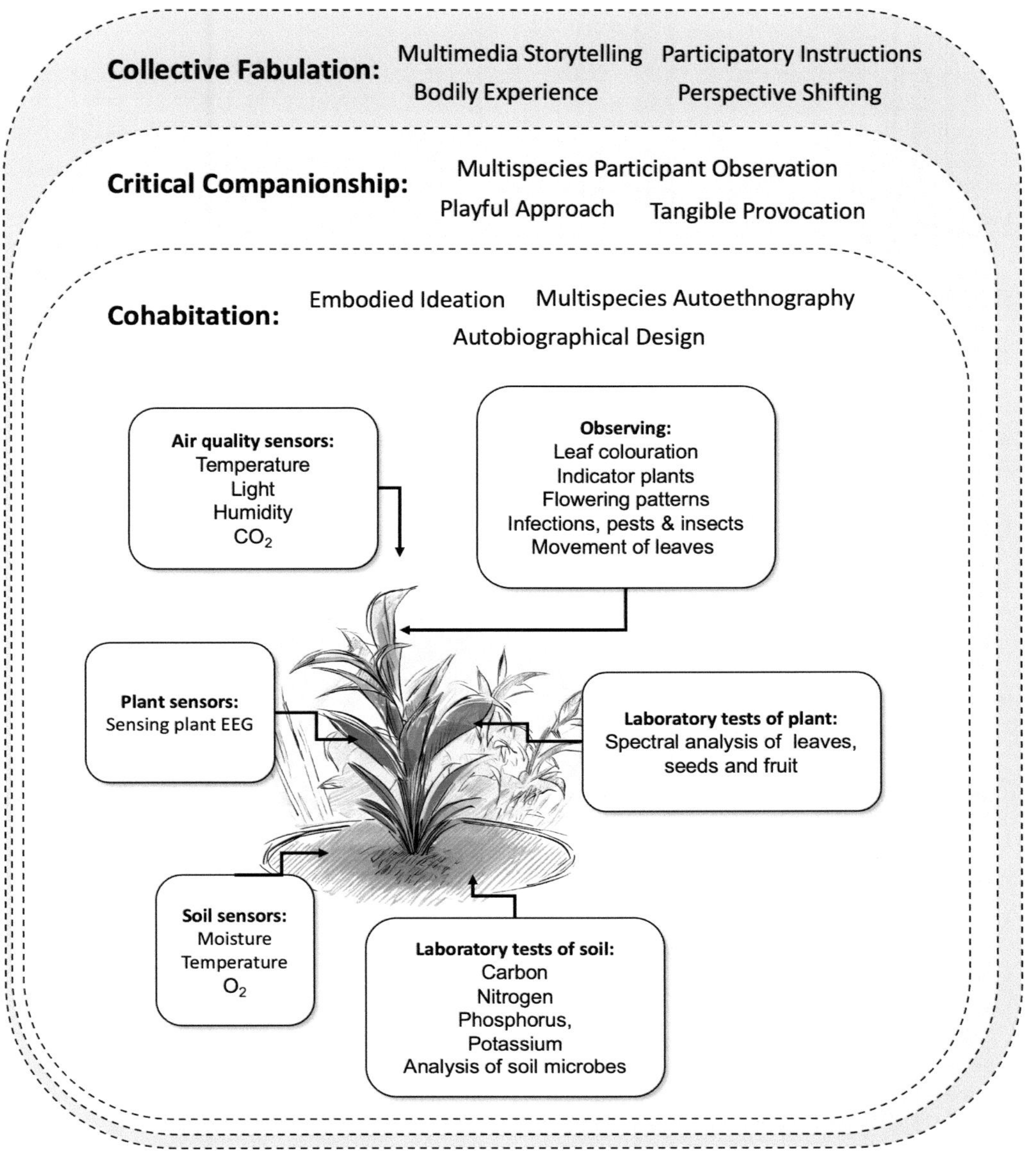

FIGURE 2.3 Ways of knowing plants. The three modes of *Collective fabulation*, *Critical companionship* and *Cohabitation* are supported by concrete methods, techniques and tools for knowing plants.

References

Allen, W. (2003). Plant blindness. *BioScience*, *53*(10), 926.

Armstrong, J., Lakind, A., Adsit-Morris, C., & Saeter, R. (Eds) (2021). *Becoming-Feral*. Objet-A Creative Studio.

Bardzell, J., Bardzell, S., & Light, A. (2021). Wanting to live here: Design after anthropocentric functionalism. *Proceedings of the 2021 CHI Conference on Human Factors in Computing Systems*, 1–24.

Bos, A. P. (2010). Aristotle on God as principle of genesis. *British Journal for the History of Philosophy*, *18*(3), 363–377.

Budinger, K., & Heidmann, F. (2019). Our symbiotic life: An exploration of interspecies relations. *Proceedings of the 2019 on Designing Interactive Systems Conference*, 1349–1362.

Desjardins, A., Tomico, O., Lucero, A., Cecchinato, M. E., & Neustaedter, C. (2021). Introduction to the Special Issue on First-Person Methods in HCI. *ACM Transactions on Computer-Human Interaction*, *28*(6), 37:1–37:12.

Fastnacht, T., Aispuro, A. O., Marschall, J., Fischer, P. T., Zierold, S., & Hornecker, E. (2016). Sonnengarten: Urban light installation with human-plant interaction. *Proceedings of the 2016 ACM International Joint Conference on Pervasive and Ubiquitous Computing: Adjunct*, 53–56.

Haraway, D. (2003). *The Companion Species Manifesto: Dogs, People, and Significant Otherness* (Vol. 1). Prickly Paradigm Press.

Jackson, M. M., Byrne, C., Freil, L., Valentin, G., Zuerndorfer, J., Zeagler, C., Logas, J., Gilliland, S., Rapoport, A., Sun, S., Peet, D., Lau, A., Han, X., Alcaidinho, J., & Starner, T. (2018). Technology for working dogs. *Proceedings of the Fifth International Conference on Animal-Computer Interaction*, 1–5.

Karban, R., Shiojiri, K., Ishizaki, S., Wetzel, W. C., & Evans, R. Y. (2013). Kin recognition affects plant communication and defence. *Proceedings of the Royal Society B: Biological Sciences*, *280*(1756), 20123062.

Keune, S. (2021). Designing and Living with Organisms Weaving Entangled Worlds as Doing Multispecies Philosophy. *Journal of Textile Design Research and Practice*, *9*(1), 9–30.

Kleinberger, R., Cunha, J., Vemuri, M. M., & Hirskyj-Douglas, I. (2023). Birds of a feather video-flock together: Design and evaluation of an agency-based parrot-to-parrot video-calling system for interspecies ethical enrichment. *Proceedings of the 2023 CHI Conference on Human Factors in Computing Systems*, 1–16.

Kohn, E. (2013). *How Forests Think: Toward an Anthropology Beyond the Human*. Univ. of California Press.

Liu, J., Byrne, D., & Devendorf, L. (2018). Design for collaborative survival: An inquiry into human-fungi relationships. *Proceedings of the 2018 CHI Conference on Human Factors in Computing Systems*, 40:1–40:13.

Loh, S., Foth, M., & Santo, Y. (2024). The more-than-human turn in human-plant interaction design: From utilitarian object to living co-inhabitant. *International Journal of Human-Computer Studies*, *181*, 103128.

McClements, D. J., & Grossmann, L. (2021). A brief review of the science behind the design of healthy and sustainable plant-based foods. *Npj Science of Food*, *5*(1), Article 1.

Messing, J. (2001). Do plants have more genes than humans? *Trends in Plant Science*, *6*(5), 195–196.

Moreira, X., Petry, W. K., Hernández-Cumplido, J., Morelon, S., & Benrey, B. (2016). Plant defence responses to volatile alert signals are population-specific. *Oikos*, *125*(7), 950–956.

Nijs, G., Laki, G., Houlstan, R., Slizewicz, G., & Laureyssens, T. (2020). Fostering more-than-human imaginaries: Introducing DIY speculative fabulation in civic HCI. *Proceedings of the 11th Nordic Conference on Human-Computer Interaction: Shaping Experiences, Shaping Society*, 1–12.

Oogjes, D., & Wakkary, R. (2022). Weaving stories: Toward repertoires for designing things. *Proceedings of the 2022 CHI Conference on Human Factors in Computing Systems*, 1–21.

Pany, P., Meier, F. D., Dünser, B., Yanagida, T., Kiehn, M., & Möller, A. (2022). Measuring students' plant awareness: A prerequisite for effective botany education. *Journal of Biological Education*, *0*(0), 1–14.

Parsley, K. M. (2020). Plant awareness disparity: A case for renaming plant blindness. *PLANTS, PEOPLE, PLANET*, *2*(6), 598–601.

Puig de la Bellacasa, M. (2016). *Ecological Thinking, Material Spirituality, and the Poetics of Infrastructure*.

Rosner, D. (2020). *Critical Fabulations*. MIT Press. https://mitpress.mit.edu/9780262542685/critical-fabulations/

Sandelin, E. (2024). *Design and Grace: An Ahuman Odyssey* KTH, School of Architecture and the Built Environment (ABE), Architecture. Konstfack, Art, Technology and Design. ORCID iD: 0000–0002–1601–6264

Sanders, D. (2019). Trapped in time: Lingering with “plantness”. *Plants, People, Planet*, *1*(2), 64–66.
Schussler, E. E., & Wandersee, J. H. (2001). Toward a theory of plant blindness. *Plant Science Bulletin*, *47*(1),), 2–9.
Sheikh, H., Gonsalves, K., & Foth, M. (2021). Plant(e)tecture: Towards a multispecies media architecture framework for amplifying plant agencies. In *Media Architecture Biennale 20* (pp. 87–99). Association for Computing Machinery.
Smith, N., Bardzell, S., & Bardzell, J. (2017). Designing for cohabitation: Naturecultures, hybrids, and decentering the human in design. *Proceedings of the 2017 CHI Conference on Human Factors in Computing Systems*, 1714–1725.
Smith, N., & Qaurooni, D. (2020). The inclusion zone: Grounded speculations in Chernobyl. *Proceedings of the 2020 ACM Designing Interactive Systems Conference*, 1929–1942.
Søndergaard, M. L. J., & Hansen, L. K. (2018). Intimate futures: Staying with the trouble of digital personal assistants through design fiction. *Proceedings of the 2018 Designing Interactive Systems Conference*, 869–880. https://doi.org/10.1145/3196709.3196766
Steer, C., Robinson, S., & Jones, M. (2015). Growth, change and decay: Plants and interaction possibilities. *Proceedings of the 33rd Annual ACM Conference Extended Abstracts on Human Factors in Computing Systems*, 2037–2042.
Thellier, M., & Lüttge, U. (2013). Plant memory: A tentative model. *Plant Biology*, *15*(1), 1–12.
Tomico, O., Wakkary, R., & Andersen, K. (2023). Living-with and designing-with plants. *Interactions*, *30*(1), 30–34.
Ulmer, J. B. (2017). Posthumanism as research methodology: Inquiry in the Anthropocene. *International Journal of Qualitative Studies in Education*, *30*(9), 832–848.
Vannini, P. (2015). *Non-Representational Methodologies: Re-Envisioning Research*. Routledge.
Ziewitz, M., & Singh, R. (2021). Critical companionship: Some sensibilities for studying the lived experience of data subjects. *Big Data & Society*, *8*(2), 20539517211061122.

3

BIOMENSTRUAL

Designing with the More-Than-Human Body

Nadia Campo Woytuk and Marie Louise Juul Søndergaard

Introduction

The human body is a vibrant site for more-than-human design. From the bacteria in our guts to the food we eat, to the water we drink and the air we breathe, our human bodies entangle a myriad of more-than-human agencies across multiple scales. Posthumanism typically foregrounds these entanglements, expanding our understanding of 'what it means to be human' to include things like technologies, the environment and plants or animals, among other examples that can be found in this book. This has led designers to shift their focus in the design processes from solely human needs towards including the needs of the rest of the more-than-human world (Forlano, 2017b; Wakkary, 2021). Thus, the purpose of more-than-human design is not necessarily only to remove the human from design, but to recognise and emphasise the ways that humans are already entangled in more-than-human worlds – as Jasmine Ulmer reminds us, "how we live, eat, drink, breathe, commute, hear, see, smell, touch, sense, and experience life are inextricable from our local ecologies" (Ulmer, 2017).

In this chapter, we present a case for how *menstruating* is inextricable from local ecologies. Menstruation and menstruating bodies are particularly potent examples of how our bodies are entangled with the rest of the world. Menstruating bodies are *leaky*, thus challenging the boundaries of the human body in a very obvious and "material" way. Viewed under a posthuman lens, menstruation challenges anthropocentric views of the body as a single contained individual with clear bodily boundaries. The human body is thus reframed as a 'more-than-human body', a body that shapes and is shaped by more-than-human worlds. The case we present, Biomenstrual, is an example of how to design with the more-than-human body, both materially and conceptually (Søndergaard & Campo Woytuk, 2023). The design project Biomenstrual is made up of a variety of artifacts and involves diverse methods, including material explorations with biomaterials and local species like moss, prototypes of menstrual pads, collages, recipes, 3D printed tools, speculative and alternative practices of menstrual care, and participatory design workshops (see Figure 3.1).

Given the imperative of addressing how design contributes and is complicit in maintaining the current state of social injustice and climate emergency (Escobar, 2018), the Biomenstrual project is deeply concerned with challenging dominant unsustainable practices of menstrual care.

DOI: 10.4324/9781003467731-5

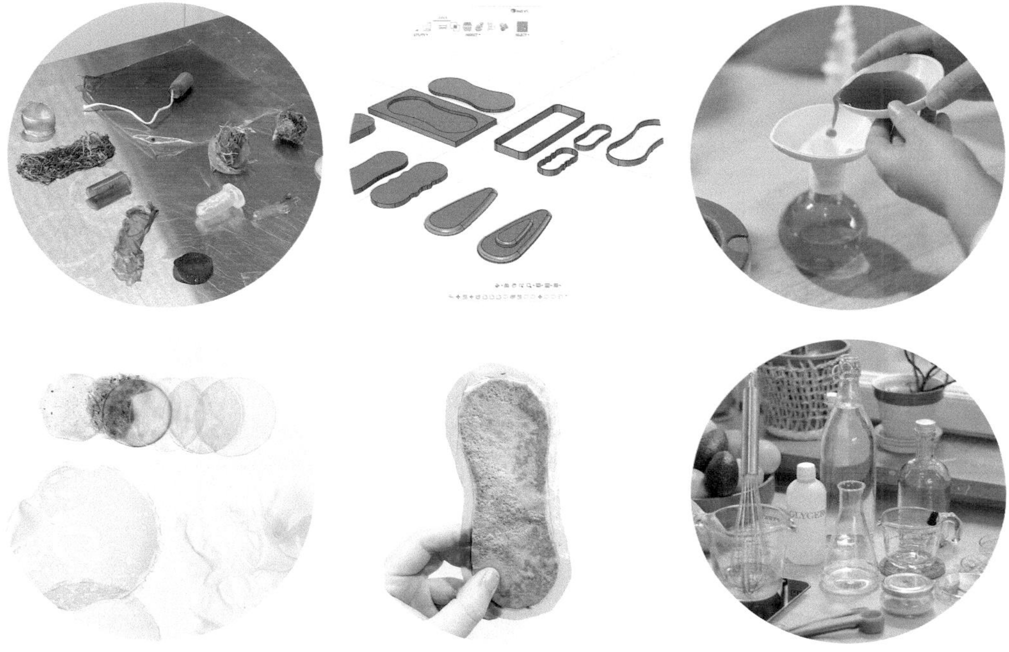

FIGURE 3.1 The Biomenstrual project (from left to right, top to bottom): Biodegradable menstrual artifacts made in a participatory workshop, 3D modelling moulds for menstrual pads, a prototype of a menstrual blood container, bioplastic samples, a biodegradable menstrual pad made with wheat gluten and agar bioplastic, and a spread of kitchen tools used for making bioplastics, which intend to reframe current menstrual care practices as Do-It-Yourself practices.

At the same time, we take an explicit feminist stance to counter patriarchal oppression on menstruating bodies, actively avoiding perpetuating shame and stigma experienced by menstruators worldwide. In Biomenstrual, we draw from posthuman feminism (Braidotti, 2022) to convey the urgency of addressing menstrual health from both a more-than-human and feminist lens, acknowledging how the struggles of patriarchal and capitalist oppression go hand in hand.

As we have seen and will see throughout this book, more-than-human design acknowledges that we cannot fully remove a human perspective, including the human body, positionality and identity. Yet instead, we can bring different ways of knowing into design, beyond human knowledge, and also highlight the existing interdependence of humans with the rest of the more-than-human world (Ávila, 2022). In taking a more-than-human design perspective, we provide an alternative to how designers can work critically and mindfully with the human body and open up for how designers already centring human needs might also extend and connect their practice to the needs of more-than-human species and worlds.

While it is not only humans that menstruate (other animal species that menstruate include some dogs, apes and bats), typically menstruation is considered an aspect of human health. Products designed for menstruation target human needs; needs for collecting or absorbing the menstrual fluids so it is possible to move around and do everyday activities. We use menstruation to showcase how the human body is already entangled in more-than-human worlds, how we can design for more caring ways of our bodies becoming part of the environment how – importantly – the body, both of the human and the designer, is not merely a neutral space but a human corporeality that interacts with and is shaped by the more-than-human world based on its intersecting identities (Homewood et al., 2021). We will start by showing existing examples of designing with 'the more-than-human body' as a way to introduce this concept. We then present the motivation and design process behind the case of Biomenstrual. Finally, we discuss the feminist posthumanist theoretical concepts that we have worked with, including Donna Haraway's figure of the "cyborg" and Stacy Alaimo's concept of "trans-corporeality", and give open-ended prompts on how to design with the more-than-human body.

The More-Than-Human Body in Design

While not all explicitly feminist, or even more-than-human, we see many examples of involving the human body in designs that relate with more-than-human worlds. Examples by authors in this book include Danielle Wilde's work on designing sustainable food practices that promote a healthy gut microbiome, also for people who struggle with serious chronic gut diseases (Wilde, 2022). The project shows that even when considering "only" the human body, we are already dealing with a rich multi-species nature within our bodies, which is continuously shaped by what enters and leaves our body. Boer et al. further explores this relation of our gut and brain through the design of a device that supports the self-tracking and cultivation of gut microbiome (Boer et al., 2020). Also, the vaginal microbiome has been explored through design work, such as by Giulia Tomasello in "Future Flora" (2016) and Serina Tarkhanian in "Insurgent Skinships" (2023). These are examples of how even the most intimate of the human body challenges assumptions of an individual contained human. While personal, these materialities of human corporality, or non-human microbiomes hosted in/on our human body, are far from private: they extend beyond and become transformed by local and global exchanges of (non)human bodies. Karey Helms' design work with breastfeeding experiences shows how healthy bacteria is exchanged between mother and child, and her biodesign material explorations of proteins found in human milk and cow milk questions how we relate to human-animal milk production (Helms, 2021). Similarly reflecting on human-cow relations of reproduction, the "Happy Cow" project by Ece Tan (2022) critically speculates on how milking, examination and insemination tools could be designed for cows to have more dignified and pleasurable experiences in industrial farming, with inspiration from human sex toys; again pointing to the corporeality of more-than-human bodies and what non-human bodies leak through our porous skin or are physically inserted in our animal body. Through drawing on her long-term experience with adapting and living with networked and AI-augmented insulin pumps and glucose monitors to manage Type 1 diabetes, Laura Forlano has extensively theorised cyborgian human-technology relations (Forlano, 2017a), which can be considered yet another example of the more-than-human body: the human body as always and already shaped by technology and science. Likewise, projects have explored how human bodies are always and constantly shaped by the environment, including environmental hazards such as air pollution, which might be sensed by digital technologies like, for instance, in the AirKit (Mahajan et al., 2021). Lastly, we see examples of the human body being used for embodied and sometimes performative ways of relating with non-human bodies, such as water, mushrooms and rocks. As a form of "embodied environmental speculation", Heidi Biggs designed an interactive pair of pants that would change in length to map future sea level rises directly on a person's body while biking around an area that will likely be under water in the future (Biggs & Desjardins, 2020). Jen Liu designed a wearable finger sensor that requires the wearer to physically insert their fingers into soil – and get their hands, and tech, dirty – to get a digital moisture measurement; hereby bringing the body into play in human-fungi-technology relationships (Liu et al., 2018). And Petra Lilja has used body performances, including practices of walking and drawing with minerals on the human body, as embodied ways of relating with rocks, fossils and minerals, in order to "embody an understanding of the common ground between the bodies of the human and the mineral" (Lilja, 2021). These multifaceted and diverse examples show the scale with which we can design for the more-than-human body; from the smallest scale of bacteria in the gut microbiome, to technologies that go under our skin, and performative experimentations of relating our bodies with local ecologies and changing environments.

Case: Biomenstrual

Biomenstrual is a design project where practice and theory mingle. Through material explorations, physical prototypes, dissemination material, collages, exhibitions, talks and workshops, the project bridges menstrual care, sustainability and feminist posthuman theory. Biomenstrual began in 2021 and involved many discussions, experiments and sketching sessions with multidisciplinary collaborators. The project splits into various threads, some involving material scientists, some involving feminist and more-than-human design researchers, some involving design students, artists and people bursting with curiosity.

We, Nadia and Marie Louise, feminist interaction designers, spent almost two years together discussing, wondering, cooking biomaterials and carefully assembling exhibitions and workshops. As often happens in feminist research, our own personal lives and experiences have been entangled in the project. Our standpoint as menstruating women afforded us access to our own bodily experiences, knowledge about the menstruating body and the ability to *imagine* otherwise (Stoetzler & Yuval-Davis, 2002). We view our research contributions through a feminist epistemological lens, where acknowledging the *situatedness* of the knowledge generated through the research is vital (Haraway, 1988). The research questions and design artifacts in Biomenstrual did not appear out of thin air: design decisions happened based on our own menstrual experiences, our and our society's values and associations with menstrual blood, the availability of local materials or the audiences and locations where we shared the work (Campo Woytuk & Søndergaard, 2022).

The project began in Stockholm, Sweden, and continued across the Nordics following Marie Louise's move to the west coast of Denmark and eventually to Oslo, Norway. Our remote collaboration during the Covid-19 pandemic consisted of many video calls in our kitchens-turned-into-labs, mail correspondence, and visits to Stockholm. Across the two years we exhibited the project in three locations: a solo exhibition in the co-designed public square Frihamstorget in Stockholm, a curated menstrual art exhibition "*MENS!*" in Halmstad, Sweden, and participation in the exhibition "*Remedy: counter responses to the design of care*" at the Oslo School of Architecture and Design, Norway.

The motivation for this project was twofold: we wanted to reframe menstrual blood as something valuable, and we wanted to design alternatives to single-use menstrual products that contain high quantities of plastics. We will talk a bit about the first now.

Motivation: Fertile Menstrual Blood and Polluting Menstrual Products

Menstrual blood, also called *menses*, is the blood and tissue expelled by the uterus during the bleeding phase of the menstrual cycle. Menstrual blood is a particularly interesting bodily fluid; it is stigmatised and "gendered", often associated with dirt and disgust, defying the boundaries of femininity when it is not concealed (Persdotter, 2022). Dominant narratives in Western societies adopt this perspective of 'menstrual concealment', therefore contributing to how we manage, and have managed in the past, our menstrual experiences (Vostral, 2008). Whether it's about blood leaking, mood swings caused by premenstrual syndrome or painful cramps, menstruation is framed as something to be ashamed of, something to hide. Furthermore, scholars have pointed out how menstruation in design often gets framed as 'a problem to solve', further pushing menstruation, and in particular, menstrual blood, into the category of useless things, of waste (Homewood, 2018; Søndergaard, 2018). Menstrual products (products that are used for absorbing, collecting, or disposing of menstrual blood, for instance tampons, sanitary pads, menstrual cups, or period-absorbent underwear) are complicit in furthering these stigmas. The corporeal function of products is often concealed under terms like "feminine/female" or "hygiene/sanitary" products, which reproduces ideas of menstrual blood as dirty and unsanitary. Hence, *menstruating*, and particularly, *menstrual blood* is situated within a problematic and ongoing history of stigmatization, casting aside menstruating bodies as disgusting, which contribute to shame and harm worldwide.

In our project, we looked at how menstrual blood is involved in more-than-human entanglements. Bodily fluids, such as saliva, urine, sweat, menstrual blood or breastmilk, blur the boundaries of the human body. The question of where the human bodily boundaries lie has been particularly pondered by the feminist posthumanities, many inspired by Donna Haraway's question: "Why should our bodies end at the skin?" (Haraway, 1987). This question fosters reflection on the essence of being human – where are the boundaries between 'the self' and 'the other'? Where 'the other' might be another human, a technology, an animal, a plant, the soil, the sea, the air, or the planet itself. Particularly, menstrual blood is an example of a bodily material that feminist scholar Julia Kristeva has called 'abject' – neither 'part of me' nor 'part of something else' (Kristeva, 1982). In its very material essence, menstrual blood, with its clumps of tissues and cells expelled from the human uterus, leaves the body and becomes something else, no longer part of that original body. This was one of the starting points for Biomenstrual: what happens to the blood after it has left our bodies? What other species, technologies and ecosystems does it 'touch'? What if it provides nourishment to others, rather than being considered a waste?

Having collaborated in working with intimate technologies previously, we were curiously drawn to this material experience of menstruation and how we might design with and around this fluid. Menstrual blood has a long history of featuring in activist art and feminist scholarship, where creators have advocated for reframing and reclaiming its value. Feminist artists and activists have used menstrual blood as a medium for their art, incorporating it as ink or paint, embedding it into resins and textiles. This particular material engagement was familiar to us – as menstruators and menstrual cup users ourselves, we were used to seeing and touching our own menstrual blood every month or so – yet it opened up for new and unfamiliar experiences and opportunities to learn about our bodies and how they interact with the world around us.

One such material practice we observed was using menstrual blood, often collected with a menstrual cup, as a fertilizer for houseplants and home gardens. Navigating a diverse panorama of online communities, from feminist activists to 'menstrual coaching' influencers to menstrual product brand accounts, we saw that many people found this act both an empowering way to reframe the value of menstrual blood and a practical way to supply nutrients to their wilting monsteras. Making their way amidst misogynistic insults and reactions of visceral disgust, we saw that people on social media ask for guidance and share tips on how to gather their blood, dilute it and pour it into their houseplants. Upon further investigation, we found that, unsurprisingly, little scientific research has been conducted on determining the qualities of menstrual blood that are nutritious for plants. At the same time, “bloodmeal” is a well-documented and commodified fertilizer consisting of dried pig’s blood, which provides nitrogen to the soil. Why was a pig’s blood a very common and socially accepted fertilizer, yet a human’s menstrual blood was not? We reflected on how the use of pig’s blood reinforces harmful logics of the human as the dominant species and speculated that the lack of scientific knowledge in the topic of human menstrual blood as fertilizer likely stems from patriarchal stigmas, and again from this framing of menstrual blood as disgusting and useless. Following Michelle Murphy’s question “why should reproduction end at our bodies?” (Murphy, 2013), we speculated, if not for human reproduction, what non-human reproduction of life could menstrual blood support?

This spurred many initial questions on the properties, stigmas and taboos of menstrual blood: what nutrients in menstrual blood are particularly nutritious to plants? Would feelings of disgust be enhanced or suppressed when doing this practice? As well as questions on how a menstruating person might go about doing this fertilization: what tools or technologies might be used to collect and give the blood to the plants? and questions on the other species involved in this process: how would the plants themselves, but also the worms or microbes living in the soil be harmed or nourished? Our design process set out to explore these questions by examining this act of fertilization, as well as the act of collecting blood, which required a close examination of existing menstrual products.

The second motivation for this project revolved around ways to resist plastic-heavy period products. Disposable menstrual products account for several hundred thousand tons of waste every year, often ending up in landfills and clogging sewer systems. The majority of tampons and menstrual pads are single use, and contain high quantities of plastic materials in their packaging, applicators, strings and even in the absorbent parts (Borunda, 2019). Menstrual pads, especially, contain superabsorbent polymers, small plastic microbeads that absorb and trap fluid (Capezza et al., 2020). Many of these products contribute to the aforementioned narratives of concealment by sanitizing menstrual blood and in turn incorporating more plastic material, which can contain traces of pesticides and carcinogens capable of disrupting hormones (Clancy, 2021). For instance, thin ‘silent’ plastic wrappers or plastic applicators on tampons seem like a good idea for privacy, discreteness and in some cases hygiene, but this veil of privacy is often contributing to narratives of concealment. The silent wrapper assures nobody ‘hears’ the existence of a menstruating body, and the applicators guarantee the menstruator does not have to touch their own seemingly ‘dirty’ blood. Many myths about menstrual blood as toxic and polluting still permeate our society, when in fact there is nothing harmful present in menstrual blood, aside from blood-borne diseases that are carried in regular venous blood as well (Clancy, 2023). Overall, many of the effects of the dominant practices in the industry are harmful to both the lives of people who menstruate and for the environment, an ‘ecofeminist’ (Mies & Shiva, 1993) challenge we wanted to work with in Biomenstrual.

Design Process

Biomenstrual was designed and imagined around a circular process (Figure 3.2): plants with absorbent qualities are gathered and turned into design materials for fabricating menstrual products, the products are used, absorbing menstrual blood, then disposed of via biodegradation and composting, providing nourishment for more-than-human species in the soil and contributing to plant growth.

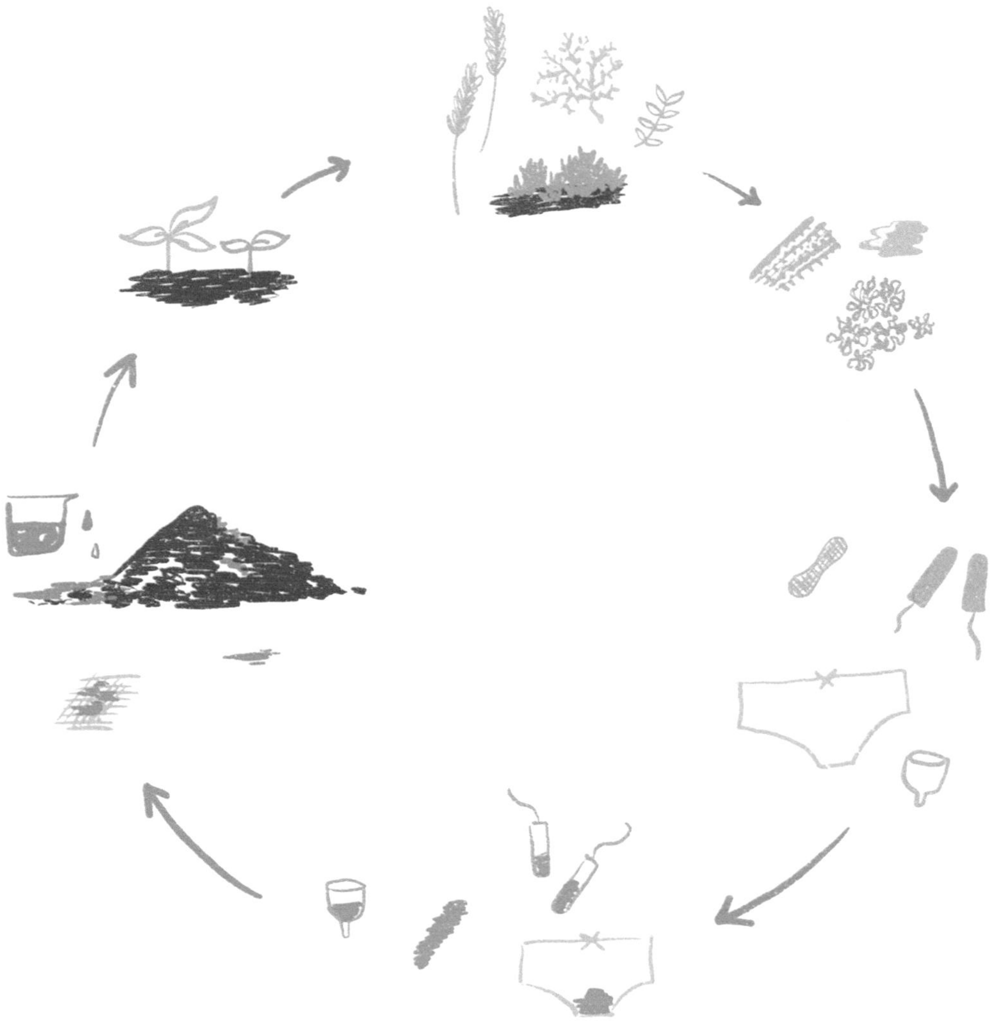

FIGURE 3.2 The Biomenstrual process imagined. Growing and harvesting materials, crafting menstrual products, collecting blood, and composting them to nourish the soil and aid new growth.

We followed a Research-through-Design methodology, where we prototyped and created artifacts, iterated, reflected, and formulated new research questions and design directions along the process (Gaver, 2012; Redström, 2017). We began the project by prototyping biodegradable menstrual products, where our process in exploring materials was heavily influenced by existing histories, practices and relations between menstrual blood and more-than-human species.

One main character in our project was *sphagnum moss*, a species of moss local to the Nordics, often found in wetlands and bogs. Curiously, sphagnum moss has been used across cultures as an absorbent material: it is capable of absorbing 25 times its weight in water, it is antiseptic, and has a low (acidic) pH value, making it difficult for bacteria and fungi to grow in its environment.

It is no wonder then that sphagnum moss was used for absorbing bodily fluids. For instance, sphagnum moss was used in bandages for wounds by British nurses during the First World War. Sure enough, it was also used for absorbing urine and menstrual fluids, taking the form of pads, diapers and stuffing for baby cradles. Indigenous communities including Sámi people, Nenets people, Native Americans, and Maori, have used dried sphagnum moss to absorb their bodily liquids. In the USA, Sfag-na-kins (1920s) and Johnson & Johnson (1970s) tried to capitalise on the use of sphagnum moss for commercial menstrual pads, yet these business endeavours never completely took off. Much of our knowledge on mosses, we gained from Indigenous scholar Robin Wall Kimmerer, who poetically phrased that the best human uses of mosses were in the hands of women during their 'moon time', referring to the bleeding phase of the menstrual cycle (Kimmerer, 2003). Inspired by and paying homage to these stories and practices, we foraged for sphagnum moss and used it as an absorbent material for menstrual pads.

Meanwhile, we got in touch with other materials and their properties through creating collaborations with material scientists, attending biomaterial courses, visiting both labs and forests, determining overall biological and chemical properties of materials that we, as designers, could leverage in our prototypes. Our collaborator and material scientist, Antonio Capezza, introduced us to a biofoam made from wheat gluten, which, similarly to the moss, demonstrated absorbent properties. We encapsulated both gluten foam and sphagnum moss between agar-based bioplastic, a common recipe for making plastic-like biodegradable sheets (Figure 3.3). We were also intrigued by the pH properties of our materials, and how these would affect and touch the intimate area of the menstruating body. While an acidic vaginal pH is often helpful for maintaining the vagina and vulva free of infections, this acidity can be thrown off by things like diet, penetrative sex, or even menstruation. Here we found the opportunity to connect the vaginal pH with the pH of our materials: we used the pH sensitive dye found in red cabbage to colour our bioplastic,

FIGURE 3.3 Biomaterial menstrual pads. Left: A menstrual pad prototype made with sphagnum moss and bioplastic. Right: We created prototypes that included seeds embedded within the bioplastic or nested between the sphagnum moss and observed how they sprouted from this medium when regularly sprayed with water.

indicating different colours according to different pH values of what comes in contact with it. We also were amazed to find that the pH value of sphagnum moss (3.5-4.5) perfectly matched the pH value of the vagina and leveraged that in our designs.

We imagined and fabricated various material composites into a familiar pad shape, as a way to bring these new and imaginative materials closer to a recognizable and everyday object used during menstruation. These pads presented varying degrees of biodegradability, and opened up questions on the next phase of the imagined circular lifecycle: what happens when these pads biodegrade in soil? To engage with this question, we performed small experiments, allowing the pads to decay, embracing accidental mould of the biomaterials, and imagining what more-than-human species are part of this process (Figure 3.3). Through speculative collages, we illustrated how macro and micro decomposers, such as earthworms, bacteria and fungi, come together and break down the biomaterials and the blood (Figure 3.4). The worms, drawn to the nutritious gluten in the pads, break down the materials, leaving a fertile and rich soil behind, similar to composting. Tardigrades, also called water bears, that often live in sphagnum moss, would cohabit with the human menstruator as they wear the pad. These speculations were meant to be open-ended, encouraging curious questioning and rethinking our feelings and associations of disgust in connection to menstrual blood, insects, microbes and what counts as 'dirt' in general.

Having assembled these initial prototypes and speculations, we invited others to collaborate and collectively imagine and respond to the questions our artifacts were generating (Figure 3.5).

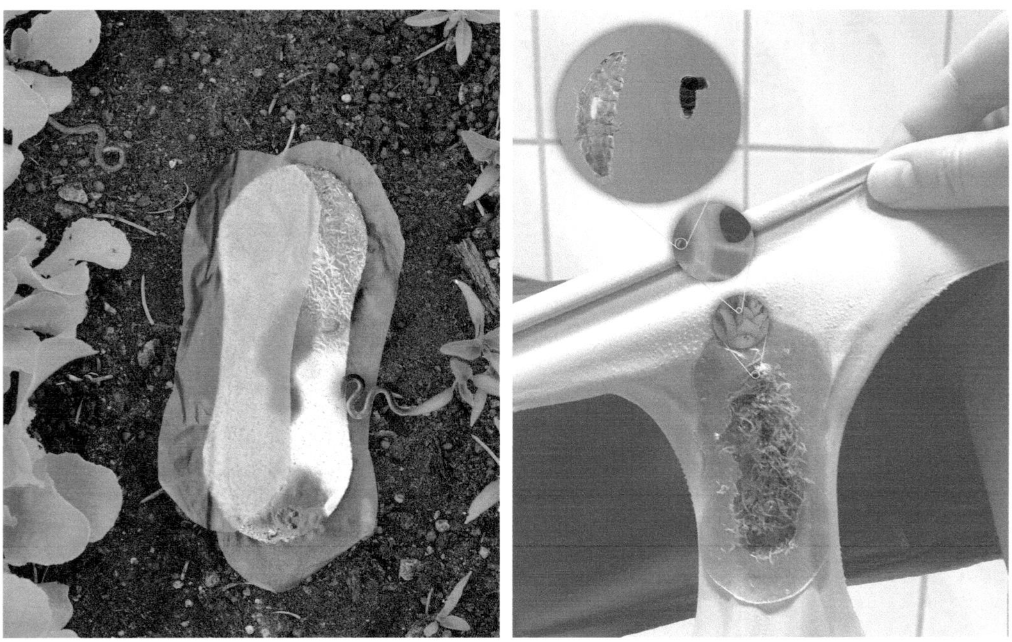

FIGURE 3.4 The breakdown of biomaterials and blood. Left: This speculative collage imagines how worms and fungi might biodegrade a used menstrual pad in a garden patch. Right: In this speculative collage, we zoom in to imagine what species might already be living on the materials we use in the pads, such as a tardigrade living in the sphagnum moss.

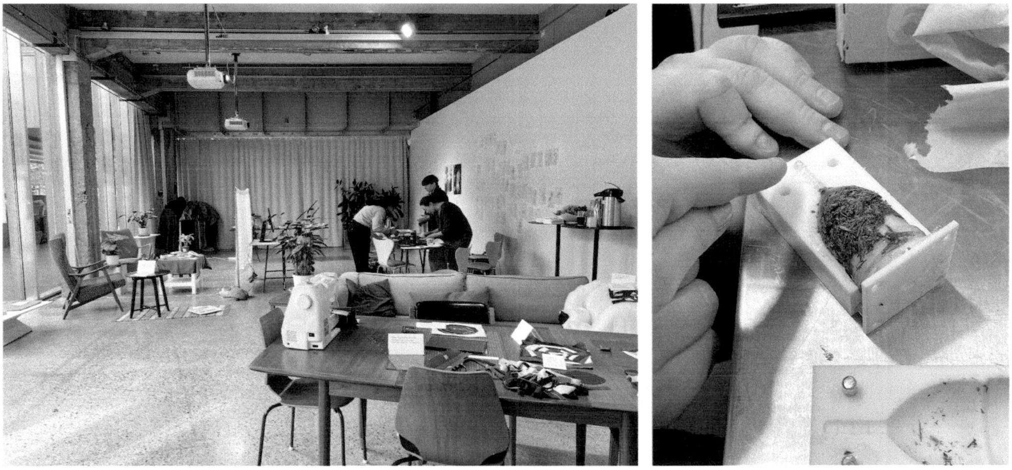

FIGURE 3.5 Workshops with biomaterials. Left: A workshop setting at Oslo's School of Architecture and Design, where participants are busy crafting bioplastics with a portable hotplate and kitchen tools. Right: A speculative menstrual cup moulded with moss, made during a workshop where participants experimented with biomaterials.

What new and alternative shapes and interactions could be designed, beyond pads, when introducing biodegradable materials? How can soil fertility be sensed and understood, once these products have decomposed? What can we know about the health of the human body, through the fertility of the soil? We opened a solo exhibition and discussed our approaches with artists, designers, researchers, menstruators and curious passers-by, and later hosted three design workshops. In the workshops, participants created not only new forms and shapes for menstrual products, but new practices of caring for their menstruating bodies (Søndergaard & Campo Woytuk, 2023). Together with designers and researchers exploring alternative approaches to agriculture, soil regeneration and energy production, we used different quantitative and qualitative approaches to sensing soil, reflecting on how we might use sensory approaches to notice more-than-human activity in the soil (Poikolainen Rosén, 2022; Søndergaard & Campo Woytuk, 2023).

Designing with the More-Than-Human Body

In Biomenstrual we have used design practice to explore how the human menstruating body is materially, symbolically, and politically situated within local ecologies and entangled with other non-human bodies, be it animals, insects, plants, plastics, waters, or soils. With our collection of design objects, we have aimed to promote how we can design more caring and sustainable menstrual practices for both the human body and the planet, while simultaneously challenging patriarchal and capitalist oppression of human and non-human bodies. We were motivated by critiquing dominant anthropocentric and unsustainable menstrual care products while reconfiguring the meaning of menstrual blood from a "disgusting" corporeal waste product to a fertile life-sustaining liquid. In the following sections we further discuss the ethical and theoretical dimensions of Biomenstrual, which serve as examples of what to consider when designing for the more-than-human body.

Ethical Dimensions of Biomenstrual

When presenting Biomenstrual we often get the question "have you tried the pads yourself?", which acts as a great starting point to discuss the practical limitations and implications of our work. While our designs are real material prototypes that could be used as they are, and women have indeed been using sphagnum moss for menstrual care practices throughout history (Kimmerer, 2003), they are also speculative and they may not fit every context and human body. Thus, it is still important for our designs to be ethically accountable, proposing more joyful, pleasurable and inclusive futures, rather than harmful ones. One way we aim to do so is by openly transmitting the knowledge we have gathered, through our open-source dissemination and workshops, despite not containing the answers to all questions. For instance, we know that one of the materials we designed with – sphagnum moss – is antiseptic and has a similar pH-value as a healthy vagina (pH 3.5–4.5), which we assume would nourish a healthy vaginal microbiome, yet we have not clinically tested this. Furthermore, we see the prototypes as design objects that dare to propose alternatives, opening up for future work in fields beyond or adjacent to design, such as medicine, biology or material sciences, who might be equipped with knowledge and techniques to help us engineer the pads into a usable and safe version.

As a further tension, our designs may also demand additional labour from menstruating people, for example in foraging or crafting the pads. Also, the additional awareness for the environmental impact of menstrual products might add to the stigma and emotional burden that menstruating people already face. Yet by doing this, we hope to shed light on how we currently 'manage' our menstrual cycles under the constraints of capitalist efficiency-driven societies: what if we did indeed have more time to attend to our bodies' cyclical changes, and manage our menstruation in more visible and collective ways? The speculative nature of our project allows for imagining these types of societal shifts, while also 'staying with the trouble' of current menstrual practices.

Biomenstrual also opens up for important conversations on material extractivism. Using off-the-shelf biomaterials might risk approaching more-than-human design from a problematic angle: using these non-human species for instrumental human needs instead of engaging deeply with the non-human's needs. In Biomenstrual, we ran into a tension between using accessible grocery store products for the pad prototypes and a slower process of foraging plants and mosses from the forest and building up knowledge on their histories and biological properties. The materials used in our project ranged from more off-the-shelf commodities (such as agar powder, or wheat gluten, which is not available for purchase but is used in the agricultural industry as a co-product of making flour) to more forgeable local species, such as the sphagnum moss. These materials exist at vastly different scales, made evident when brought into conversation which each other by manifesting through the same artifact (for instance a pad with both agar/gluten and moss). While we engaged with the cultural histories and needs of the moss, it became very clear that we wouldn't advocate for sphagnum moss pads to be commodified and scaled at an industrial level, as this would risk depleting peatlands, which serve a vital role in maintaining the climate. At the same time, exploring scaled-up and commodified materials like agar powder allowed us to make prototypes without having to develop the biochemical knowledge and expertise of its extraction process from seaweed. In tune with our feminist commitments, we envision both accessible practices as well as practices that allow us to attend to the locality of more-than-human species. This might mean that practices differ across ecologies, where different materials and species might be involved depending on availability across the globe.

As a final ethical tension, while the Biomenstrual project draws on and seeks to honour and celebrate ecofeminist relations between menstruating bodies and nature, there is a risk that by emphasizing the connection between menstruation and nature, the gendering of menstruation as an intrinsically female experience, and women as "closer" to nature, comes to surface, furthering essentialist associations that we indeed do not ascribe to. However, we, as humans, are not separate from nature, and our Biomenstral has indeed succeeded – for some participants – to reconfigure values of menstruation away from connections to womanhood and reproduction of humankind, towards an ecosocial awareness of making kinship beyond humans. We believe that overall, these practical, material, symbolic and political reflections of the human menstruating body as lively, vibrant and entangled with local ecologies allow us to know, experience and care for, not just the menstruating body, but all human bodies, *and* the local ecologies in new ways. Menstrual hygiene products that contain plastic generate trash and pollute and harm the environment. But menstrual blood can also be nutritious and benefit the environment, as we have explored in this project. Whether our designs are potentially polluting or nourishing for the human corporeality and all other bodies that menstrual fluids touch and transform, they show us that the menstruating human body is intimately and inextricably linked with local ecologies, and that the health of the human body depends on the health of the environment.

Theoretical Dimensions of Biomenstrual: Posthumanism and Feminism

In Biomenstrual, posthuman philosophies have taught us how the human body extends into the environment. We find that this theoretical framing is also useful beyond cases of menstruating, when designing with any aspect of the more-than-human body.

Posthuman philosophies manifest very literally when looking at the boundaries of the body, and bodily fluids are an evocative example. Animal residual flows fertilize gardens and agricultural fields. A birthing parent's vaginal microbiome sustains the immune system of their baby after a vaginal birth. Breastfeeding exchanges healthy bacteria between parent and baby, and breast milk provides nourishing food for growth. Premarin, short for pregnant mare's urine, is an oestrogen drug where oestrogen is extracted from horses' urine to be used as oestrogen hormone replacement therapy for peri-menopausal women (Haraway, 2012).

The human body also extends into the environment in many less tangible or materially visible ways: through the air we breathe, through hormones and antibiotics that are flushed out with the toilet water, through designed objects we use and then discard whether this is single-use plastics, clothes, or technology. Some of these are related to everyday practices: just think of water and air pollution, food, body overheating, or endocrine disruptors found in many beauty, household and industrial products. And some of these relations are long-term, manifesting in ways that we do not yet know but can only anticipate or start to see proof of. For instance, emerging scientific studies show that the menstruating body is disrupted by environmental pollutants in the long term in ways we cannot perceive in human time (Clancy, 2021). Fertility decreases and more irregular menstrual cycles are just some of the adverse effects. Environmental pollutants can even cause disorders like endometriosis and fibroids, and in the worst cases cancers such as ovarian cancer (Clancy, 2021). Therefore, it is urgent to make this link between human (menstruating) bodies and the environment, to attend to and prevent sustained and long-term harm caused by the Anthropocene.

The concept of 'the more-than-human body' problematises and challenges Western boundaries, dichotomies or distinctions between human corporeality and ecologies, and between human and other animal species. In Biomenstrual, we found new materialist and posthumanist perspectives useful for describing these more-than-human bodily relations through theoretical concepts that have inspired and transformed our design practice. Donna Haraway's "cyborg" is a material reality and a metaphor for thinking about human-machine-organism hybrids in a post-gender world, where–similar to her "companion species" concept–it is more crucial that humans commit to make kin (relate to other species) than to reproduce humankind (Haraway, 1991; 2008). This concept has inspired our reconfiguration of menstrual blood away from it being a 'failed attempt at reproduction' or the uterine lining that could have held an embryo, towards it being a nourishing fluid that can fertilize depleted soils. Another concept that has inspired our thinking about the urgency of attending to polluting aspects of the menstruating body is Stacy Alaimo's "trans-corporeality", which highlights how human bodies, as inherently toxic bodies, are products of the Anthropocene; and thus human health, social justice and environmental justice cannot be severed (Alaimo & Hekman, 2008). Trans-corporeality begins with the human body in order to disrupt Western human exceptionalism by pointing to how "all creatures, as embodied beings, are intermeshed with the dynamic, material world, which crosses through them, transforms them, and is transformed by them" (Alaimo, 2018). Following this, the human is not a contained body, disembodied or removed from the world, but embedded in local ecologies. In our design case, embedded in coastal and urban ecologies in Denmark, Norway and Sweden. Furthermore, Vijaya Rettakudi Nagarajan's concept "embedded ecologies" has taught us that how we see and relate to the natural world is shaping and shaped by the cultural world, and not an idealised and romanticised relation, but one that is "contested, historicized and culturally specific" (Nagarajan, 1998). Learning from historic and cultural practices with mosses, as inspired by Kimmerer's "Gathering moss" (Kimmerer, 2003), has for instance led us to inquire into Nordic historical menstrual care practices. While there is surprisingly little archival material, we did find stories and images of women knitting and crocheting their own menstrual pads, and importantly we found documentation of Indigenous knowledges of Sámi people using sphagnum moss as an absorbent material in diapers and pads. These practices are of course situated within a history of Scandinavian colonial erasure of Sámi culture and knowledge, so we need to navigate these spaces with care, making sure we–as non-Indigenous designers–are not furthering colonial mindsets. In line with Kimmerer, Indigenous scholar Zoe Todd reminds us that the human-animal dichotomy is a Western construct and that Indigenous communities have always been relating with nature (Todd, 2016), hereby bringing an important postcolonial critique to posthumanism, which is crucial to listen to in Western more-than-human design.

Lastly, our more-than-human design practice cannot be severed from feminist scholarship, and we have been deeply inspired by Rosi Braidotti's critical and feminist take on posthumanism. Braidotti foregrounds that posthumanism must by necessity connect with, and acknowledge pre-existing connections with, intersectional feminist theories that have paved the way for posthumanism (Braidotti, 2022). A feminist posthumanism challenges the universal (decentred) Man and connects the posthuman conditions with power and oppression and questions of race, gender, sexuality, disability and class, which, for instance, highlights how marginalised groups are disproportionately affected by the climate crisis (Braidotti, 2022; Søndergaard et al., 2022). Before we started practicing more-than-human design, we practiced feminist design for several years, so for us these inherent layers of feminism and posthumanism are evident and something we hold close to heart throughout our design practice. Designing with the more-than-human body implicitly brings feminist principles in contact with more-than-human design, such as a focus on embodiment and emotion, on positionality, on the fleshy, messy parts of the body and on women's care work and labour. These are principles key to Data Feminism (D'Ignazio & Klein, 2020) and Feminist HCI (Bardzell, 2010) and promoted throughout the Feminist Designer book (Place, 2023). Justice-oriented and humanistic designers might find it challenging or counter-productive to engage with more-than-human design because of its Western logics of posthumanism and assumptions of who/what is "human", which is not a single category that can be decentred. We see the feminist principles going hand-in-hand with more-than-human design, and a necessity for more-than-human design to critically engage with feminist critiques of patriarchy, capitalism and colonialism's responsibility in current extractivist practices in design and the environmental crisis.

We hope these theoretical concepts can inspire *more* more-than-human designers to engage with intersectional feminist issues that necessarily flow through all design decisions of what life is made possible or excluded, supported, or abandoned, through our designs.

Conclusion

We hope this chapter finds its way into the hands of curious designers and researchers coming both from traditions of feminist, embodied or somaesthetic design (Höök, 2018), as well as posthuman thinkers and scholars already thinking from a perspective of multispecies or more-than-human care. We hope you find yourself asking how to design for and with the more-than-human body in your own work, and that you find that there is a diverse abundance of fascinating, unexpected and unexplored relations between humans and non-humans. We end with a few prompts for continued reflections on how to design with the more-than-human body:

- How can I embrace and draw on my own positionality (body, identity, location) in designing with the more-than-human?
- What non-human species are part of my body? What non-humans leak into and are absorbed by my body, and what ecologies (local and global) does my body extend into?
- What curious (ignored, forgotten, or yet-to-come) relations do I find if attuning human corporeality and the self to local ecologies?
- How does it feel to materially and symbolically bring non-humans close to the body through the design process and in final design objects?

References

Alaimo, S. (2018). Trans-corporeality. *Posthuman Glossary*, London, Bloomsbury, 435–438.

Alaimo, S., & Hekman, S. J. (2008). *Material Feminisms*. Indiana University Press.

Ávila, M. (2022). *Designing for Interdependence: A Poetics of Relating*. Bloomsbury Visual Arts.

Bardzell, S. (2010). Feminist HCI: Taking stock and outlining an agenda for design. *Proceedings of the SIGCHI Conference on Human Factors in Computing Systems*, 1301–1310. https://doi.org/10.1145/1753326.1753521

Biggs, H. R., & Desjardins, A. (2020). High water pants: Designing embodied environmental speculation. *Proceedings of the 2020 CHI Conference on Human Factors in Computing Systems*, 1–13. https://doi.org/10.1145/3313831.3376429

Boer, L., Bewley, H., Jenkins, T., Homewood, S., Almeida, T., & Vallgårda, A. (2020). Gut-tracking as cultivation. *Proceedings of the 2020 ACM Designing Interactive Systems Conference*, 561–574. https://doi.org/10.1145/3357236.3395588

Borunda, A. (2019, September 6). How tampons and pads became so unsustainable. *Environment*. https://www.nationalgeographic.com/environment/2019/09/how-tampons-pads-became-unsustainable-story-of-plastic/

Braidotti, R. (2022). *Posthuman Feminism*. Polity.
Campo Woytuk, N., & Søndergaard, M. L. J. (2022). Biomenstrual: More-than-human design of menstrual care practices. *Temes de Disseny*, *38*, Article 38. https://doi.org/10.46467/TdD38.2022.116-131
Capezza, A. J., Lundman, M., Olsson, R. T., Newson, W. R., Hedenqvist, M. S., & Johansson, E. (2020). Carboxylated Wheat gluten proteins: A green solution for production of sustainable superabsorbent materials. *Biomacromolecules*, *21*(5), 1709–1719. https://doi.org/10.1021/acs.biomac.9b01646
Clancy, K. (2021, August 2). How endocrine disruptors affect menstruation. *American Scientist*. https://www.americanscientist.org/article/how-endocrine-disruptors-affect-menstruation
Clancy, K. (2023). *Period: The Real Story of Menstruation* (1st ed.). Princeton University Press.
D'Ignazio, C., & Klein, L. F. (2020). *Data Feminism*. The MIT Press.
Escobar, A. (2018). *Designs for the Pluriverse: Radical Interdependence, Autonomy, and the Making of Worlds*. Duke University Press.
Forlano, L. (2017a). Data rituals in intimate infrastructures: Crip time and the disabled cyborg body as an epistemic site of feminist science. *Catalyst: Feminism, Theory, Technoscience*, *3*(2), Article 2. https://doi.org/10.28968/cftt.v3i2.28843
Forlano, L. (2017b). Posthumanism and design. *She Ji: The Journal of Design, Economics, and Innovation*, *3*(1), 16–29. https://doi.org/10.1016/j.sheji.2017.08.001
Gaver, W. (2012). What should we expect from research through design? *Proceedings of the 2012 ACM Annual Conference on Human Factors in Computing Systems – CHI '12*, 937. https://doi.org/10.1145/2207676.2208538
Haraway, D. (1987). A manifesto for Cyborgs: Science, technology, and socialist feminism in the 1980s. *Australian Feminist Studies*, *2*(4), 1–42. https://doi.org/10.1080/08164649.1987.9961538
Haraway, D. (1988). Situated knowledges: The science question in feminism and the privilege of partial perspective. *Feminist Studies*, *14*(3), 575. https://doi.org/10.2307/3178066
Haraway, D. (1991). *Simians, Cyborgs, and Women: The Reinvention of Nature*. Routledge.
Haraway, D. (2008). *When Species Meet*. University of Minnesota Press.
Haraway, D. (2012). Awash in urine: DES and Premarin® in multispecies response-ability. *Women's Studies Quarterly*, *40*(1/2), 301–316.
Helms, K. (2021). Entangled reflections on designing with leaky breastfeeding bodies. *Designing Interactive Systems Conference 2021*, 1998–2012. https://doi.org/10.1145/3461778.3462048
Homewood, S. (2018). Reframing design problems within women's health. *DRS Biennial Conference Series*. https://dl.designresearchsociety.org/drs-conference-papers/drs2018/researchpapers/42
Homewood, S., Hedemyr, M., Fagerberg Ranten, M., & Kozel, S. (2021). Tracing conceptions of the body in HCI: From user to more-than-human. *Proceedings of the 2021 CHI Conference on Human Factors in Computing Systems*, 1–12. https://doi.org/10.1145/3411764.3445656
Höök, K. (2018). *Designing with the Body: Somaesthetic Interaction Design*. The MIT Press.
Kimmerer, R. W. (2003). *Gathering Moss: A Natural and Cultural History of Mosses* (1st ed). Oregon State University Press.
Kristeva, J. (1982). *Powers of Horror: An Essay on Abjection*. Columbia University Press.
Lilja, P. (2021). Tracing matters of scale by walking with minerals. *Nordes*, *1*(9), Article 9. https://archive.nordes.org/index.php/n13/article/view/608
Liu, J., Byrne, D., & Devendorf, L. (2018, April 19). Design for collaborative survival: An inquiry into human-fungi relationships. *Proceedings of the 2018 CHI Conference on Human Factors in Computing Systems*. https://doi.org/10.1145/3173574.3173614
Mahajan, S., Gabrys, J., & Armitage, J. (2021). AirKit: A citizen-sensing toolkit for monitoring air quality. *Sensors*, *21*(12), Article 12. https://doi.org/10.3390/s21124044
Mies, M., & Shiva, V. (1993). *Ecofeminism*. Zed Books.
Murphy, M. (2013). Distributed reproduction, chemical violence, and latency. *S&F Online*, *Life (Un) Ltd: Feminism, Bioscience, Race*. https://sfonline.barnard.edu/life-un-ltd-feminism-bioscience-race/distributed-reproduction-chemical-violence-and-latency/
Nagarajan, V. (1998). *The Earth as Goddess Bhū Devī: Toward a Theory of "Embedded Ecologies"*, in Lance Nelson (Ed.), *Folk Hinduism. In Purifying the Earthly Body of God: Religion and Ecology in Hindu India*. State University of New York, 1998 (pp. 269–295).

Persdotter, J. (2022). *Menstrual Dirt: An Exploration of Contemporary Menstrual Hygiene Practices in Sweden*. Arkiv förlag. https://doi.org/10.13068/9789179243685

Place, A. (Ed.). (2023). *Feminist Designer: On the Personal and the Political in Design*. The MIT Press.

Poikolainen Rosén, A. (2022, June 13). Relating to soil: Chromatography as a tool for environmental engagement. *Designing Interactive Systems Conference*. https://doi.org/10.1145/3532106.3533503

Redström, J. (2017). *Making Design Theory*. The MIT Press.

Søndergaard, M. L. J. (2018). *Staying with the Trouble through Design: Critical-feminist Design of Intimate Technology* [Ph.D, Aarhus University]. https://doi.org/10.7146/aul.289.203

Søndergaard, M. L. J., & Campo Woytuk, N. (2023). Feminist posthumanist design of menstrual care for more-than-human bodies. *Proceedings of the 2023 CHI Conference on Human Factors in Computing Systems*, 1–18. https://doi.org/10.1145/3544548.3581083

Søndergaard, M. L. J., Kannabiran, G., Chopra, S., Campo Woytuk, N., Gamage, D., Alabdulqader, E., McKinnon, H., Winschiers-Theophilus, H., & Bardzell, S. (2022). Feminist voices about ecological issues in HCI. *Extended Abstracts of the 2022 CHI Conference on Human Factors in Computing Systems*, 1–7. https://doi.org/10.1145/3491101.3503717

Stoetzler, M., & Yuval-Davis, N. (2002). Standpoint theory, situated knowledge and the situated imagination. *Feminist Theory*, *3*(3), 315–333. https://doi.org/10.1177/146470002762492024

Todd, Z. (2016). An indigenous feminist's take on the ontological turn: 'Ontology' is just another word for colonialism. *Journal of Historical Sociology*, *29*(1), 4–22. https://doi.org/10.1111/johs.12124

Ulmer, J. B. (2017). Posthumanism as research methodology: Inquiry in the Anthropocene. *International Journal of Qualitative Studies in Education*, *30*(9), 832–848. https://doi.org/10.1080/09518398.2017.1336806

Vostral, S. L. (2008). *Under Wraps: A History of Menstrual Hygiene Technology*. Lexington Books.

Wakkary, R. (2021). *Things We Could Design: For More than Human-Centered Worlds*. The MIT Press.

Wilde, D. (2022). Shitty food-based world-making: Recasting human|microbiome relationships beyond shame and taboo. *Futures*, *136*, 102853. https://doi.org/10.1016/j.futures.2021.102853

4

TRYING OUT SHIT!

Experimental Approaches for Relating with Microbes

Danielle Wilde and Tau Lenskjold

Introduction

From a more-than-human vantage point, this chapter seeks to explore designerly ways of relating to microbes – the plethora of critters that co-constitute us as embodied and social beings but are most often invisible in both human experience and design practices. The aim is to experiment, through design, with collective and experiential ways of making tangibly present the microorganisms that live *around*, *on* and *within* us; that make us at once hosts and living assemblages – biomolecular networks that encompass a host and its microbes, what biologists refer to as *holobionts* (Bordenstein & Theis, 2015) – and to consider how designers might engage proactively with these relations.

Designing with microbes is a new frontier for design, often relegated to biodesign. However, there are many ways this design work can be approached. There is the low-hanging fruit of educating designers to work with microbes.[1] We contend that this work must not only be technical but also philosophical – engage with the ethics of microbial participation in design, examine issues of power and bias – and recognise that it is not always so. In this chapter, we highlight the value of taking an experimental approach to the challenge. The work draws on participatory research through design, which converges methods from participatory design, embodied design and research through design, to critically and speculatively open up experimental spaces of inquiry (Wilde, 2022). In this chapter we position this approach against more teleologically and instrumentally driven applications of *zōē*, or biological life, in design, meaning understood in terms of the purpose they serve, e.g., in developing new materials.

Participatory research through design (pRtD) uses embodied methods – thinking through moving, making and doing collaboratively, co-creatively – to focus on process rather than outcome, and to engage *zōē*, or more-than-human life through (and as part of) *bios*, the social modes of human living. In the case we use as our example, we use pRtD to attain better awareness of the importance of microbial flourishing, through design actions, and offer alternative paths to fostering resilience when living with the effects of human-microbial discord.

An orientation towards designerly explorations of what Rees (2020; 2018) terms the *microbiocene* – "the world as it emerges from the perspective of the virus" (Rees, 2020;

DOI: 10.4324/9781003467731-6

2018) – has led to a growing body of work in the humanities and social sciences, which examines the importance of microbes in almost every aspect of human and societal constitutions. In this chapter we aim to describe and discuss the role design research may play in making such more-than-human, nature-entangled perspectives present in people's lives. The research draws on knowledge in biology, ecology, the medical sciences and the social sciences, as well as design research, to inform and develop new approaches to engaging the more-than- human through pRtD. The chapter's theoretical foundation is framed by the notion of "probiotic participation" (Lenskjold & Wilde, 2022), building on Lorimer's (2020) dictum of "using life to manage life". As our case will demonstrate, our research supports the assertion that an embodied engagement with the materiality of what is at stake can have a transformative impact on the relationships that people develop with challenging circumstances. This assertion holds true in relation to external challenges such as adjusting to climate change (e.g., Wilde et al., 2022), as well as internal challenges such as discomforts and disorders related to the gut or skin (Lenskjold & Wilde, 2022). Our research demonstrates that being aware of this potential of embodied engagement enables designers to support people to understand, anticipate and act upon futures different from the diagnostics and prognostics offered by scientific and medical knowledge regimes and practices (Wilde, 2020). In the Shit! project, we continue that probiotic exploration, with an inward view towards ourselves as symbiotic human-microbiome assemblages, or holobionts (Margulis & Fester, 1991).

Human relationships with our microbiomes are particularly difficult to grasp because our microbiomes are simultaneously part of us and consist of a multitude of species living on and inside us, in continual relation with the external world. This symbiotic relationship is not only important for our well-being but is fundamental to our individual and collective existence. For example, the gut microbiome has been declared "our second brain" (Ochoa-Repáraz & Kasper, 2016); it is said to shape our physical health, our emotional and mental well-being, our cravings, desires, moods and more (Carpenter, 2012). Yet, our awareness and understanding of our gut microbiome is mostly black-boxed and directed towards its effects (which may take the form of an ominous rumble, a tummy ache, qualities of defecation, etc.). With this chapter we aim to demonstrate that design research methods, in particular those that fall under the umbrella of pRtD, hold a potential to materialise, and thus make both tangible and discursively manifest our inter-species relationships with our microbiome and the socio-material importance of these relations; and that such a discourse is essential if designers are going to design with microbes.

Our research centres on methodological inquiry, and builds on previous work, where we examined the possibilities of the experimental design workshop as central method and approach for conducting pRtD (Wilde & Lenskjold, 2023). Building on that work, the contribution of this chapter is threefold:

1) we lay a foundation for thinking biologically as designers, by considering existing approaches, and the possibilities of participatory research through design as probiotic intervention;
2) we revisit and delineate the methodological repertoire of the Shit! project as it has unfolded until now, to open up our inquiry into where to go next;
3) we explore the project's future methodological direction by turning our attention to probiotic interventions in everyday life by means of self-experimentation kits.

This work contributes to a growing research programme (Wilde, 2022; Lenskjold & Wilde, 2022; Wilde & Lenskjold, 2023) that has as its aim to forge new – psycho-socio-politically engaged – directions for designing with microbes, expanding biodesign to incorporate an ecological purview that cuts across human subjectivities, social relations and environmental concerns (Guattari, 2000). We begin this journey by taking a step back to consider designing and biology.

Towards Probiotic Participation

Following calls for a radical transformation of design in light of the omnipresent environmental crisis, growing interest among designers and design researchers has been levelled at decentring the human (Forlano, 2016; Wakkery, 2021), informed by "a more-than-human intellectual zeitgeist catalysed by the Anthropocene" (Lorimer, 2020, 4). For the most part, these interests seem to have been directed either outwards, toward the world as a multispecies *pluriverse* (Escobar, 2018), or inwards towards microbial life as decontextualised biodesign experiments in laboratory settings (e.g., Kim et al., 2022), or by examining body-microbiome interactions as a controllable interface (Bell et al., 2023)

Increasingly, we find experimental exceptions to this binary positioning, for example, in the work of a cohort of designers that have emerged from KTH in Sweden (see: Campo Woytuk et al., 2020; Helms, 2022; Helms & Fernæus, 2021; Helms et al., 2021; Juul Søndergaard & Campo Woytuk, 2023) and our own work with the Shit! project. However, despite these concerted forays, we argue that the space of human-microbiome interplay remains largely overlooked in the design imaginary.

In order to open up space for reflection on kits for self-experimentation and creative imagining, we examine the motivations of design work that engages with microbes and reflect on the philosophical implications of these – largely instrumental – entanglements. Rather than provide an exhaustive review, we aim to characterise the field, using a few carefully selected examples. The purpose is to open a conceptual space; lay a foundation for discussion of our case and the experimental and methodological development of the Shit! project; encourage reflective consideration of the potential of undertaking design experimentations into human-microbial relations by means of what we have termed 'probiotic participation'.

Anthropologist Stefan Helmreich asserts that the mere biological fact that "we humans are sideways mash-ups Frankensteins – made up of a welter of teeny microbial friends and enemies" (2014, p. 53) has political consequences. He argues for a shift towards "symbiopolitics" (p. 56), which insists on the entangled existence and mutual dependence of many life forms at different scales. We argue that such political consequences should call us to examine *with* and *for whom* we design, first and foremost. Is it humans? The ecologies within which we are intimately entangled and on which we are wholly reliant? Or, perhaps – to differing degrees in each design endeavour – both?

BioDesign is a burgeoning movement at the bleeding edge of design practice that has drawn inspiration from two related, albeit very different, fields: BioArt and synthetic biology (e.g., Anker, 2021; Chieza et al., 2019; Ginsberg et al., 2017; Liu, n.d.). BioArt merges biology and art, using a range of scientific methods and materials – such as biotechnology, genetic engineering and living organisms – as artistic mediums to engage with the politics of human microbial entanglement, raising questions through philosophically and scientifically engaged artistic practice. Synthetic biology, in contrast, applies engineering principles to biology, to redesign and reengineer organisms to have new abilities for various useful purposes. By definition, synthetic biology is an instrumentally engaged field, whereas BioArt is philosophically engaged, even as it engages the instruments of science. While many of the techniques used in both fields are still relatively new to design, BioArt – which, at its foundation, is an artistic practice – demonstrates the accessibility of these scientific methods. With the parallel promise of material advances (as discussed in, e.g., Chieza et al., 2019 and seemingly endless social media posts about, for example, synthesising spider silk[2]), their potential is becoming both visible and of increasing interest to designers.

There are myriad examples of BioDesign in studio practice and, increasingly, in commercial practice, ranging from speculative approaches, for example, Vincent Callebout Architects' Dragonfly Metabolic Farm for Urban Agriculture (Vincent Callebout Architects, 2009) and in speculations catalogued by Dunne and Raby (2013); developmental prototypes such as those crafted and grown by Faber Futures, who work with bacteria to dye textiles and achieve other design effects "at the crossroads of design and biotechnology" (n.d.); work led by researchers and students at MIT, programming natto (fermented soybeans) to make dynamically responsive textiles for dancers (Yao et al., 2015); and the architecturally inspired experiments of Neri Oxman et al., undertaken with bacteria and insects, silkworms and bees. Through to products in the making, such as Giula Tomassello's Future Flora,[2] and fully-fledged products, such as those catalogued by Denend in their book on BioDesign projects (2015); and the work of industry pioneers working developing, growing and otherwise working with new bio-materials, e.g., bioMATTERS (n.d.), and ecoLogicStudio (n.d.).

In many cases, the use of microbes in developing these products is less developed than processes used to physically transform, chemically synthesise or otherwise modify materials for use to attain certain desirable properties. For example, creating design artefacts using industrially developed chitin, which can be harvested from the shells of invasive crabs (Muzzarelli, 2013), and the aforementioned spider silk – whether it be herded at great effort,[3] or synthesised.

When considering that biodesign has developed or taken inspiration from biochemical engineering, it is perhaps not surprising that the dominant approach to the integration of natural science techniques into design practice is an instrumental approach. Designers might ask questions such as: How is it useful to me, as a designer? or: What kinds of attributes does it have that I can leverage in the service of my designing? With such questions, the political edge and philosophical intent of BioArt seems largely subsumed by well-meaning intentions to develop exciting new (perhaps more sustainable?) materials, forms, products, systems and services for (and through) designing.

With regard to the concerns of this chapter, a more recent turn towards biological life in design is emerging. This turn engages politically and philosophically with the question of our microbial entanglement, exploring experimentally what it means to design with and for microbes, and specifically for humans as holobionts. As discussed in Wilde & Lenskjold (2023), Pauline Yurman (2022), for example, speculates on the impact of the more-than-human by using bodily fluids, such as breastmilk or urine, as a drawing medium. Through speculation, this work connects to the fluid speculations of Karey Helms, Marie Louise Juul Søndergaard, Nadia Campo Woytuk et al., whose research collectively focuses on leaky, pregnant, menstruating bodies, body-environment relations, the notion of transcorporeality across design research and HCI (Campo Woytuk et al., 2020; Helms et al., 2021; Helms & Fernæus, 2021; Juul Søndergaard & Campo Woytuk, 2023) and the speculative ethics of such forays (Helms, 2022). Their work collectively explores how and why to design self-reflexively with and for bodily fluids. In complement, Boer et al. (2020) look to design research methods to externalise experience and reflect on the self through tracking, cultivation, and sampling of the gut microbiome from inside the mouth. We see such externalisations as elements in unfolding processes of connectivity.

We find further inspiration in: Haldrup et al.'s (2022) commitment to approaching other species as co-creators of experiences and knowledge; van Gaalen's (2021) examination of how fictions might effectively foster multispecies cooperation; burgeoning efforts to bring together HCI researchers to discuss shared and diverging visions, perspectives and experiences, of more-than- human-design in HCI (Coskun et al., 2022); and to exchange methods for designerly engagement with "nature", where nature is approached as an open and ambiguous idea that

> embraces diverse kinds of more-than-human entanglements, including (but not only): farming, companion species, microbiomes, body ecologies, forests and other large-scale landscapes (e.g., oceans); or cohabitation in houses … [indeed] all possible relevant vectors of nature-related design: multispecies, cohabitation, posthuman sustainability, posthuman care.
>
> (Tomico et al., 2023)

And yet, despite these concerted forays, the space of human-microbiome interplay remains largely overlooked in the design imaginary.

Drawing from the broader territories of biological design and emergent research into human-microbial relations, our approach is conceptually undergirded by human geographer Jamie Lorimer's (2017; 2020) notion of probiotic intervention. Lorimer defines probiotic interventions as "interventions that are explicitly framed in contrast to modern, antibiotic approaches to health and environmental management" (2017, p. 28). What Lorimer outlines as "a probiotic turn" in the Global North, is located across vastly different spatio-temporal scales – from the reintroduction of apex species in nature conservation to bolster biodiversity to purposely ingesting hookworm larvae as "gut buddies" in DIY helminth therapy (p. 43).

Lorimer's expansive notion of probiotic intervention calls for a reassessment of human-environment relations and their attendant political ramifications in late modernity (p. 23). Sifted through the traditions and practices of embodied and collaborative design research, the idea of probiotic interventions forms a conceptual understanding that is guiding the Shit! project's theoretical development under the rubric of *probiotic participation*. It is, to be sure, an inquiry with a narrower scope, but it similarly engages human-microbiome relations transversally and as part of a meshwork crossing multiple scales and material, social and political dimensions. But, importantly, always with a specific focus towards forms of participation.

The examples from design research and theory presented in this chapter foreground the impact of microbial life and its attended material impact on human lives. Within this landscape, the research lens of probiotic participation serves a dual purpose: first, it explores ways of making microbial life's vital importance present and interactionable in our everyday lives by working with and through mundane practices and materials – to foster and nourish what we provisionally may call a *probiotic mindset* and to probe its effects; second – acknowledging the tradition of pRtD and seeking to maintain its collaborative ethos while simultaneously investigating experimental ways of extending participation, or "making kin" (Haraway, 2016) through design – it interrogates, through practice, what participation might mean when extended to encompass nonhuman species.

We now present the development of the Shit! project and the methods employed. This includes a diverse range of experiments aimed at breaching the "invisible" psycho-socio-political barrier that stands between each of us, human-holobionts, and our gut microbiomes. This case has been discussed in other ways elsewhere (Wilde, 2022; Lenskjold & Wilde, 2022; Wilde & Lenskjold, 2023). Here, we bring a new reading to the material by focusing on the methods themselves as forms of inquiry, considering what each method affords in terms of our larger agenda of breaking new ground in nature-entangled design research practice.

Initial Exploratory Methods

Acknowledging the strong historical orientation of design towards the human body, societal and socio-technical entanglements, the Shit! project began as a way of looking at the human body as a multispecies event – as holobiont, drawing on Margulis and Fester (1991). From this perspective, our initial discussions revolved around the symbiotic relationships with our microbiomes and how these relationships increasingly serve as a site for experimental design. From the inward looking understanding of the body as multiple, we expanded our purview "trans-corporally" (Alaimo, 2010) to consider and include other relationships that humans have with microbes at different scales: in our homes, workplaces, social contexts and as the traversal of substantial matter from agrarian ecologies, building infrastructures of bodily waste handling (Laporte, 1993).

Many research collaborations begin from discussions of shared interests and ongoing work followed by a jolt of excitement when a new line of correspondence suddenly appears. The Shit! project began this way. From our disparate yet shared interests in food, multispecies design and the field of pRtD, we quickly zoomed in on our symbiotic, yet oblique, relationships to our gut microbiome. Our mutual interest was in part fuelled by our vastly different personal experiences with our gut; one of the authors has a history of living with irritable bowel syndrome (IBS) while the other had never been troubled with gut issues.

Beginning from the intuition that an inquiry into the microbiome could start from everyday experiences, we devised an experiment informed by autoethnography – a method for approaching cultural norms and expectations through a systematic analysis of personal experiences (Ellis, Adams, & Bochner, 2011). To engage with our gut as researchers, we began documenting the consumption of sustenance and bodily excretions through photos accompanied with annotations of emotional states of the "transformation and passages from kitchen counter to toilet bowl and from mouth to anus" (Wilde, 2022). The panel of consecutively logged and catalogued images resulting from one of the author's experiments (Figure 4.1) demonstrates three important findings: first, it helped us establish a foothold for engaging with human-microbial relations by extrapolating the simplistic idea of documenting input/output events. It did not enable us (as one could have hoped) to discern or obliquely identify the working microbiome as a cipher of recurring patterns able to be correlated with specific behaviours. But rather, alerted us to mundane yet complex threshold events and practices taking place outside the body itself (Paxson, 2023). Second, and closely related, it enabled us to recognise the materials involved (especially foodstuff and fæces) as valuable experimental tokens for subsequent design experiments. And, third, the psychological and social ramifications of addressing abject matter and the cultural taboos associated with bodily waste, in turn stressed the ethical and political implications of the area of research. To read more on this, and an associated autoethnographic text, please refer to Wilde (2020).

Following the autoethnographic experiments, we opened the project for collaborative and public engagements – a Kitchen Research Lab in the Wild was developed and conducted with Design Research Masters students as part of the Danish annual National Research Week (Forskningens Døgn). Wearing lab coats and operating from two ad-hoc research stations, with an adjacent video screen, running a looped exposition about the human gut microbiome, we invited

FIGURE 4.1 Autoethnographic registration: This compilation of images, shot consecutively over time, depicts what went into and what came out of the author's body. Images include toilets, plates, menus, packaging, foods, vitamins, snacks and other things that the author ingested or excreted.

visitors to engage in participatory actions and dialogue around food, fæces and multi-species concerns. At the first research station the attendees were introduced to and treated with foods made by, with or containing non-human organisms (sourdough bread, kimchee, sauerkraut, local honey) and subsequently invited to draw their gut anatomy, using food-based paints. At the second research station, guests were presented with food items made from identical ingredients, with similar tastes, but formed differently – either as aesthetically pleasing food or to resemble fæces. The aim was to explore people's reactions and willingness to eat and reflect on their bodily and cognitive responses, and more broadly to ascertain people's interest in the presence of beneficial microbial life forms. The most significant takeaway of these experiments was that a confrontation with sensorial experiences and bio-socially constructed taboos led many of the visitors to share other facets of their lives related to food and the gut, and most notably their struggle with different forms of dysbiosis – the imbalance and malfunctioning of the microbiome in or on the body.

Bringing Shit to Those for Whom it Matters – From Facilitated Inquiry to Seeded Discovery

Following initial experiments, our attention turned to the Danish Colitis and Crohn's Patient Association (Colitis-Crohn Foreningen, 2021; CCF). The contact came via a student member of the research team with a history of chronic ulcerative colitis. The student was a member of the CCF and acted as an important intermediary. In the first meetings with members of the CCF board of directors, we discussed our research interests and aims and compared these with their relations to endocrinology, pharmacology and related medical research areas, and ultimately deemed our approach to provide an interesting alternative.

We were subsequently invited to develop and execute two one-day workshops as part of the association's yearly summer camp that serves to bring the community together in a safe space, and generally focuses on social activities. The first workshop involved 32 adult participants (26 years and over), including 16 sick people, each accompanied by a healthy family member, who participated alongside them. The second, involved 16 young adults (aged 19–30), all of whom are ill. Most of the participants (in both workshops) knew each other well. To illustrate the severity of chronic Colitis and Crohn's disease, one member of the adult group had died since the previous year, a loss that highlights the vulnerability of the participants and the important role of the summer camp as a site for mutual support.

For the same reason, the building of trust with members of the CCF and careful considerations of how to intervene into the safe space was of utmost importance in our preparations. To build rapport, we arranged with the CFF management to arrive at the venue the evening before the workshop, sleep over and have dinner, breakfast and lunch together with the participants.

In comparison with our previous research activities, "the workshop" was agreed upon as a format of engagement which was well known to the stakeholders. For the research team, the workshop, as method moulded in the tradition of participatory design (Wilde & Lenskjold, 2023), offered a blueprint for collective speculation and reflection in action, and methodological explorations of ways to breach the black box that surrounds human-microbiome interactions, and questions pertaining to the status of symbiopolitics in design unfurled through collaboration and self-experimentation.

Each workshop was conceived as four consecutive 45-minute activities, with breaks for coffee and lunch in between. The activities followed a carefully scripted narrative arc designed as a steady progression of confidence and trust. Each activity reconfigured embodied design around, with and through food to create a fertile space in which participants could find ways to address vulnerability and taboo while in the process engendering, experiencing and imagining the gut microbiome as actor and potential ally in their struggle for wellbeing. The following photo-montages combine images from both events (Figures 4.2, 4.3 and 4.4) show (1) the progression from drawing their own gut, (2) collaboratively building and performing the food, (3) making and eating cakes formed as fæces to confront taboos and (4) finally reflection on the experience.

FIGURE 4.2 Draw your gut (Workshop 1). Working alone, but together – side by side – workshop participants drew their gut anatomy from memory using food-based paints (spirulina, cacao, turmeric mixed with water, beetroot juice). In the adult group, half the participants have chronic gut diseases; the other half are healthy family members. In the activity, we did not distinguish between who was sick and who was healthy. However, the participants asked for this information. On analysis, the health of each participant was not discernible from their portrait. This outcome surprised sick and healthy alike.

FIGURE 4.3 Build your gut (Workshop 2). Participants, working in four groups, performed the mechanics of their gut – i) mouth to stomach; ii) stomach; iii) small intestine; iv) large intestine and rectum – using food-based elements to play the roles of food and biological processes (e.g., apple cider vinegar stood in for stomach acid, and juice for digestive juices). Top: a passing manœuver from group (i) mouth to stomach, to group (ii) stomach (Workshop 1). Bottom, left: the large intestine of a person with ulcerative colitis (the beetroot juice on the stockings indicates blood on the intestinal walls) (Workshop 1). Bottom right: preparing food for the mouth (Workshop 2).

FIGURE 4.4 Shitty cakes (Workshop 3). The primary task in this activity was to model your fæces. The materials used were organic chocolate cake (made from pre-biotics, pro-biotics and polyphenols), various "superfoods" and spices (bee pollen, açai berry, dried and powdered sea buckthorn, freeze-dried raspberries, chili flakes, cacao nibs; edible gold powder). Top left: beginning the task (Workshop 1). Top right: a poo-cake; Bottom left: a participant compares their model stool with the Bristol Stool Chart, a standardised model used by doctors in Western medical contexts to discuss stool characteristics with patients (Workshop 2). Centre right: a participant eats their poo-cake (Workshop 1). Bottom right: the table set up, with "superfoods" in the foreground.

The experiment documented here fostered dialogue, broke down barriers, and highlighted the shared realities of human experience. Notably, participants shared very private information and were able to have fun while doing so. In the reflection session, they revealed strained relationships with doctors and a reluctance to discuss issues beyond the medical realm. Communicating their illness with relatives (including those who were present) was also challenging. Most participants experienced a change in attitude toward their future lives with gut disease due to their involvement, and reiterated this change in a follow-up written survey. The workshop activities and subsequent dialogue helped them to breach the communicative status quo and open up about the intimate struggles of their daily experience, by sharing highly personal, material details about their (literal and metaphorical) 'shit', which led to them discovering that they are perhaps not so different from others – healthy or sick.

Our ambition of making the microbiome matter for the participants was less evident in the ensuing results. While the microbiome was addressed directly (e.g., illustrated in an animated video with which we began the day) and woven into activities (e.g., through the use of probiotic ingredients and as pieces of plastic token objects in the performance of the gut system) the participants remained preoccupied with the newly afforded experiences of sharing their struggles. This suggests a lack of insight, on our part, into the depth of challenge faced by the participants, and how to address their challenges, which perhaps overshadow any possibility of exploring new, unknown territory. In brief, with this experiment, we did not quite manage to open up the symbiopolitical ramifications of human-microbiome relations, in terms of prompting a new or heightened interest in probiotic potentials or the microbiome as such. Routes to enable probiotic participation, we learned, are among other things contingent on circumstances, and research agendas do not always translate easily to participants embroiled in the day-to-day struggle of coping with the consequences of dysbiosis.

The then director of the CCF board described the workshop as groundbreaking: "no-one is approaching our illness in this way, your work changes everything" (personal correspondence), and participants contacted us afterwards to ask for the cake recipes,[4] so they could do the workshop activities at home with their families, thus taking ownership of the method. Despite the participants' lack of conscious engagement with their microbiome, their embodied interactions with pre- and probiotic foodstuff formed microbial exchanges throughout the workshop, as they held and passed ingredients on to each other, made and digested shitty cakes. In the desire to bring home activities, in particular recipes that explicitly highlight the health benefits of microbial entanglements, a new potential for engendering the microbiome is brought to the fore, this time as part of making cakes (and modelling fæces with them) together as a family. This possibility at once highlights the difficulty of designing for a microbial mindset and the strength of participation as a vehicle for addressing the microbiome in different settings and under different circumstances.

This development prompts us to reflect on the nature of the activities scripted into the workshop, their ability to give access to the microbiome, and what it might take to reorient the scripts for home use. The possibility of people conducting this work at home raises many questions for us as design researchers. Collective, collaborative, self- experimentation in facilitated workshops allows the spectre of performativity to protect people from their vulnerabilities. There is little possibility to ensure that scripts for at-home enactment can do the same. The Shit! workshop has been carefully designed to scaffold care, and the ability to be increasingly, radically, open and vulnerable. It is unclear to us whether our workshop participants – let alone someone who had not been in the facilitated space – would understand how to do this, or even recognise that this is what the workshop supported them in becoming. In the next section, we consider the possibility of kits for at-home self-experimentation, beginning with related work, before moving on to the notion of kits for creative experimentation.

Self-experimentation Kits as Method for Microbiome Relating

We understand kits, broadly, as collections of physical and/or virtual objects, and instructions, scripts, or scripted things (Akrich, 1992), gathered together and designed to support participants to conduct certain kinds of activities. Following this definition, kits are stabilised networks, sent off to be used in a specific way, able to reach into a different domain or context of action than is usually accessible to a researcher. This aspect of kits differs fundamentally from workshops that are by definition anchored in a (heterotopic) place or location. Other than Cultural Probes (Gaver, Dunne & Pacenti, 1999) and their (many) derivatives (e.g., see Mattelmäki, 2006) kits for self-experimentation barely exist outside of medical or artistic contexts. Though, as we discuss below, with emerging work that engages with bodily fluids in particular (e.g., Campo Woytuk et al., 2020; Helms, 2022; Helms & Fernæus, 2021; Helms et al., 2021; Juul Søndergaard & Campo Woytuk, 2023), and our own work in this space (Lenskjold & Wilde, 2022) this situation is changing. To consider kits as a methodology, we reflect on participatory kits for: *creative imagining;* and for *self-experimentation*, discussing related work and reflecting on how the emerging landscape of self-experimentation is opening up new possibilities for Design. Our objective is to identify characteristics that can assist in developing self-experimentation kits for multispecies engagement, in ways that speak to design research communities.

Kits for Creative Imagining

Roque et al. (2021) provide a thoughtful review and discussion of kits in the context of culturally responsive computing. They emphasise that the possibilities and limitations of kits emerge as people create with them "in relation to [their] experiences, histories, and futures". (Tzou, et al., 2019). This builds on Tim Ingold's understanding of material interaction as a generative and improvisational process between the material and the maker (Ingold, 2010), recognising that "what is defined as "material" and how relations to particular materialities are narrated and enacted are shaped by knowledge systems" (Barajas-López & Bang, 2018). As Donna Haraway reminds us:

> It matters what matters we use to think other matters with; it matters what stories we tell to tell other stories with; it matters what knots knot knots, what thoughts think thoughts, what descriptions describe descriptions, what ties tie ties. It matters what stories make worlds, what worlds make stories.
>
> (2016, 12).

Our interest in kits and participation lies at the intersection of embodied engagement with/and the materiality of what is physically and conceptually at stake.

At the intersection of HCI and experimental design research, Cultural Probes (Gaver, Dunne & Pacenti, 1999) support guided exploration by participants, often through the use of ambiguous or absurd prompts. Their intent is to gather inspirational material about how people experience themselves and their relationships with the world, bracketed by the research concerns; the design researchers then use this material as inspiration for developing designs. The method builds on artistically informed social movements such as The Situationist International (Plant, 2002), and art movements such as Fluxus (Friedman, 1998; Higgins, 2002), for whom scripts and instruction sets serve as a key methodology.

> [Friedman's] works are deliberately playful – absurd even – and describe a series of actions that [leave] their actual realization in the hands or minds of the viewer. By operating in

language as instructions, descriptions, and documents, they slip between past, present, and future, operating in a suspended state of possibility with endless potential for variation and interpretation.

(Kalmar Art Museum, n.d.)

Where Cultural Probes differ from these instructional texts is that the performance of the instructions is meant to result in something that is sent back to the researchers. Our interest in kits, and self- experimentation, aligns with the subversive intent of Cultural Probes, and of Events. Our objective is not to gain inspirational input for design, rather it is to prompt particular embodied experiences that invite participants to reflect on and reposition themselves in relation to a core concern – their relationship with the microbiome. A key question is whether the need to return something to the designer is necessary for the enactment of the instructions? Or if putting the instructions in the world is sufficient. Or whether their embodiment in a kit matters in fundamental ways?

Kits for Self-experimentation

Opportunities for at-home-testing expand the reach of medical care into the home. Many devices and protocols are used by the quantified self-community and people who suffer from chronic bowel conditions. Today, anyone can send a sample off to have their stool analysed, or purchase a product that supports breath-based gut analysis. In many parts of the world, when a person turns 50, they receive a kit to test for markers of colorectal cancer or its precursors. The kit contains clear and simple instructions; the packaging is medical, which generates trust – the aesthetics of the materials and graphics are familiar to most from hospital or doctor visits; and the instructions describe a clear and easy-to-follow medical protocol, understandable across languages. The aim of this at-home-testing is to increase the success of early detection of colorectal cancer, which is a major cause of death in developed societies, and costs healthcare systems billions (Favoriti et al., 2016). All human actors involved in this process have something at stake and are motivated to succeed in the process.

This normalisation of self-testing opens a pathway for at-home-testing that might be more experimental. If we consider the Shit! workshops, they are praised for taking a new approach that does not aim to technically or medically solve the issues that chronically ill people are facing, but rather open up new ways of thinking through embodied engagement with the materiality of what is at stake.

The ethics of promoting at home, self-experimentation are delicate. In the Shit! project, we must consider the ethics of working with vulnerable populations; multispecies collaboration; transgressing taboos and potentially, biosafety guidelines. Certainly, if people are tempted to literally work with their intimate bodily fluids – take samples of their fæces, or grow bacteria that they harvest from within or without their bodies – concerns about bio-safety and ethics are well-founded (for a discussion, see Ledford, 2010; Schmidt, 2008). Considering our subject matter, we must therefore pay careful attention to these material matters as we consider the relocation of scripts from the controlled social settings of workshops to the home. No less important, we must be mindful that we are working with vulnerable individuals, whose conditions may cause them to overlook caution, as they consider how to interpret the provided scripts. For our scripts will not be neutral. Here we might turn to Peter- Paul Verbeek (2008) who urges a consideration not only of the moral decisions and responsibilities of designers, but the morality of the artefacts (in this case, scripts) themselves. Our move to kits is therefore not yet resolved.

Repositioning Methods to Drive an Inquiry

As a new space in design research, nature-entanglements requires new methods. In this chapter, we delve into methods as catalysts of such an inquiry. The methods are not instrumental tools – they defy instrumentality, even if they yield measurable outcomes. This stance is a critical departure from, for example, self-tracking, or behaviour change efforts, where the focus may be on the microbiome, but is determined to measure what, arguably, cannot be measured.

Furthermore, the purpose is not to seek inspiration for design, but rather to cultivate embodied experiences that might lead to insights for design research participants. The role of the imagination, of inventing possibilities and finding epistemic freedom in the absurdity of the tasks that are set, cannot be overestimated. By inviting participants to shift their worldviews, these methods open the possibility of shifting practices. Equally, by shifting practices, they open up the possibility of shifting worldviews.

The methodology – whether enacted through workshop, kit or instruction sets – challenges the mode of designer as "one who develops things" (i.e., products, systems, services), transforming it to one who facilitates change, spearheaded by others. This is a fundamental tenet of participatory research through design, with its focus on process and change-making, and we suggest it is what makes the Shit! Project experiments so compelling.

The methods described are not the outcome of an *a priori* methodological strategy, but rather (in hindsight) a continuous repositioning around the central theme of the workshop. Through this repositioning process, our focus shifted from open-ended inquiries to establish the project's foundation, to more structured approaches, centred on the established workshop format. At each stage, the development of methods has been influenced by specific and contextual circumstances, as well as the opportunities that emerged along the way. This process can be likened to a strategic repositioning of available opportunities, propelling our inquiry forward.

At the heart of everything lies the workshop – whether the traditional workshop or the more unconventional "Research Lab in the Wild". These workshops serve as spaces for experimentation, where different methods and techniques are applied. Our consistent thread has been to explore participation with, through and around living organisms, guided by a probiotic mindset and methodical openness – or to put it more bluntly, a penchant for trying out shit. As we transition from workshops to kits in the next phase of the project, we bring a heightened awareness of continuous repositioning. Our experiments will adapt and evolve based on participant feedback and their lived experiences. This deliberate consideration ensures that our investigations remain dynamic and responsive to the evolving context.

Questions

- How can participatory design methods – specifically, facilitated through food-related activities – be leveraged to foster novel experiences of human-human gut microbiome relating?
- How may experiences of multispecies liveliness help to bolster resilience and a sense of joy or hope among individuals living with chronic gut health challenges?

Author Note

Danielle Wilde and Tau Lenskjold are joint authors of this chapter and made equal contributions.

Notes

1 E.g., by participating in https://waag.org/en/project/biohack-academy/; https://textile-academy.org/about/; www.biodesignchallenge.org; www.badaward.nl; https://igem.org or through degree courses such as www.materialfutures.com/.
2 https://gitomasello.com/Future-Flora.
3 E.g., Voice of America (2017) A Simple, New Way to Spin Spider Silk in the Lab: www.youtube.com/watch?v=wK52CJXaCy4; SciShow (2019) Why We Can't Make Spider Silk: www.youtube.com/watch?v=UX2LHcLxjio; The Thought Emporium (2020) Grew Real Spider Silk Using Yeast: www.youtube.com/watch?v=2hf9yN-oBV4; SciShow (2024) We Finally Made Synthetic Spider Silk: www.youtube.com/watch?v=rLtOcr6gMaQ;
4 For the Shit! workshops, we modified cake recipes to be better suited to developing healthy gut bacteria, by swapping out foods such as what flour and eggs, and replacing them with inulin-rich foods such as chickpea flour and flax seeds.

References

Akrich, M. (1992). The de-scription of technical objects. In W. Bijker & J. Law (Eds), *Shaping Technology/Building Society: Studies In Sociotechnical Change*, The MIT Press, 205–224.

Alaimo, S. (2010). *Bodily Natures: Science, Environment, and the Material Self.* Indiana University Press.

Aleksejeva, J., Biedermann, P., Gavriliuc, I., Orlovate, O., & Wilde, D. (2019). Crafting bioplastics: Materially reconfiguring everyday food practices. RTD2019 Research Through Design Conference. Article 23.

Anker, S. (2021). Epistemic practices in Bio Art. *AI & Soc*, 36, 1389–1394. https://doi.org/10.1007/s00146-021-01152-w

Barajas-López, F., & Bang, M. (2018). Indigenous Making and Sharing: Claywork in an Indigenous STEAM Program. *Equity & Excellence in Education*, 51(1), 7–20. https://doi.org/10.1080/10665684.2018.1437847

Bell, F., Ramsahoye, M., Cofe, J., Tung, J., Alistar, M., & ACM (2023). µMe: Exploring the human microbiome as an intimate material for living interfaces. In *Designing Interactive Systems Conference* (DIS '23), July 10–14, 2023, Pittsburgh, PA, USA: ACM, 2019–2033. https://doi.org/10.1145/3563657.3596133

bioMATTERS (n.d.) www.biomatters.org/

Boer, L., Bewley, H., Jenkins, J., Homewood, S., Almeida, T., & Vallgårda, A. (2020). Gut-tracking as cultivation. In *Proceedings of the 2020 ACM Designing Interactive Systems Conference*, 561–574. Eindhoven Netherlands: ACM. https://doi.org/10.1145/3357236.3395588.

Bordenstein, S. R. & Theis, K. R. (2015) Host biology in light of the microbiome: ten principles of holobionts and hologenomes. *PLoS Biology*, 13(8), e1002226.

Denend, L. (2015). *Biodesign*. Cambridge University Press

Carpenter, S. (2012). That gut feeling. *Monitor on Psychology*, 43(8), 50.

Chamberlain, A., & Crabtree, A. (2020). Research 'In the Wild'. I: Chamberlain, A., Crabtree, A. (Eds) *Into the Wild: Beyond the Design Research Lab. Studies in Applied Philosophy, Epistemology and Rational Ethics*, vol. 48. Springer, Cham. https://doi.org/10.1007/978-3-030-18020-1_1

Chieza, N., Ginsberg, A. D., Lee, S., & Agapakis, C. (2019, 1 February). Design with science. *Journal of Design and Science*. Retrieved from https://jods.mitpress.mit.edu/pub/issue4-agapakis-lee

Campo Woytuk, N., Juul Søndergaard, M. L., Ciolfi Felice, M., & Balaam, M. (2020). Touching and being in touch with the menstruating body. In *Proceedings of the 2020 CHI Conference on Human Factors in Computing Systems* (CHI '20). Association for Computing Machinery, New York, NY, USA, 1–14. https://dl.acm.org/doi/10.1145/3313831.3376471

Coskun, A., Cila, N., Nicenboim, I., Frauenberger, C., Wakkary, R., Hassenzahl, M., Mancini, C., Giaccardi, E., &Forlano, L. (2022, April). More-than-human concepts, methodologies, and practices in HCI. In *CHI Conference on Human Factors in Computing Systems Extended Abstracts* (pp. 1–5). New Orleans, LA, USA.

Dunne, Anthony, & Fiona Raby. *Speculative Everything: Design, Fiction, and Social Dreaming*. MIT Press, 2013.

Ecological studio (n.d.) www.ecologicstudio.com/
Ellis, C., Adams, T. E., & Bochner, A. P. (2011). Autoethnography: An overview. *Historical Social Research/Historische Sozialforschung*, 36(4 (138)), 273–290.
Escobar, A. (2018). *Designs for the Pluriverse: Radical Interdependence, Autonomy and the Making of Worlds*. Duke University Press.
Favoriti, P., Carbone, G., Greco, M., Pirozzi, F., Pirozzi, R. E. M., & Corcione, F. (2016). Worldwide burden of colorectal cancer: A review. *Updates in Surgery*, 68(1), 7–11. https://doi.org/10.1007/s13304-016–0359-y
Faber futures (n.d.) https://faberfutur https://faberfutures.com/
Forlano, L. (2016). Decentering the human in the design of collaborative cities. *Design Issues*, 32(3), 42–54. https://doi.org/10.1162/DESI_a_00398
Friedman, K. (Ed.) (1998). *The Fluxus Reader*. John Wiley & Sons.
Gaver, B., Dunne, T., & Pacenti, E. (1999). Design: Cultural probes. *Interactions*, 6(1), 21–29. https://doi.org/10.1145/
Ginsberg, A. D., Calvert, J., Schyfter, P., Elfick, A., & Endy, D. (2017). *Synthetic Aesthetics: Investigating Synthetic Biology's Designs on Nature*. MIT Press.
Guattari, F. (2000). *The Three Ecologies*. Athlone Press.
Haldrup, M., Samson, K., & Laurien, T. (2022). Designing for multispecies commons: Ecologies and collaborations in participatory design. In *Proceedings of the Participatory Design Conference 2022*, vol. 2, 14–19. Newcastle upon Tyne, United Kingdom: ACM. https://doi.org/10.1145/3537797.3537801.
Haraway, D. (2016). *Staying with the Trouble: Making Kin in the Chthulucene*. Duke University Press.
Helmreich, S. (2014). Homo microbis: The human microbiome, figural, literal, political. *Thresholds*, 42, 52–59. https://doi.org/10.1162/THLD_A_00076
Helms, K. (2022). A speculative ethics for designing with bodily fluids. In *CHI Conference on Human Factors in Computing Systems Extended Abstracts* (pp. 1–11).
Helms, K., & Fernaeus, Y. (2021). Troubling care: Four orientations for wickedness in design. In *Proceedings of the 2021 ACM Designing Interactive Systems Conference* (pp. 789–801).
Helms, K., Juul Søndergaard, M. L., & Campo Woytuk, N. (2021). Scaling bodily fluids for utopian fabulations. In *The 9th Bi-Annual Nordic Design Research Society Conference: Matters of Scale, 2021*. https://doi.org/10.21606/nordes.2021.23.
Higgins, H. (2002). *Fluxus Experience*. Univ of California Press.
Juul Søndergaard, M. L., & Campo Woytuk, N.(2023). Feminist posthumanist design of menstrual care for more-than- human bodies. In *Proceedings of the 2023 CHI Conference on Human Factors in Computing Systems* (CHI '23). Association for Computing Machinery, New York, NY, USA, Article 38, 1–18. https://doi.org/10.1145/3544548. 3581083
Kalmar Art Museum (n.d.). 92 Events. www.kalmarkonstmuseum.se/en/exhibition/92-events/
Laporte, D. (2002). *History of Shit*. MIT Press.
Ledford, H. (2010, 6 October). Garage biotech: Life hackers. *Nature*, 467(7316), 650–652. https://doi.org/10.1038/ 975467650a
Lenskjold, T., & Wilde, D. (2022). Shitty stories: Experimenting with probiotic participation through design. In Lockton, D., Lenzi, S., Hekkert, P., Oak, A., Sádaba, J., Lloyd, P. (Eds), *DRS2022: Bilbao*, 25 June – 3 July, Bilbao, Spain. https://doi.org/10.21606/drs.2022.521.
Liu, A. (n.d., February 19). Untitled (Labor of Love). *Ani Liu.* https://ani-liu.com/untitled-labor-of-love
Liu, J., Byrne, D., & Devendorf, L. (2018). Design for collaborative survival: An inquiry into human-funghi relationships. *Proceedings of the Designing Interactive Systems Conference*, 779–788.
Lorimer, J. (2017). Probiotic environmentalities: Rewilding with wolves and worms. *Theory, Culture & Society*, 34(4), 27–48. https://doi.org/10.1177/0263276417695866
Lorimer, J. (2020). *The Probiotic Planet: Using Life to Manage Life*. University of Minnesota Press. https://doi.org/10.5749/j.ctv19cw9vg
Margulis, L., & Fester, R. (1991). *Symbiosis as a Source of Evolutionary Innovation*. MIT Press.
Mattelmäki, Tuuli (2006). *Design Probes.* Aalto University.
Muzzarelli, R. A. A. (1977) *Chitin*. Elsevier Science & Technology.

Ochoa-Repáraz, J., & Kasper, L. H. (2016). The second brain: Is the gut microbiota a link between obesity and central nervous system disorders? *Current Obesity Reports*, 5(1), 51–64.

Oxman (n.d., February 19). Untitled (Labor of Love). *Oxman Nature X Humanity* https://oxman.com/

Paxson, H. (Ed.) (2023). *Eating Beside Ourselves: Thresholds of Foods and Bodies*. Duke University Press.

Plant, S. (2002). *The Most Radical Gesture: The Situationist International in a Postmodern Age*. Routledge.

Rees, T. (2018). *After Ethnos*. Duke University Press. https://doi.org/10.1215/9781478002284

Rees, T. (2020). From the anthropocene to the microbiocene. *Noema*, June 10, 2020. www.noemamag.com/from-the-anthropocene-to-the-microbiocene/.

Roque, R., Tamashiro, M. A., McConnell, K., & Granados, J. (2021). Opportunities and limitations of construction kits in culturally responsive computing contexts: Lessons from ScratchJr and family creative learning. In *Interaction Design and Children* (pp. 246–256). ACM. https://doi.org/10.1145/3459990.3460728

Schmidt, M. (2008). Diffusion of synthetic biology: A challenge to biosafety. *Systems and Synthetic Biology*, 2(1–2), 1–6. https://doi.org/10.1007/s11693-008-9018-z

Tomasello, G. (n.d.). *Future Flor.* https://gitomasello.com/Future-Flora

Tomico, O., Altarriba Bertran, F., Keune, S., Buruk, O., Wilde, D., & Wakkary, R. (2023). Designerly ways of engaging with nature: Exploring the methodological landscape of more-than-human design research. In *26th International Academic Mindtrek Conference* (Mindtrek '23), October 03–06, 2023, Tampere, Finland. ACM, New York, NY, USA, 4 pages. https://doi.org/10.1145/3616961.3616993

van Gaalen, S. (2021). Functional fictions for multispecies cooperation. *Interactions*, 28(2), 14–15. https://doi.org/10. 1145/3450506

Verbeek, P. (2008). Morality in design: Design ethics and the morality of technological artifacts. In P. Kroes, P. E. Vermaas, A. Light, S. A. Moore, & P. Verbeek (Eds), *Philosophy and Design: From Engineering to Architecture* (pp. 91–103). Springer.

Vincent Callebout Architects (2009) *DRAGONFLY Metabolic Farm for Urban Agriculture*. New York, U.S.A. https://vincent.callebaut.org/object/090429_dragonfly/dragonfly/projects

Wakkary, R. (2021). *Things We Could Design: For More Than Human-Centered Worlds*. MIT Press.

Wilde, D., & Lenskjold, T. (2023). Shit! towards an experimental multiple-perspective approach to human- microbiome relations. *Human-Computer Interaction*, 1–28. https://doi.org/10.1080/07370024.2023.2276527

Wilde, D. (2015). Embodying material ideation. *Proc. Participatory Innovation Conference*, 386–397.

Wilde, D. (2020, Summer). Design research education and global concerns. *She Ji Journal of Design, Economics and Innovation*, 6(2), 170–212.

Wilde, D. (2022). Shitty food-based world-making: Recasting human|microbiome relationships beyond shame and taboo. *Futures*, 136, 102853. https://doi.org/10.1016/j.futures.2021.102853

Yao, L., Ou, J., Cheng, C., Steiner, H., Wang, W., Wang, G., Ishii, H., & Assoc Comp Machinery (2015). bioLogic: Natto cells as nanoactuators for shape changing interfaces. CHI '15: Proceedings of the 33rd Annual ACM Conference on Human Factors in Computing Systems, 2015, 1–10. https://doi.org/10.1145/2702123.2702611

Yurman, P. (2022). More-than-human fluid speculations. In D. Lockton, S. Lenzi, P. Hekkert, A. Oak, J. Sádaba, & P. Lloyd (Ed.), *DRS2022: Bilbao*, 25 June – 3 July, Bilbao, Spain. https://dl.designresearchsociety.org/drs-conference-papers/drs2022/researchpapers/121/

5

DESIGNING WITH BODIES OF WATER IN THE HYDROCENE

Thomas Laurien, Anna Schröder, Inna Zrajaeva and Christoph Matt

Opening the Tap?

*The puz-
zling num-
eral coinci-
dence that
most animal
bodies and
the surface
of our dear
planet all
consist of
roughly two-
thirds water.
There are
certainly a
lot of wa-
tery bodies
on this wa-
tery planet -
there is cer-
tainly a lot
of water in
this compost
...*

DOI: 10.4324/9781003467731-7

For people living in so-called modern societies, thinking of water in everyday life often points to images of water as "pure matter" (H_2O), as something tapped from a seemingly endless reservoir designated for human use – as *aqua nullius*,[1] empty, passive and non-relational. For these people water is something that flows in anonymous hidden-away systems of pipes, and is most often taken for granted; this kind of water can be called *modern water* (Linton, 2010). This understanding of water also correlates with underwater illiteracy – the inability or disinterest in engaging with life and space below the water surface.

In the current dominating political systems, water-related tensions like water scarcity and droughts, polluted waters, melting glaciers, flooding and rising sea levels, are predominantly being looked at as matters of the water crisis. In the context of the global ecological crises, we authors acknowledge and experience this water crisis, which we see rooting in the social relation to modern water (Linton, 2010). The social imagery of what water is weighs in with its crisis, in other words, as Jamie Linton points out, "water bears the traces of its social relations, conditions, and potential" (Linton, 2010, p. 7).

When dealing with these tensions and realities in the context of design practices with posthumanist intentions, we argue that they instead need to be understood as matters of *relational water* – water entangled with human and more-than-human life, water with agency, water with histories and futures, water *as* life. Building on scholars like Veronica Strang (Strang, 2004), Astrida Neimanis (Neimanis, 2009), Jamie Linton (Linton, 2010), Cecilia Chen (Chen, 2013), and Robin Wall Kimmerer (Kimmerer, 2013) our design works grasp these issues on the paradigmatic level of relational water.[2]

As implied in the title of this chapter, we find the concept of *bodies of water*, coined by Astrida Neimanis, both inspiring and productive (Neimanis, 2009; 2017). The concept can enable and strengthen multispecies empathy, and an expanded idea of care since both human bodies and a vast number of more-than-human bodies share that we all consist of water, and are totally dependent on water for our survival. We also want to emphasise that bodies of water, or water bodies, not only are spaces of species assemblages but also assemblages of more-than-human (infra-)structures. Water bodies, and waterscapes, are habitats in the literal sense, lifeworlds for a myriad of human and more-than-human beings to hold on to – be it glaciers, lakes, wetlands, rivers, puddles, or the seas. When water is perceived and treated as relational, as opposed to just a resource to (over)use, or as threats on different scales, it challenges the current dominating mechanistic and techno-fixated paradigm. Within this paradigm there is a constant strive to gain as much of a complete understanding of all water bodies as possible, in order to control them – an approach and pedagogy that might need to be let go of (Verlie, 2021). Taking the Anthropocene as a point of departure for the climate, water and biodiversity crises, Bronwyn Bailey-Charteris casts up the idea of the *Hydrocene* as a curatorial theory of critical engagement with water bodies, to recognise their agency, and to produce situated knowledge (2021, p. 431–445).

We – four designers, design educators and design researchers, with different professional backgrounds and practices – acknowledge that each watery body is a unique *place of uncertainty* (Bailey-Charteris, 2021, p. 432). In this chapter, we will share some moments and insights from different watery cases, where we have tried to be true to this premise. In the following sections: Watery meeting, Watery making and Watery telling, we will share our experiences from our different engagements with water: Thomas Laurien through working with the Saläng brook in Borås,[3] Anna Schröder through working with the shore of the Öresund,[4] Christoph Matt through working with three different sites: a Parco delle Cave creek in Milano, the River Lutz and the Rote Wand glacier in the Großes Walsertal,[5] and finally Inna Zrajaeva through working with the Öresund strait[6] (see Figure 5.1).

FIGURE 5.1 A collage of the four design practices. A: Flow Feelers searching for an opening. B: Capturing reflections from a seaweed noticing walk. C: Memories of the extinct Rote Wand glacier. D: Becoming an ocean carer.

Source: All photographs by the authors.

In "poetic field notes", or *interludes* (Åsberg et al., 2020, p. 3) we speak through the voice of one merged "designer-I" and share our knowledge of what it can mean to design with bodies of water. They show how we translate concepts from watery literature into watery doing in our design practices as a means to explore and experience relations and the agency of water.

Watery Meeting

Meeting the Saläng Brook in Borås

How to meet and greet a body of water who was put underground a long time ago, forced to flow a long distance through hidden and now forgotten pipes? How to create a sense of proximity between human beings on the ground, and a culverted brook buried beneath asphalt, gravels and soil? We, a group of men in a study circle named "Flow Feelers" decide that we want to get to know precisely such a body of water – water deprived of the manifold relations it once had. We also want to experience if and how a small, long-neglected body of water maybe can help us participants in the men-only circle to see the connections and ideas that make up the web of oppression of women, indigenous peoples, queer minorities, other-than-human animals, forests, mountains – and water. And even more importantly, we start the circle with the hope (and subversive idea) that we then can bring these insights into our ordinary professional and everyday realities, thus maybe challenging patriarchal structures from within. We look for traces of the brook in the archives and find old maps and photographs with clues here and there. At an early stage, we also visit the urban area where the brook most likely is hidden, and we then try to read the topography. How would the water flow had it not been silenced and buried? We also try to use our bodies as tools to connect. We put our ears to the ground hoping that we will hear a watery mumbling from below. We place ourselves in a line across a playground where we know that the brook once ran; we close our eyes and try to imagine a physical meeting with the water (see Figure 5.2A). Eventually, we even jump into the murky water of the city river to try to find the hidden mouth of the culvert with our bare hands, using our spread fingers as flow feelers.

Meeting the Öresund Shore in Malmö

Cold hands stuffed deep into their pockets, a group of six people shares memories of seaweed at the urban beach of Malmö, where the land of Swedish Scania meets the waters of the Öresund strait. Stories of siblings torturing each other with disgusting seaweed, and culinary adventures in foreign cultures come up. Without clear instruction I send the group out to notice what the sea has left on the shoreline, to notice what they usually overlook, and to follow their curiosity along the following bit of the coast. There is no right or wrong way to encounter seaweed during this walk, no correct order to do things. What is usually a walk of fifteen minutes down the beach stretches to a whole hour of rummaging around, picking up and inspecting washed-up seaweed. Wondering. How did this weed end up so far outside the water? Is it still alive? … Why are some attached to stones (see Figure 5.2B)? People find their own ways up to the dunes, and down to the waterline, some duck under piers as we pass them. By now the group has stretched out, and initial uncertainty has yielded wondrous scouting. Hands are carrying bewildered stones, some jacket sleeves wet from inspecting shallow waters.

Meeting the Parco delle Cave Creek in Milano

The Sympoietic Society art collective meets with the urban ecosystem that connects ways of water and humans in the city of Milan. We are invited as a collective to exchange knowledge with citizens about the complex, and often invisible, infrastructures of collection, transformation and distribution of water. In the "Collecting Wor(l)ds" workshop we, local experts, the collective and myself, take participants on a guided walking tour in the ecosystem of "Parco delle Cave". A recycling paper based watery blue fanzine logbook designed with a fluid typography system, conceptualised and produced within the local community beforehand, provides instructions with information about the entangled aquatic system of human and more-than-human actors. An exercise section proposes a practical experiment based on noticing walking in the aquatic ecosystem as an art practice. While we walk through the urban area with a barely noticeable creek at its heart, participants explore the habitat, listening to the experts and write down words in a water glossary referring to the urban waterscapes surrounding us. The glossary lists the entire alphabet with blank placeholders, where the participants can fill in their associations as keywords easily by only using a pen (see Figure 5.2C). As the participants are locals and internationals, they are adding to the alphabet in their native language, where for example "W" stands for water in English and "A" for Acqua in Italian. Exactly this process makes all the difference as the word "water" has manifold meanings depending on the language. We keep on collecting different points of view, perspectives, and perceptions in relation to the knowledge, which is conveyed by the local experts throughout the walk. Once the glossary is fully filled, it builds the foundation for further reflections on how to meet water. This collection process carefully considers the agency of language with indications of ecological, social and economic associations, and connects words with watery worlds.

Meeting the Öresund strait in Malmö

On a grey October morning, a group of high school students are reading stories about Blue Mussels, European Eels, Herring and other species that inhabit the waters of the Öresund. Occasionally they look up from their handout and turn to gaze out the window where just a couple of metres away the sea is silently waiting at the shoreline. At the beginning of the day, each student was bestowed with a Öresund inhabitant through a little ritual. As I am looking at their concentrated faces, I wonder if asking them to identify with a species changes the way they look at the information. One boy, clearly unhappy with his pairing, objects to being assigned to the Common Cockle, as his classmates giggle. The written profiles introducing them to their matched species are quite different from the ones they know from biology books (see Figure 5.2D). They describe the Herring as the founding fathers of Malmö, the Blue Mussels as the relentless workers who purify the brackish sea, and the Eel as an influential political agent able to end the career of hungry Swedish politicians. When they're done, I tell them that now they bear the responsibility of representing the interest of their ocean "actor". Not only that, but the students

FIGURE 5.2 Watery meeting. A: Practising "Standing brook". B: Lifting stones with seaweed-strength, washed up far outside the water. C: Collecting words and connecting worlds via the water glossary. D: Ocean species' cards.

Source: All photographs by the authors.

will work in groups of four, teaching them about the relationships that define life in the Öresund. After all, I tell them, that no species can survive alone, especially not us humans.

Watery Making

Making with the Saläng brook in Borås

One moist spring evening the "Flow Feelers" group decides to walk to the hillside where the brook starts as a tiny rivulet. Earlier we have been reflecting on the importance of rituals, but also on the experience of a *lack* of rituals – rituals that maybe could connect us better with the natural world. Kennet, one of the study circle leaders, has made some preparations beforehand so that we can make a ritual together that evening, or rather prototype it since we don't really know how to go about it. The idea is to acknowledge and, in some sense celebrate, the water cycle since one can't engage oneself with a brook or a river without also starting to reflect on the sea, and then, how the water as clouds comes back again to start the cycle anew. We walk up the hill and find a clearing in the forest. Kennet has collected raindrops from leaves, now stored in a tiny jar. He has also brought some big maple leaves, plucked from a twig that a storm gave him the other day. One at a time, we get a maple leaf in our hands, and by shaping it into cups we can receive some water from the jar (see Figure 5.3A). There is only a draft of the continuation of this making, and it says: "Now go away in private and return these drops to the water cycle". After a while, we gather again and share thoughts about what we have just experienced.

Making with the Öresund Shore in Malmö

Bright sun shines on a large table, on which a handful of co-visitors have spread pieces of seaweed, some seagrass, and some beach trash – all gathered during one hour of attentive walking along the coastline. With a view of the beach section we have just left behind, I introduce the technique of cyanotype, a procedure to produce contact photographs with sunlight. The participants are not the first ones to engage with seaweed through this technique, inspiration comes from English 19th-century botany.[7] We adapt the technique for our reflection process. Shortly after, people rake their fingertips through the loose pile of gathered seaweed to find the one specimen they want to print. Which one looks particularly interesting? Once all arrangements are prepared, we carry the table from the shade into the sun. Now we wait and see how the seaweed evaporates water under the sun's warmth, and how the seaweed changes colour, as it dries. Some long moments later the prints are ready to be washed. Each person sways their paper in tap-water-filled trays, and pale beige turns dark blue. The white seaweed silhouettes reveal delicate structural details. Silhouettes on other prints burst out of their outlines, the sea salt has interfered with the chemical process. As we hang the prints to dry in the sea breeze people point out the different forms, sizes, and seaweed kinds once more (see Figure 5.3B). Together, we notice how much the prints and our hands smell like seaweed, like the ocean – in wonder and discomfort.

Making with River Lutz in Thüringerberg

The workshop "Waterscape" within the Nature Design Camp is an open invitation to critically examine the anthropocentric perspective on bodies of water and how we can overcome it. Within

FIGURE 5.3 Watery making. A: Prototyping a water ritual. B: Hanging seaweed cyanoprints to dry. C: Co-designing with River Lutz. D: Seeking new perspectives on the eel.

Source: All photographs by the authors.

an art hike and on-site printmaking workshop, we hike with the participants through the forest of an alpine valley, down to the river Lutz. During the hike, I prepare participants for watery making, without providing any directions, only instructing them to follow the soundscape of the river. Taken from the expertise and interview with regional biologist Anna Weber, I pass on the knowledge about the current state of the Lutz and an ongoing political debate taking place in the habitat. Once we are at the river, we start the process of co-designing with the water, where artworks are created with colour pigment and the rivers' water as a medium. I ask everyone to find a place at the riverbed where they feel they belong. In a first 60-second design sprint, we acknowledge the river as a more-than-human subject and focus only on one river specific element. We reproduce what we see onto water-sensitive paper, without having a look at the actual canvas. In the following 60-second design sprint, I ask the group to expose their drawing directly to the rivers' current (see Figure 5.3C). No hesitation from the group, just a surprising and revealing moment. One participant takes out the soaked canvas only to encounter an altered portrait, where the river directs its own self-portrait. A participant starts to reflect on how nature often is depicted in the visual arts as background, an object instead of being acknowledged as a subject. With this critical thought in mind, we slowly hike back and discuss hands-on practices to overcome the human/nature dualism within artistic methods, eventually to actively make a difference with and for more-than-human actors.

Making with the Öresund Strait in Malmö

For the students, it might be hard to imagine how the needs of Eelgrass and Bristle Worms relate to their lives. A clearer example is needed, so we ask the students to consider an expansive new coastal housing development project in the city's harbour area. We ask them what their ocean actor might want these new harbour neighbourhoods to look like. Maps of the "Nyhamnen" neighbourhood and of the Öresund area that we have prepared beforehand are handed out to the groups. Soon the room is filled with animated chatter, and ideas are being discussed. But the transformation from restless high school students into a cautious Green Crab or an excitable Arthropoda will require more coaxing. So we take them out to meet the ocean. We go out to search along the shoreline, looking through the clear, nearly still water for glimpses of life and subtle activity in the sand. We get to know how the Öresund meets the city and we spot the ships that routinely pass in the distance and under the Öresund bridge. Most of the students grew up here, they have seen this ocean a thousand times. It is their home. But by looking through the lidless eyes of the Herring, or the photoreceptors of the Bladderwrack, they might begin to rethink what it means to call the sea your home (see Figure 5.3D).

Watery Telling

Telling about the Saläng Brook in Borås

After three semesters, the study circle "Flow Feelers" has come to an end, but there is one more intervention to make, one more experience to have. There is now knowledge in the group about where the culverted brook flows, where these pipes are then connected to other pipes that carry rainwater, and where exactly all this mixed water eventually merges with the body of water in the city river. I propose a *walking culvert* activity, as a pendant to the *standing brook* experience in the playground at the very beginning of the study circle. It is intended to be a respectful closure of some kind, for now. I have prepared a liquid mixture of water, wheat flour and blue food colourant, and after having handed out brushes to the participating flow feelers, the group starts to move. We start from the point where the brook disappears underground. One of us paints a big blue dot on the wooden box that protects the "hole". We then move on, and every ten metres or so, someone in the group bends down to the ground to paint a dot: "You are here, and here, and here". (see Figure 5.4A). This slow and peculiar performative procession attracts views. People walking by stop and curiously ask what we are doing, and so the activity gradually transforms into a watery *telling-in-action*. The brook, who has been socially dead for such a long time, is now little by little being reconnected with human beings again.

Telling about the Öresund Shore in Malmö

Standing in a circle, the first spoken words are the participants' memories of seaweed: being tortured with it by their siblings, and culinary strangenesses, some share how disgusting they find to touch seaweed while swimming, which for some is a fear they rarely overcome (see Figure 5.4B). Later, we stroll along the coastline, where some fingers and tips of shoes are submerged in the calm water which barely moves. The most curious thoughts seem to come up during the aimless glance at the wet sand and the stare out to sea. Telling what comes to mind seems slow and in irregular waves while hands or feet are kept busy. The casual conversation that splashes around the group contains rather questions than answers when smaller groups of participants come together for a short moment to share their discoveries and thoughts with each other before they disperse again shortly after. Right after the walk, and again one week later I ask my co-visitors about their memories of the seaweed excursion, and what they took out of it. Only in this retrospective, the participants formulate what had wavered in between us at the shore: the excursion had shifted their perspective towards being surprised, curious and playful. Being there with the seaweed, at the shore, hands-on, had allowed them to explore an alternative relationship with seaweed. Years later I find myself coming back to tell about this watery telling: excited about one of the participants, who had joined the excursion to counter their fear of seaweed and who I had met a few weeks later on a snorkelling tour, then in full curiosity and excitement about what was under the surface.

Telling about the Rote Wand Glacier in Sonntag-Stein

Loose textile pieces flutter in harsh alpine winds, they belong to the structure of a sculpture that is surrounded by a handful of local children and their parents. The "Glacier Ghosts" storytelling performance shares fictional and non-fictional tales tied to environmental education and the imagination of the generation of tomorrow. The regional tailor Claudia Ebster has brought to life how short stories from Finland, Italy and Austria describe those glacier ghosts: manifold beings in white (being alive), blue (being melting) and black (being extinct). Each colour tells us about the current state of the glacier or its habitat. Within one hour we walk with participants to three different stations, where different narrators share a collage of fictional stories about the glacier ghosts living in glacier habitats. While these stories unfold, Isadora, Cilia and Sanna, all collaborating performance artists of the "Sympoietic Society" art collective, take turns wearing one sculpture after the other like a costume attached to their bodies. They use their human vessels to enact the current state of glaciers, incorporating movements of being static, moving and flowing. These intuitively enacted stories create a relatable connection to the Glacier Ghosts sculptures, as if they were alive. In the final act of the performance, we lead the participants to a carefully curated outdoor gallery, where 75 postcards from Finland, Italy and Austria are being displayed (see Figure 5.4C). These postcards were once blank templates, now with their added content they reveal a holistic story of how children from various glacier-related habitats have used their imagination in a series of workshops, to create drawings and short stories of alternative futures of glaciers and their living creatures. This combination of rational and speculative storytelling builds a collective memory and vision of more-than-human beings in the past, present and future.

Telling about the Öresund Strait in Malmö

When we get back to the workshop room, it is time for the students to tell their stories. "You have 30 more minutes and then you need to present your proposals from the perspective of our ocean creature". Suddenly things get serious, no more jokes about the names of molluscs or side conversations. The students storm the supply table where we have laid out paper, scissors, glue and other craft materials to build their prototype. "It's not ready!" a girl protests, offended by my request to take a photo of her prototype. Any previous concerns I had about whether they were listening or absorbing anything about their ocean actors were quickly put to rest as they went to work. Showing their ideas with their hands, drawing, cutting and glueing, and electing who will speak for each group. We ask them to build a presentation set up outside, next to the water, because even though they were told to present to us and Alex Cammaroto, the marine biologist from the Öresund pedagogy centre, I want them to feel that the water is also watching (see Figure 5.4D). The first groups present timidly, but Alex taking the character of a talent show

FIGURE 5.4 Watery telling. A: Practising "Walking culvert". B: Sharing seaweed memories by the coast. C: Glacier Ghosts postcards with kids' drawings and stories. D: Students representing proposals in front of the Öresund strait.

Photograph 5.4A by Kennet Öhlund. Remaining photographs by the authors.

host brings just the right amount of humour and seriousness to the matter. Soon the groups start to relax, bringing wit and humour to their presentation. The ocean actors come alive. A platform for people to use: humans above water and blue mussels underneath. An oath for every new resident in Nyhamnen to respect their ocean neighbours. Regulations on how high the building of Nyhamnen can be as to not shadow the light-sensitive eelgrass meadows. It is remarkable how thoughtful each of the proposals is. How easily the students have gone from not being aware of the existence of a creature, to caring and understanding after a couple of hours at the shore.

Closing the Gap – Kinning

Needless to say, designing with *you* – water – can be done in numerous ways. We would however argue that one core element of watery design practices is a focus on relations, and not any kind of relations, but relations between human subjects and more-than-human subjects – instead of between human subjects and nonhuman objects (or resources). For this to happen we actively need to subjectify you who we are trying to engage with. We see our watery design practices as part of the artistic tide of the Hydrocene; they too can flow-on with a "potential for more radical ways of relating to water ethically" (Bailey-Charteris, 2021, p. 431). Ursula K. Le Guin has captured this transformative move in the imperative: "One way to stop seeing trees, or rivers, or hills, only as natural resources, is to class them as fellow beings – as kinfolk. I guess what I'm trying to do is to subjectify the universe because look where objectifying it has gotten us" (Le Guin, 2014, p. M16). We also need to become better at listening to, and learning from, indigenous persons (scholars and thinkers, designers, and activists) who are linked to worldviews where this understanding of kinship is millennia-old and central to this day; we all need to learn or relearn "the grammar of animacy" (Kimmerer, 2013, pp. 49–59). In processes where this understanding is taken into account, the designer can then become a facilitator of relation building, and someone who can encourage reflections on kinship between me and you. You, brook and river, you, sea, glacier, and shore – and you, all wondrous creatures who know these bodies of water as your home. We (seek to) acknowledge you. We (seek to) cherish sameness when appropriate, without forgetting the importance of underscoring and respecting differences (Sachs Olsen, 2022, pp. 320–321). Through watery meeting, making and telling, we seek to be affected by you. We try to understand you as entities and communities with your own agencies.

Encountering you creates familiarity; familiarity can transform into affinity; affinity can spark a sense of kinship. Kinship in this sense is not static; it is a process. When done in intentional ways, building on indigenous worldviews and languages, in English this process could be called *kinning*. Melissa Nelson says:

> Using *kin* as a verb reminds us that kin is always alive. It's a movement and it's a flow and it's a process, kind of like how wind and air flow. Or water, again getting back to that river as kin, as 'kinning.' So *kin* as a verb is permeable and it's movable, and it requires some action or intention.
>
> (Hausdoerffer, 2021, p. 140)

We want to immerse ourselves in you. Bron Taylor articulates the importance of immersive encounters with the more-than-human, and how this is part of the kinning process:

> Direct, visceral, sensory experiences in nature – including experiences of awe and wonder at the beauties, mysteries, and sometimes terrors of nature, especially through personal encounters with nonhuman organisms – are a common pathway to kinship sentiments and ethics. Kinship sentiments and ethics involve perceptions of being related to someone, with corresponding feelings of caring and responsibility toward one's relations.
>
> (2021, p. 31)

But, encountering you, water, also creates strangeness. We adjoin Neimanis' tactic of estrangement and invite to keep things strange when designing with bodies of water (Neimanis, 2023b). We also try to learn how to become good visitors to unfamiliar waters by acknowledging ourselves as water bodies with inner, watery workings that we do not and will never fully be able to understand (Neimanis, 2023a).

We are watery as we capture our liquid practices in this writing. As we do, we let the uncertainties spill out so that some tacit experience can trickle further. We are watery as you flow through our doing, watering down rigid theory and method protocols, and rinsing out intentions of mastering you. Like water meeting the shore and creating a fluid space in between, is the line between theory and practice one to dissolve as well? You are unsteady, and take a different shape in each of our practices, which allows us to grasp your fluidity.

We are watery as we stay afloat, as we are treading deep water,
as we sink into you, and let you sink in, sea arm, lake, pond,
basin, bay, brook, beck, rivulet,
canal, channel, culvert, creek, delta,
estuary, fjord, flood, glacier, gulf,
spring, lagoon, lake, sea, ocean, puddle,
pond, pool, river, aquifer, spring,
strait, swamp, torrent, wash, wetland, cloud,
fog, mist, rain, snow, mixed water, modern water,
process water in the industries, tapped water,
bodies of water, water in bodies.
waterbodies
water bodies
watery bodies
bodies of water
bodies.

Questions

- How can your practice engage with bodies of water in relational ways, embracing more-than-human life and the agencies, histories and futures of water?
- What is the current social imaginary of the body of water you design with/for/as?
- What kind of conversations and interactions are being enabled by a subjectification of a body of water?
- How can design play a role in watery kinning processes?

Notes

1 Aqua Nullius, Terra Nullius, and similar concepts, derive from colonial strategies to "clear" land and water from existing relations, as ways to justify colonisation: "There is nothing here, and no one owns it".
2 In contrast, for design work that directly relates to urban water management and sustainable use of water see van Asseldonk et al., 2017.
3 In the project "Flow Feelers – A study circle in applied ecofeminism for men", initiated and led by Thomas Laurien and Kennet Öhlund, a group of men explored ecofeminist and posthumanist concepts in relation to a hidden away urban culverted brook (see Flow Feelers, 2023; Laurien et al., 2023; Skimmer och härvor, 2024).
4 In the project "Getting in Touch with Seaweed" Anna Schröder explored alternative, and radically non-exploitative, relationships between humans and marine seaweed species within participatory design research (see Schröder, 2022).
5 In the project "Fiumana", the "Sympoietic Society" art collective engaged with artists, scientists and the public to investigate urban bodies of water in Milano. In the "ICE * In Case of Emergency" project by the Sympoietic Society, research, experiments, and interventions were conducted at glaciers, their habitat and with its communities in Finland, Italy, and Austria. In the "Nature Design Camp" project, the Dear Deer artist duo consisting of Isadora Alves and Christoph Matt hosted an interdisciplinary art and science symposia with the community of the UNESCO Biosphärenpark Großes Walsertal, exploring the relationship of water and humans from diverse perspectives (see Sympoietic Society, 2024a; Sympoietic Society, 2024b; Nature Design Laboratory, 2024).
6 The project "Ocean Care" proposes an approach to reimagine different human actions like urban development, food or the collection of ocean data from the perspective of marine species (see Zrajaeva, 2022; Feral Malmö, 2023).
7 English botanist Anna Atkins was the first person who illustrated a book with photographs in 1843. Her cyanotypes of British algae marked the beginning of a new era in botanical visual documentation.

References

Bailey-Charteris, B. (2021). Revealing the hydrocene: Reflections on watery research. *Przegląd Kulturoznawczy, 2 (48)*, 431–445. https://doi.org/10.4467/20843860PK.21.031.14351

Chen, C., MacLeod, J., & Neimanis, A. (Eds) (2013). *Thinking with water*. McGill-Queen's Univ. Press.

Feral Malmö (2023). *Cockles, Herring, and Eel Grass Weigh in on Malmö's Nyhamnen Development* [Online]. https://designandposthumanism.org/2023/01/10/cockles-herring-and-eel-grass-weigh-in-on-malmos-nyhamnen-development/

Flow Feelers (2023). *Flow Feelers* [Instagram]. www.instagram.com/flowfeelers/

Hausdoerffer, J. (2021). Playful kinning: An interview with Melissa Nelson. In G. Van Horn, R. W. Kimmerer, & J. Hausdoerffer (Eds), *Kinship—Belonging in a world of relations. Vol. 2: Place* (pp. 136–147). Center for Humans and Nature.

Kimmerer, R. W. (2013). *Braiding Sweetgrass: Indigenous Wisdom, Scientific Knowledge and the Teachings of Plants* (First paperback edition). Milkweed Editions.

Knechtel, J. (Ed.) (2009). *Water*. MIT Press.

Laurien, T., Jönsson, L., Lilja, P., Lindström, K., Sandelin, E., & Ståhl, Å. (2023). An emerging posthumanist design landscape. In S. Herbrechter, I. Callus, M. Rossini, M. Grech, M. De Bruin-Molé, & C. John Müller (Eds), *Palgrave Handbook of Critical Posthumanism* (pp. 1–23). Palgrave Macmillan. https://doi.org/10.1007/978-3-030-42681-1_42-2

Le Guin, U. K. (2017). Deep in admiration. In A. Tsing, H. Swanson, E. Gan, & N. Bubandt (Eds.), *Arts of Living on a Damaged Planet: Ghosts and Monsters of the Anthropocene* (pp. 15–21). University of Minnesota Press. https://www.jstor.org/stable/10.5749/j.ctt1qft070

Linton, J. (2010). *What is Water? The History of a Modern Abstraction*. UBC Press.

Sachs Olsen, C. (2022). Co-creation beyond humans: The arts of multispecies placemaking. *Urban Planning*, 7 (3), 315–325.

Nature Design Laboratory (2024). *Nature Design Camp*. Online. https://naturedesignlaboratory.com/ [Online]. (forthcoming)

Neimanis, A. (2009). Bodies of water. In *Water* (pp. 82–91). MIT Press.

Neimanis, A. (2017). *Bodies of Water: Posthuman Feminist Phenomenology*. Bloomsbury Academic, an imprint of Bloomsbury Publishing Plc.

Neimanis, A. (2023a). Hydrofeminism. In R. Braidotti, E. Jones, & G. Klumbytė (Eds), *More Posthuman Glossary* (pp. 64–66). Bloomsbury Academic.

Neimanis, A. (2023b, December 12). *Mullets, Breathing (An Ethics of Letting Be Strange)*. [Talk at Symposium]. With and Against the Current: Exploring Hydrofeminism(s), Copenhagen, Denmark. https://artsandculturalstudies.ku.dk/Calendar/2023/with-and-against-the-current/

Schröder, A. M. (2022). *Getting in Touch with Seaweed: Exploring a Non-Exploitative Relationship With a More-Than-Human Actor Through Design Research* [Master Thesis, Malmö University]. www.diva-portal.org/smash/record.jsf?pid=diva2%3A1677428&dswid=9113

Skimmer och härvor (n.d.). *Flow Feelers*. www.skimmerochharvor.net/pageff

Strang, V. (2004). *The Meaning of Water*. Berg.

Sympoietic Society (2024a). *ICE * In Case of Emergency*. https://incaseofemergency.earth/

Sympoietic Society (2024b). *Fiumana*. https://sympoieticsociety.com/

Taylor, B. (2021). Kinship through the senses, arts, and sciences. In G. Van Horn, R. W. Kimmerer, & J. Hausdoerffer (Eds), *Kinship – Belonging in a World of Relations. Vol. 1: Planet* (pp. 30–47). Center for Humans and Nature.

van Asseldonk, Alissa & Scheepers, Renee & Raijmakers, Bas. (2017). *Let's talk water: The Readership in Strategic Creativity at Design Academy Eindhoven*. Eindhoven: Design Academy Eindhoven.

Verlie, B. (2021). *Learning to Live with Climate Change: From Anxiety to Transformation* (1st ed.). Routledge. https://doi.org/10.4324/9780367441265

Zrajaeva, I. (2022). *Ocean Care: Building a relationship of care between the sea and humans with ocean data* [Master Thesis, Umeå Institute of Design]. https://umu.diva-portal.org/smash/record.jsf?pid=diva2%3A1678338&dswid=7436

Åsberg, C., Holmstedt, J., & Radomska, M. (2020). *Methodologies of Kelp: On Feminist Posthumanities, Transversal Knowledge Production and Multispecies Ethics in an Age of Entanglement* (pp. 11–23).

6

WEATHERING WITH STORMS AND GROUNDS AS A MORE-THAN-HUMAN DESIGN PRACTICE

Encountering Winds, Soils and Rocks

Delphine Rumo and Gloria Lauterbach

Introduction

Listen. Patiently. *Weathering with* is a process of alteration over time that focuses on the in-betweenness of human and more-than-human bodies. It is an invitation for changes and adaptations. This chapter encourages more-than-human design practice to meet and critically *weather with* current anthropo-influenced phenomena related to storms and grounds in urban environments, including uncontrollable wind occurrences and the irreversible depletion and degradation of soils and rocks. Storm and ground phenomena are interconnected through irrevocably destabilised carbon cycles: the anthropo-influenced equilibriums that contribute to accelerated climate change, extreme weather patterns, and biodiversity and geodiversity loss. We bring into dialogue and discuss two design practices challenged by encounters with these phenomena.

We explore a conceptual take on *weathering with* that can translate into opportunities to lean into post-anthropocentric worlds. Our dialogue contributes to a form of more-than-human design practice that takes guidance from a posthumanist political ecology. We use the term 'more-than-human' as a reminder that earthly phenomena can have causal powers and capacities on their own, influencing design processes (Clark, 2011; Laurien et al., 2022). A posthumanist political ecology approach allows us to emphasise the interdependencies and entanglements of humans and all earthly phenomena. In our encounters with storms and grounds, it is foremost listening that creates responses, broadening and altering our design practices. In doing so, we are simultaneously challenging established ideas related to storms and grounds and being challenged by our design perspectives.

We *weather with* not merely as a way to move beyond human-centred perspectives by including storm and ground phenomena in design (Tarcan et al., 2022). *Weathering with* also helps us to critically reflect on our practices, slowing down and recalibrating our senses to more-than-human worlds. As we *weather with*, we contribute to design practices that explore pluriversal methodologies and knowledge-making to re-imagine and enact futures toward post-anthropocentric design (Hermansen et al., 2023; Kambunga et al., 2023; Lauterbach, 2022; Oliveira Serpa, 2023; Rumo, 2023; Veselova, 2023). Our task is to probe ways of creating spaces where different worldviews can form while resisting solutionist design strategies.

DOI: 10.4324/9781003467731-8

This contributes to design research and practice that demand a critical redesign of ethical ontological assumptions within the field of design, rejecting human exceptionalism.

In this chapter, we understand *weathering with* as an exercise to explore the space in between us and the storms and grounds we work with. *Weathering*, terminologically, brings a variety of exciting conceptual imprints. Neimanis and Hamilton (2018), for instance, understand the concept of *weathering* as an awareness of how human bodies are interlinked with their environments in a world threatened by polycrises. With this conceptualization, the authors go beyond the understanding of the concept of *weathering* as a mere metaphor and engage with the idea as a situated practice (Neimanis and Hamilton, 2018). No one enters the process of *weathering* the same, no one *weathers* alike. As nothing *weathers* identically, to *weather* is a situated practice that encapsulates a sociopolitical dimension (Neimanis and Hamilton, 2018). We expand the concept of *weathering* to *weathering with* in the following to underline that the concept is always an entangled and reciprocal practice.

We understand the practice of *weathering with* as taking place in steps and with very small nuances. To us, it is a timeless exercise, forbearing and non-directional. It neither fits into a general mould nor does it follow a predefined script. The way we *weather with* our phenomena is inspired by the concept of 'dialogue' coined by the philosopher Martin Buber (2002) and the idea of inward explorations practised by science-fiction author Ursula Le Guin (2023). On the one hand, Buber's humanist dialogical approach supports our *weathering with* practice, helping us to *weather with* because of the other and simultaneously, alongside the other, similar to posthumanist theories (Neimanis and Walker, 2014). In our design practices, we walk towards the other while focusing on the in-between of the weathering bodies (Buber, 2002). Le Guin's practice, on the other hand, means for us *weathering with* through ceaseless introspective explorations of inward spiralling as we reflect on storm and ground phenomena and re-imagining our own situatedness and design activities. Together, Buber's and LeGuin's approaches visualise two different but complementing directions of acting, thinking and engaging that help us build a bridge between our (still too often anthropocentric) design approaches and posthumanist ecologies.

The chapter focuses on a particular challenge for more-than-human design practice: how to meet with the intricate web of large-scale, omnipresent and all-encompassing phenomena that permeate our surroundings. The first part is written in the first person, describing two individual design practices from two design researchers and practitioners. Our working definition of *weathering with* brings forth design processes that are backward, inward, circular and reciprocal. It is an invitation to slow down and lean into patient transformations while acknowledging power dynamics that influence design interventions. As we carefully attend to timescales exceeding human imagination and engage with forces beyond human control, we explore dialogical and introspective design practices of alterations. We explore a practice of *weathering with* storms and grounds that navigate the complex and nuanced interplays between broader ecosystems and individual engagements, slowly interweaving our work through the dependency of winds, soils and rocks. Considering the distinctness of the phenomena we engage with, we also individually *weather with*. We alter with storms and grounds, we explore potential forms of interaction, empathise with the phenomena, rehearse our limits, experience varying timescales, and depart from the design practices we are familiar with.

The chapter aims to answer two questions: How can *weathering with* grounds and storms inspire more-than-human design practices? How can *weathering with* help to lean into post-anthropocentric imaginaries, embodiments and engagements? In this chapter, we bring our practices into dialogue as we explore *weathering with* grounds and storms.

A Reading of Storms and Grounds

A Reading of Storms, Gloria's Meeting with Niklas

Back then, in 2015, it was in Zurich, Switzerland that I first encountered my more-than-human conversational partner[1] when a storm, known by the name of Niklas, ripped off a big chunk of copper from a local church roof (Figure 6.1). This roof piece, as it crashed into the street, resembled more of a statue monumentalizing weather phenomena than an emergency in need of an immediate response. It crashed there to be met: its colour, resulting from the oxidation process, revealed information about the air quality of the City of Zurich. The shapes pressed into the material carried insights about the force of the respective storm. The incident itself revealed challenges around building techniques in times of constantly increasing extreme weather phenomena. Mostly, to me, the monument allowed for a brief moment to imagine an entirely different worldview, one in which we live *with* the earthly phenomena we are part of.

After this occurrence in 2015, I started to reflect on the societal benefits of *weather monuments* compared to seeing a *roof debris* only as a disturbance or threat in need of immediate removal. From this moment onwards, to me, the way we deal as societies with more-than-human phenomena, such as the storm event in Zurich, reveals a mindset of looking at, of operating the world instead of being genuinely curious about more-than-humans and openly looking *for* them. In such a world, human safety, human needs and human infrastructures always have priority over guardianship, cohabitation and the urgently needed mediation between humans and more-than-humans. In a first step, this storm occurrence did help in laying bare the politics of power I am part of. In a second step, the alternative postures I accessed through the monument enabled me to imagine urban environments from a new point of view.

For the field of more-than-human design, my work can be seen as a blueprint for how to exchange single bricks of storylines to create new narrations. If we transform a storm debris into a weather monument and then observe where we stand, the following questions (and many others) arise: how does a world in more detail look like that monumentalises storms? What relevance does the weather has in that lifeworld? What is the relationship between humans and more-than-humans in such a worldview?[2]

It became my interest to work on keeping these imaginaries alive. Consequently, this meant building a rework of the storm monument as encountered in Zurich, Switzerland, and to plant it in Espoo, Finland. This rework, 'Kreuztstrasse', became the artistic output of my doctoral thesis through which I have been able to research also in practice how else cooperation between humans and more-than-humans may look like (Figure 6.1).

To read storms and to cooperate with them required me to expose myself to storms, to attend and observe them more carefully than I may have ever done before and to follow emerging, deepening leads. The making of storm recordings (video and sound) for instance as well as the process of learning how to build a roof myself – one way to understand what holds and keeps materials together that are constantly exposed to the weather – allowed me to plant the questions and imaginaries in the lifeworld I currently stand. Not surprisingly, 'Kreutzstrasse' also contributed to this process by revealing the capability of continuously generating new imaginaries and responses through the link that has been created between two locations (Zurich, CH and Espoo, FI) and their asynchronous weather patterns. Expressible hence as a *dialogical work* it stirs for me the question if a lush connection can be established between different occurrences, times, materials and realities.

FIGURE 6.1 Storm and ground phenomena in Zürich and Helsinki, including a roof piece ripped off a church by storm Niklas in 2015 in Zürich, the making of the rework of the storm monument in Otaniemi in 2018, a bedrock soon to be blasted by explosives, and piles of grounds on a construction site in Laajasalo, Helsinki, in 2023.

Source: Stadtpolizei Zürich.

For the field of more-than-human design, the work and working process allow me to ponder what *dialogical* could mean for the field and how a redefinition may contribute to creating spaces for more caring encounters. And, further, the discussed work can be understood as an exercise how we as design researchers can open up spaces where phenomena stay present and unfold on their own terms to be observed and attended.

A Reading of Grounds, Delphine Sitting on a Bedrock

I have been meeting grounds entangled with Finnish construction since 2021. These grounds include bedrocks, boulders, gravels, clayey soils, engineered soils and contaminated soils that are blasted, excavated and piled on sites of construction, waiting to be removed, reprocessed, reused, or landfilled. In the fall of 2023, I encountered a piece of ground threatened by construction in Laajasalo, on the East side of Helsinki. The ancient landscape was part of the largest exposed Precambrian rocks in Europe (City of Helsinki, 2019). Aged over 3.5 billion years, more than millions of human lifetimes, the bedrock showcased areas of grey granite, resembling a washed-up whale (Figure 6.1). As I lingered and sat on an exposed tip of this 'whale', admiring the undulating grey and pink geological formations, I could hear the sounds of heavy machinery nearby and feel the rock beneath me shaking.

The plan for the Laajasalo area is to build new roads, followed by apartments and a tramline that will connect the island to the city centre (City of Helsinki, 2022).[3] In the last decades, the City of Helsinki's urban planning strategies have mainly focused on physical growth through construction to improve the city's financial resources and investment profile (City of Helsinki, 2021). In this economic context, massive volumes of ground become entangled with the forces of urban expansion and industrialised construction, where current building techniques lead to the alternation and removal of soils and rocks. To ensure the city's continued growth on solid foundations, the demand for urban land intensifies. In turn, urban expansion and infrastructural development put pressure on the local geology, causing shifts in the delicate balance of the underlying grounds. Soils are sealed and covered with concrete and stones, preventing exchanges of energy, water and gases and bedrocks are drilled and blasted, creating noise, gas, dust and ground vibration that disrupt local ecologies (European Environment Agency, 2021; Scalenghe and Marsan, 2009). Soon after my visit, the bedrock I was sitting on forever disappeared, blasted by explosives, and flattened for the construction project. A dominant drive for profit maximization through material growth resulted in the subjugation of the ancient rock.

While sitting on the bedrock, I paid attention to the nearby old-growth forest of Stansvik. Grounds around the bedrock had maintained the plant life of the forest, which could access nutrients, water and support for anchorage. In turn, the forest was interacting and influencing the large rocky landscape, slowly weathering the bedrock I was sitting on. Salty water and winds continuously abraded the surface of the rock, creating fine particles of sand and dust (Tuuling et al., 2011). The bedrock had worn away, slowly mineralizing younger and thinner layers of clay, silt and organic matter that contributed to the composition and fertility of the surrounding forest (Rajakaruna and Boyd, 2021). In addition, physical, chemical and biological weathering processes, including acidic rain, ice formation, the growth of plant roots and microorganisms and winds, broke down the rock into smaller particles over time, encouraging dynamic cycles of nutrients that could support life (Hack, 2020).

Walking around Lajasaalo, I also noticed how the ancient bedrock was entangled with the fast-paced lives of earthwork contractors hired by the city to construct roads and residential buildings (Figure 6.1) (Rumo, 2023). In their world, construction was bathed in an illusion of measure and control over the rock. To complexify the situation, forest activists were resisting the city's development plans, camping in the Stansvik's forest (Kangasluoma, 2023). These entangled configurations: gatherings of industrialised activities, expansion of infrastructures, intensification of capital-seeking investment, discourses of economic growth, the disappearance of bedrock, depletion of soils, forest activists, contractors, heavy machinery, geological temporalities, soils and rocks, outlined a certain quality of mediation between human and more-than-human bodies. Walking around Laajasalo, I could experience slow geological processes, moments of unequal relations of domination, and acts of resistance.

I was *weathering with* ground phenomena as I followed soils and rocks across sites of construction, meeting people closely working with grounds, driving with people from site to site, sitting on bedrocks, taking pictures, recording videos of rocky landscapes, and practising slowing down. As I was attending to fast-paced economies of construction and slow geological processes, my understanding of soils and rocks changed over time, generating spaces to engage with these phenomena otherwise. Other *weathering* processes happened through exploratory creative interventions that I describe further on page 86. Paying attention to the grounds entangled with construction helped me build ways to engage with different agencies of bodies that *know* differently: contractors, rocks, soils, machinery, design practitioners and researchers, and my gut microbiota, eroding each other over time. I found that *weathering with* grounds in that way was a process of slow exploration and negotiation.

Two More-Than-Human Design Practices

Gloria's Research Practice with Storms

Storms bring and take along, they stir, and alter materials and environments; storms enable earthly matter to be in flux (Ingold, 2007); they take place above ground, give voice to their environments as well as imprints of their strength; nothing remains silent, nothing in its present form. Storms inherently belong to the earth as they listen and react to anthropocentric practices.

In Helsinki, located on the shoreline of the Baltic Sea, winds are part of the everyday. Here, a warm, windy summer day is different from a warm, windless day. Also, the soundscape of a windy city such as Helsinki is different from a city without winds. Even though winds are invisible, they do alter their environments – and everyone part of those – quite drastically.

As "materially and temporally ambiguous", pointedly characterised by Lisa Moffitt (2023), winds greatly impact all levels of their lifeworlds: winds create and change microclimates within seconds, transport microparticles over hundreds of kilometres, destabilise structural as well as temporal components of buildings, infrastructures and environments. Winds erode and intensify (Moffitt, 2023). Winds give energy to power infrastructures and create movement; they energise, challenge and harm. And they transform through the environments they pass by. All these factors give storms and winds both a material and conceptual quality which I explore in my research and practice as a more-than-human design researcher.

Considering the force and scale of wind phenomena, their unpredictability, invisibility and hard-to-control nature makes winds, from human point of view, uncooperative. However, with the risks of extreme weather events advancing, storms will increasingly consume our environment (European Environment Agency, 2021; Kahraman et al., 2021). This increase will give them more voice, visibility and power in the near future, and humans the possibility to learn how to recalibrate anthropogenic positioning. During my research with storms I learnt that to follow such vast and ungraspable phenomena meant that I had to step back from my own expectations, past learnings and rigid sense-making automatisms. Methods such as *weak action* by artistic researcher Tuija Kokkonen (2021) as well as *rehearsing* by scholars Sabeth Buchmann and Constanze Ruhm (2013) offered me counter-approaches that allowed me to focus on the in-between of a storm and myself as two *weathering* entities. In her work, Kokkonen (2011) tests the borders of societies as performative bodies. *Weak action* enabled me to actively step back and re-evaluate the setting I was part of, to give other/s room to unfold. Buchmann et al. (2016, p. 16) work with the notion of *rehearsal* as a "zone of transition". In their work and research, a *rehearsal* is a marker of processes of renewal crossing through power structures of sociocultural settings. As an endless form of practice, and an endlessly transformative practice, to rehearse means, among others, to reflect, improvise, interrupt and make futures (Buchmann et al., 2016).

Through learning about the approaches by Kokkonen (2021) and Buchmann and Ruhm (2016) a cooperation with winds and storms seemed suddenly accessible. Together, weak action and rehearsal are methods that can help eroding our anthropocentric structures, rebalancing agency between humans and more-than-humans and, at the same time, enabling bridge-building to more-than-humans other/s. The more I practised cooperating with a storm rework, the more agents were to be included into the rehearsal and the more I had to re-adjust my own position and agency within. While working with very diverse more-than-human others such as wind, nesting birds, surrounding trees, the sea close by – all shaping the work – it was the notion of sound that started to foreground and so I gave it space to unfold. I started to work with the audio and sound recordings I had collected at the beginning of my research and explored them side-by-side with my own experiences with storms, the notes I collected in the form of a storm diary and a variety of readings about storms.

The pathway I followed led me to an attunement practice with a focus on my senses. The storm recordings captured wind sounds the moment they met their environments. The forms and shapes of these *singing environments* I have heard of many times before, be it winds blowing through construction sites or birch trees or wind stirring up water. Yet, totally failing to reproduce them made me experience a big distance from the environments I was part of. So I started to explore the idea of a 'Storm-Sing-Along' or 'More-than-Human Karaoke' to share my observations with a wider audience and to hear out others on how to build the meaningful bridges to the more-than-human world we are lacking.

The 'More-than-human Karaoke' is best described as a spatial installation that includes and merges the voices of the karaoke singers and the various environments shown in the video works (Figure 6.2). As a multi-dimensional experiment, it aims to probe a variety of perspectives and voices by exploring the lifeworld of more-than-human sounds. The karaoke format inspires collaboration and co-creation with other singers and the more-than-humans encountered in the video works. Through participatory engagement – shared sound making – the storm sceneries become easily accessible, voices of a few tend to come closer to the environments we see in front of us. This exercise is a rehearsal, it is a realization of the distance to nature we have grown accustomed to and that requires a rethink of anthropocentric practices.

'More-than-Human Karaoke' at Feral Festival (2023) and ECQI (2024).

Delphine's Research Practice with Grounds

Grounds seemingly provide a sense of stability; solid earthly surfaces, undisturbable, and timeless that are witnessing the ebb and flow of the world around them over countless ages. They seem to act as platforms and social stabilisers for the construction of buildings, highways, bridges, subways, sewers and power lines (Toland et al., 2018). But 'permanent' grounds are formed of constantly moving geo-tectonic plates and intricate sedimented layers (Meulemans and Granjou, 2020; Clark, 2011). They can change wind patterns and ingest eroded rock debris, animals and plant materials, creating deposits of porous living layers (Povinelli, 2016; Latour and Weibel, 2020).

When I started to pay attention to the grounds entangled with construction, I was drawn to the massive piles of soils and rocks shuffled in corners of the sites I visited. I practised following these piles for a few seasons from one construction site to another, and from storage areas to landfill sites, hanging out with clayey soils in snowy landscapes, and sitting on bedrocks in the summertime (Rumo, 2023). To *follow* and *hang out* with grounds was inspired by ethnographic

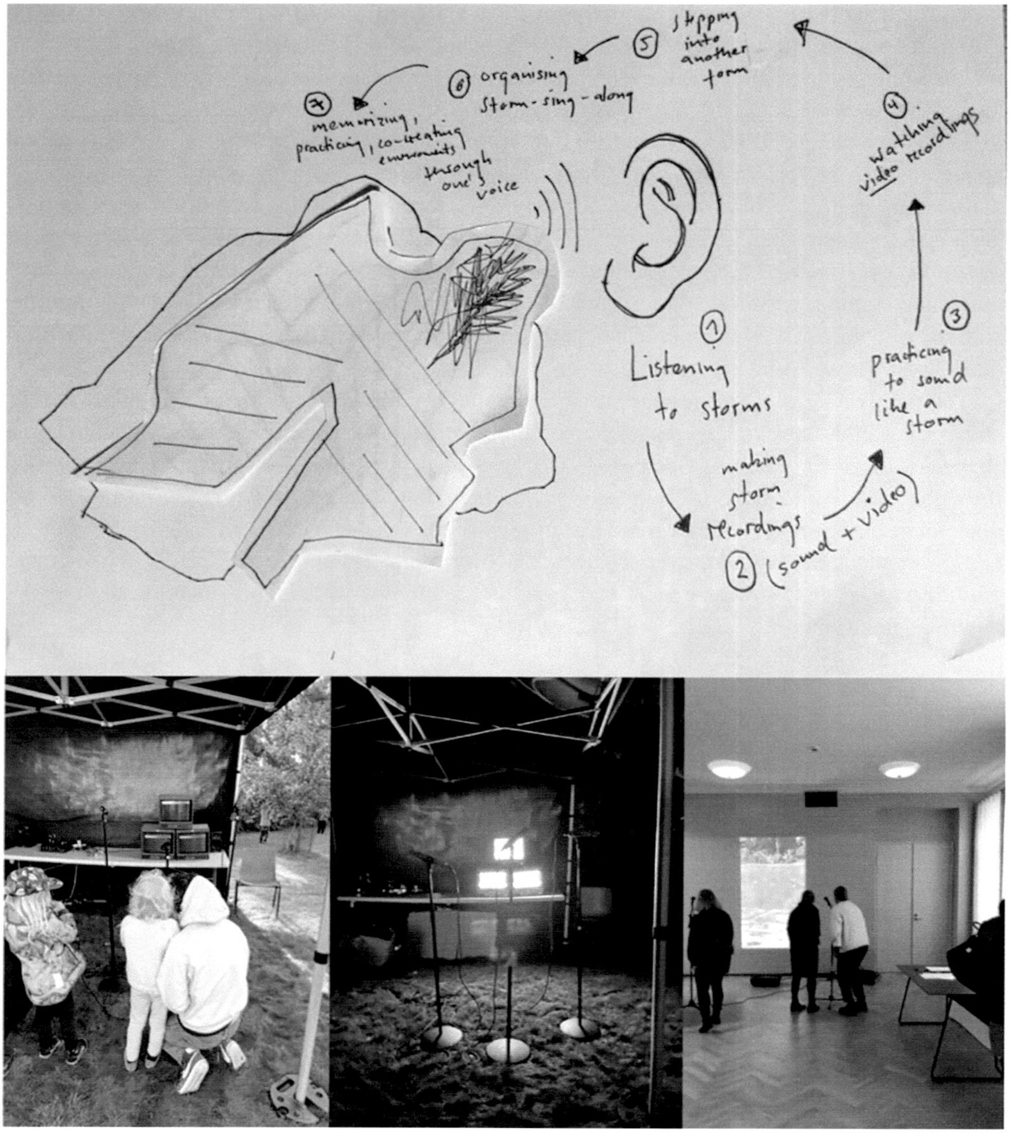

FIGURE 6.2 This sketch shows how *weathering with* may take place through 'More-Than-Human Karaoke'. Here, in this process, I am leaving open where practices lead to, whom they all include, invite, inspire and engage. Storm sounds teach us to imitate through more carefully listening to and perceiving the environments we are part of.

methods that included walking and driving interviews with contractors, engineers and city planners and participant observations where I gathered pictures and videos to help me narrate the life of soils and rocks (Geertz, 1998; Sodero, 2019). It was a snowballing process, one person referring to another, and one encounter with a certain ground opening questions that brought me to another site.

As I followed and hung out with grounds, my initial curiosity for metrics – the volumes of excavated and blasted soils and rocks, and their associated carbon emissions and biodiversity losses – became entangled with ethical considerations. I found that grounds were mainly understood as instrumental to profit maximization and urban expansion. Planners and engineers described soils and rocks as 'good' or 'bad', 'profitable' or not, and 'useful' or not. I observed how these words translated into values that shaped decision-making regarding grounds. To follow and hang out with soils and rocks helped me to erode extractive worldviews and their normative 'sustainable' solutions, which mainly focused on profit-making and material growth (Rumo, 2023).

I physically met more grounds, engaging with complex and messy entanglements of human-to-ground relations. I noticed how soils and rocks had existences that radically differed from human activities, having causal powers and capacities of their own (Clark, 2011; Rumo, 2023). Planners and engineers were well aware of the intricate relationships between grounds and urban development. However, the role of soils and rocks in re-orienting planning strategies and reconfiguring the city was often overlooked (Robertson, 2020; Rumo, 2023). Structurally recalcitrant soft and liquid clayey soils and surprisingly ungraspable rocky geologies resisted and challenged construction processes, influencing plans for urban expansion (Rumo, 2023). As I met more grounds, my perspective changed, altered by intricate stories of resistance and inner transformations intertwined with grief and joy (Rumo, 2023). The ungraspability of ground phenomena also inspired me to explore new types of designerly engagements, weaving encounters with the tangible world with introspective explorations.

I further attended to troubled geological landscapes by practising introspective acts of critical reflection. Drawing from Le Guin's (2023) practice of exploring ways to go inward, and following my gut feeling, I linked my observations of systemic land degradation with people's inner world. In the fall of 2023, I invited rocks, excavators, microbiota and gut feelings of urban dwellers to participate in the intervention 'Rocking Guts'. The intervention was presented at the event 'Feral Festival. Living-with Feral Ecologies. A Festival' during Helsinki Design Week 2023. 'Rocking Guts' included video and sound following the life of soils, rocks, and people's inner worlds, and the writing of postcards (Figure 6.3). A short film showing side-by-side activities of construction and alive human gut microbiota featured the sound of my own gut microbiotas' environments captured with stethomicrophones. The intervention was an open invitation for people living in the Helsinki region to engage in conversations about large-scale construction alongside their personal gut experience. Participants were invited to explore acts of resistance against extractive economies of urban development. Once written on postcards, gut feelings were posted to decision-makers responsible for overseeing local construction projects (Figure 6.3).

The participatory intervention 'Rocking Guts' was an exercise to *weather with* grounds collectively. I wanted to animate stories of grounds with people's embodied and participatory gut feelings (Figure 6.3). I weaved stories of ground extraction with people's gut microbiota; the collective of microorganisms living in their digestive tracts that influence their emotions and cognitive functions (Appleton, 2018). The idea was to retreat to the darkness of people's guts, to the wet and peaceful inner environments of human bodies, to seriously (re)consider the existence, agency and experiences of grounds. The intervention was an invitation to explore inward spiralling and to re-imagine ground phenomena beyond stories of extraction and destruction. Experiencing gut feelings was a way to recognise grief in the face of massive land extraction, and perhaps joy in resisting more destruction.

FIGURE 6.3 A still showing bedrock and gut microbiota from the short film 'Rocking Guts' presented at Feral Festival (2023). Participants writing postcards to decision-makers from the City of Helsinki during the interventions. Sketch showing how weathering with ground phenomena is an invitation to meet with extractive economies of urban development, gut microbiota, soils, rocks and people's inner world.

Bringing Weathering Practices into Dialogue

In our design interventions, we articulated weathering with storms and grounds as a critical practice of continuous recalibration to more-than-human phenomena through dialogical and introspective encounters (Figure 6.4). First, we explored the use of *weathering with* as an invitation to reimagine language and worldviews. Second, we *weathered with* as an invitation to embody more-than-human phenomena through sounds and other senses. Finally, we *weathered with* to open spaces for participation, inviting people inside and outside of the design field to engage in further breaking down and abrading our design position.

Weathering through Language and Worldviews

As we met storm and ground phenomena, some entrenched ideas, mindsets and worldviews began to erode. While *weathering with* grounds, we encountered troubled geological landscapes: blasted, excavated, and extracted soils and rocks. Following and hanging out with these landscapes made us aware of different temporalities: fast-paced industrialised construction and slow geological processes (Rumo, 2023). We surrounded ourselves with the possibility of being with grounds otherwise than the forms of engagements that dominate current construction techniques, industrialised activities, and economies driven by material growth and profit maximization (Neimanis and Hamilton, 2018). Well-established ideas and conventional language wore down in light of the anciency of geological landscapes and their irrevocable disappearance

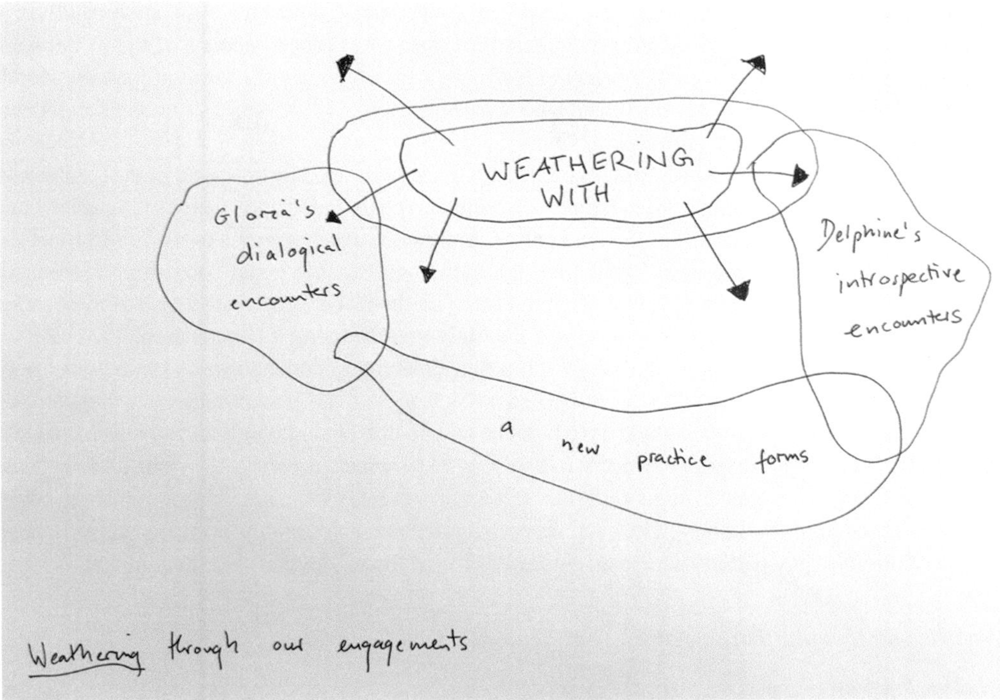

FIGURE 6.4 Our process of *weathering with* in dialogue

(Rumo, 2023). Similarly, *weathering with* storms eroded our ideas about wind patterns and their challenges for urban environments, such as the constant increase in extreme weather phenomena and the near-future limits of densification strategies.

Our *weathering with* processes enabled us to re-evaluate how to be and attend to the irreversible depletion and degradation of soils and rocks, and uncontrollable wind occurrences. To *weather with* earthly phenomena challenged our senses, memory and human presence, softening our design practices. We let ourselves be altered by grounds and storms which changed our perceptions of these phenomena and helped us lean into post-anthropocentric worlds. As we *weathered with* grounds and storms, we also engaged with new ideas of human-geologies-weather patterns interdependency and responsibility, moving beyond dichotomies such as large-small, inward-outward, life-death, biotic-abiotic, fast-slow, action-inaction, air-rocks and up-down. The practice of *weathering with* was a humbling experience, one that offered entry points into a deeper exploration of alternative languages and worldviews.

Weathering through Embodiments, Sounds and Other Senses

Weathering with also enabled us to enter into a dialogical work through embodiments, sounds and other senses. We were curious to *weather with* grounds and storms as a way to physically engage with the ungraspability and unknowingness of large-scale and omnipresent phenomena. To encounter storms dialogically (as opposed to monologically) meant that we never knew what our counterpart would 'voice' next (Lauterbach, 2022). For this type of genuine encounter to happen, we needed to give space to the other. As Buber (1983) highlights, we needed to acknowledge that a dialogue is always reciprocal, very subtle, and rare but transformative. A dialogue is reciprocal and transformative because it only takes place because of the in-between of one and the other. This makes *weathering with* a process of constantly following and responding while leaving space between those who *weather*. Buber (1983, p. 13) himself calls this practice a "waiting towards and giving room".[4]

A practice of "waiting towards and giving room" to grounds and storms could help us enter into conversations with more-than-human phenomena with as few preconceived ideas, expectations and answers as possible (Buber, 1983, p. 13). Further, in/through these encounters we got to meet the phenomena we worked with in different ways by attending, listening to, and observing more carefully the encounters themselves, with one's main focus on the in-between. In the case of *weathering with* storms, grasping the process became possible through rehearsing it (Lauterbach, 2022). As we rehearsed designing with storms, looking for missing links, interconnections, and in-betweenness, we intuitively developed a tacit and embodied knowledge of the phenomena. The intervention 'More-than-Human Karaoke' enabled us to practice how to dub or sing winds and sounds of wildlife and nature. We listened to and recorded wind sounds, experienced a variety of winds and observed how winds themselves transformed with our built environments (for example, through sharp edges versus round edges of buildings). The exercise recalibrated ourselves bodily as our senses were challenged to observe, grasp and attune to an entirely different force, an other-than-human one.

Weathering through Participatory Engagements

Weathering with processes also opened spaces for participatory explorations and further discussions about storm and ground phenomena. We *weathered with* each other's practices, discussing and discovering the interconnections between grounds and storms. We also brought our observations into conversations with other scholars and the broader public. The 'Feral Festival' was one example of an event where discussions about storms and grounds further opened up and almost continued their own lives. Through the intervention 'Rocking Guts', for instance, we *weathered with* people involved in decision making regarding the use of urban spaces. We engaged with the site of Lapinlahti, a building under threat of real estate development, and people involved in the initiative 'Kaikkien Helsinki'[5] who were dedicated to democratic participation in decisions regarding the city of Helsinki. As we shared ideas and exchanged contacts, we ran strings between projects related to urban grounds. Similarly, the intervention 'More-than-Human Karaoke' enabled us to practice *weathering with* together with various locations and people. In the installation, collaborative sound-making prompted 'singers' to develop an awareness of how they performed and attended environments both individually and collectively.

Through these discussions and participatory engagements, we facilitated interventions that opened spaces for further encounters between grounds, storms and people. Additionally, we let our design practices be altered by a broader range of people. This helped us carve more-than-human design practices that further engaged with a post-anthropocentric worlds that would interlink ground and storm phenomena with a wider range of people, sketching opportunities for change through design and non-design approaches alike.

In Conclusion, Weathering Beyond

Considering the current polycrises, and our position within an emerging field of more-than-human design, we found it important to lean into post-anthropocentric worlds. We used the concept of *weathering with* as a way to focus on the situatedness and power dynamics at play when engaging with storm and ground phenomena. We brought the concept into practice by shedding light on the embodied and political aspects of uncontrollable wind occurrences and the depletion and degradation of soils and rocks. We *weathered with* storm and ground phenomena through dialogical and introspective interventions, offering a productive way of discussing the entanglements of our research practices with these phenomena. As we *weathered with*, we also expanded on the idea of *designing with* more-than-humans, exploring ways to change the power structures we were entangled with, including established languages and worldviews.

This chapter contributes to emerging research practices that explore encounters and engagements with large-scale phenomena. Examples of creative work related to grounds include storytelling, for instance, based on the relationship between contaminated grounds and designers (Latva-Somppi, 2020), representations of altered landscapes and historical strata of places through sculptural encapsulations (Brooks, 2017), and further exploration of human-decentring perspectives through thinking-with minerals using walking (Lilja, 2021), reading stones as collaborators using sounds (Madsen, 2020), or speculative participative engagements that consider grounds as not inanimate but living phenomena (Akama et al., 2020; Gatto and McCardle, 2019). Additionally, examples of interventions that engage with storms include new positionings where patience and loss of control become part of the practice and product (Devendorf, 2019), engagements that explore how to give air phenomena active roles in architectural designs and processes (Atmos lab, 2022) and investigations of wind as sound to prompt embodied interactions with earthly phenomena (Keenan, 2023).

Bringing storms and grounds closer to our design practices prompted us to engage with temporalities and scales beyond human experiences. We had to acknowledge our fleeting human existences and the limitations of individual and collective actions related to uncontrollable wind occurrences and the depletion and degradation of soils and rocks. Our practices were changed and complexified by diverse layers of discursive, embodied, and participatory entanglements, outlining the limitations of abstracting storms and grounds in more-than-human design. It became impossible to *design with* or *without* these phenomena. They could not be *designed with*, as in *included* in our design practices but only *weathered with*. As we sketched links between storms, monuments, sounds and voices, and between large-scale construction sites and individual gut feelings, *weathering with* outlined the need for more-than-human design practice to move beyond tendencies to reductionism and solutionism (Blythe et al., 2016). This acknowledgement meant altering our design practices and slowing down, exploring the unfolding of empathetic interventions. In this process, our practices became tools for more tender encounters, helping us narrate the in-betweenness, leaving spaces for humbling interactions, as we constantly got altered by grounds and storms, power structures, and troubled landscapes (Tokarczuk, 2019).

Our approaches to *weathering with* can strengthen design practices that explore counter-narratives of established worldviews and contribute to envisioning more inclusive, post-anthropocentric ones. We invite design practitioners and researchers to explore *weathering with* as a design approach that emphasises 1) an enquiry-driven, rather than solution-focused practice, interested in asking questions rather than seeking answers, 2) a sensitivity for bridge-building towards more-than-human phenomena through slowing down and non-linear practices, switching from monological to dialogical relationships, and 3) the situated, embodied, and political aspects of design. In other words, we invite practitioners and researchers to approach *weathering with* as an individual and collective design practice that can generate spaces for pauses and interruptions to explore the in-between, dynamic, and complex realms of *designing with*.

We want to conclude this chapter with a critical reflection on our practices, considering the urgency of the polycrises. We *weathered with* ground and storm phenomena that remained ungraspable, nevertheless deeply interwoven with human lives. This opened potential routes of exploration for more-than-human design practices that can engage with complex exigencies. Additionally, we kept questioning the strength of our design interventions: How can our more-than-human design practices more seriously and vigorously engage with the urgency of increasingly uncontrollable weather patterns and ground depletion and degradation? How can our practices keep up with fast-paced environmental and societal collapses? As we *weather with* further with omnipresent and all-encompassing more-than-human phenomena that permeate our close surroundings we look forward to exposing ourselves to the urgency of the polycrises and exploring ways to strengthen our design practices by sharing and discussing them with and through other/s.

Questions

- How can more-than-human design practices more seriously and vigorously engage with fast-paced environmental and societal collapses related to ground depletion and degradation and increasingly uncontrollable weather patterns?
- How can 'weathering with' help to lean into post-anthropocentric imaginaries, embodiments and engagements?

Notes

1 In the following, there will be terms and qualities used that might be confused with anthropomorphisation of more-than-humans. To use expressions in the text such as "conversational partner" or "storms…listen and react" means here instead, on the one hand, to show the limits of our languages and structures and, on the other hand, the need to exercise how to bridge to others through eroding anthropocentrism from within.
2 May questions be the engine of the field of more-than-human design? May questions help to envision a different form of cooperation with the world and more-than-humans? May asking as many questions as possible be the first step to re-learn how else to learn with others and to change from a monological to a dialogical relationship with more-than-humans?
3 The construction of the tramline that will connect Laajasalo to the city centre started in 2021 with an estimated completion in 2027.
4 In the German translation Buber (1983, p. 13) describes this ethos as "Gegenwartende" [a waiting towards] and "Gegenwährende" [giving room].
5 In English: 'Helsinki for Everybody'.

References

Appleton, J., 2018). The gut-brain axis: Influence of microbiota on mood and mental health. *Integrative Medicine: A Clinician's Journal*, 17(4), 28–32.

Blythe, M., Andersen, K., Clarke, R., & Wright, P. (2016). Anti-solutionist strategies: Seriously silly design fiction. Presented at the 2016 CHI Conference, San Jose, 4978.

Brooks, D. (2017). *Rock, Mosquito and Hummingbird: A Prehistory of Governors Island.*

Buber, M. (1983). Ich und Du [I and Thou]. Reclam.

Buber, M. (2002). *Meetings*. Autobiographical fragments. Routledge.

Buchmann, S., & Ruhm, C. (2013). Subject put to the test. *Texte zur Kunst*, 90. www.textezurkunst.de/90/buchmann-ruhm-subjekt-auf-probe

Buchmann, S., Lafer, I., & Ruhm, C., Eds (2016). Putting rehearsals to the test. *Publication Series of the Academy of Fine Arts Vienna*. Sternberg Press.

Burnett, B., & Evans, D. (2016). *Designing Your Life: How to Build a Well-Lived, Joyful Life*, 1st edition. New York: Knopf.
City of Helsinki (2019). Soil and bedrock [online]. *Helsingin kaupunki*. Available from: https://www.hel.fi/helsinki/en/housing/plots-land-buildings/earth-bedrock/ [Accessed 7 March 2023].
City of Helsinki (2021). *Helsinki City Strategy 2021–2025: A Place of Growth*. Helsinki.
City of Helsinki (2022). Laajasalo. *City of Helsinki*. Available from: www.hel.fi/en/urban-environment-and-traffic/urban-planningandconstruction/urbandevelopment/laajasalo [Accessed 15 January 2024].
Clark, N. (2011). *Inhuman Nature: Sociable Life on a Dynamic Planet*. London: SAGE Publications Ltd.
Devendorf, L. (2019). A machine for being frustrated. Available from: http://artfordorks.com/a-machine-for-being-frustrated/ [Accessed 15 January 2024].
European Environment Agency (2021). What will the future bring when it comes to climate hazards – Overview. *Briefing* [online]. Available from: www.eea.europa.eu/publications/europes-changing-climate-hazards-1/what-will-the-future-bring [Accessed 15 January 2024].
Geertz, C. (1998).Deep hanging out. *The New York Review of Books*, 45(16): 69–72.
Hack, H. (2020). Weathering, erosion, and susceptibility to weathering. In M. Kanji, M. He, L. Ribeira e Sousa, Eds, *Soft Rock Mechanics and Engineering*. Cham: Springer International Publishing, 291–333.
Haraway, D. (2016). *Staying with the Trouble: Making Kin in the Chthulucene*. Durham: Duke University
Hermansen, M. T., Chilet, M., Ureta, C., & Marín, P., Eds (2023). *Design For More-Than-Human Futures: Towards Post-Anthropocentric Worlding*. London: Routledge.
Ingold, T. (2007). Earth, sky, wind, and weather. *The Journal of the Royal Anthropological Institute*, 13, S19–S38. www.jstor.org/stable/4623118
Kahraman, A., Kendon, E. J., Chan, S. C., & Fowler, H. J. (2021). Quasi-stationary intense rainstorms spread across Europe under climate change. *Geophysical Research Letters*, 48, e2020GL09236 https://doi.org/10.1029/2020GL092361
Kambunga, A. P., Smith, R. C., Winschiers-Theophilus, H., & Otto, T. (2023). Decolonial design practices: Creating safe spaces for plural voices on contested pasts, presents, and futures. *Design Studies*, 86, 101170.
Kangasluoma, E. (2023). Ympäristö | Mielenosoittajat leiriytyivät Helsingin Stansvikin kiistanalaiseen metsään [online]. *Helsingin Sanomat*. Available from: www.hs.fi/kaupunki/art-2000009911019.html [Accessed 20 November 2023].
Keenan, F. (2023). The drum and silk: The experience of imitating wind as sound. *Venti Journal*, 2(3). Available from: www.venti-journal.com/fiona-keenan [Accessed 15 January 2024].
Kokkonen, T. (2011). Non-humans and performance. *Journal for Artistic Research,* 0. https://doi.org/10.22501/jar.7736
Kokkonen, T. (2021, August 13). *Performance by Non-Humans*. Tuija Kokkonen. http://tuijakokkonen.fi
Latour, B., & Weibel, P. (2020). *Critical Zones: The Science and Politics of Landing on Earth*. MIT Press.
Latva-Somppi, R. (2020). Traces from the Anthropocene: Working with soil. *RUUKKU - Studies in Artistic Research*, 14, 1–26.
Laurien, T., Jönsson, L., Lilja, P., Lindström, K., Sandelin, E., & Ståhl, Å. (2022). An emerging posthumanist design landscape. In S. Herbrechter, I. Callus, M. Rossini, M. Grech, M. de Bruin-Molé, & C. John Müller, Eds, *Palgrave Handbook of Critical Posthumanism*. Cham: Springer International Publishing, 469–491.
Lauterbach, G. F. (2022). *The Open Seam. On Becoming a Knot with a Storm-sculpted Roof, on Speaking-with from Within*. Helsinki: Aalto University.
Le Guin, U. K. (2023). *Space Crone*. London: Silver Press.
Lilja, P. (2021). Tracing matters of scale by walking with minerals. *Nordes Conference Series*.
Madsen, T. M. K. (2020). *The Voices of Stones*, performance.
Meulemans, G., & Granjou, C. (2020). Soils: A new frontier for environmental knowledge and policies. *Revue d'anthropologie des connaissances*, 14 (4).
Miller, E. (2019). *Reimagining Livelihoods: Life beyond Economy, Society, and Environment*. University of Minnesota Press.
Moffitt, L. (2023). Designing with wind: Architecture's model environments. *Venti Journal*, 2(3). Available from: www.venti-journal.com/lisa-moffitt [Accessed 15 January 2024].

Neimanis, A., & Hamilton, J. M. (2018). Open space weathering. *Feminist Review*, 118 (1), 80–84.
Oliveira Serpa, B. (2023). Militant design research: A proposal to politicize design knowledge-making. *Revista Disena*, 22, 1–15.
Povinelli, E. A. (2016). *Geontologies: A Requiem to Late Liberalism*. Duke University Press.
Rajakaruna, N., & Boyd, R. (2021). Geoecology. *Oxford Bibliographies*. Available from: www.oxfordbibliographies.com/display/document/obo-9780199830060/obo-9780199830060-0125.xml [Accessed 18 December 2023].
Robertson, S. A. (2020). Re-placing soil and its mattering in more-than-human cities. *Australian Geographer*, 51(3), 307–324.
Rumo, D. (2023). Grounded circularity: The livelihoods of surplus clay. *Journal of Cultural Economy*, 16 (4), 631–639.
Scalenghe, R., & Marsan, F. A. (2009). The anthropogenic sealing of soils in urban areas. *Landscape and Urban Planning*, 90 (1), 1–10.
Shapiro, J., & McNeish, J. A. (2021). *Our Extractive Age: Expressions of Violence and Resistance*.
Sodero, S. (2019). Vital mobilities: Circulating blood via fictionalized vignettes. *Cultural Geographies* 26 (1): 109–125. https://doi.org/10.1177/1474474018792656.
Tsing, A. L. (2015). *The Mushroom at the End of the World: On the Possibility of Life in Capitalist Ruins*. Princeton: Princeton University Press.
Sundberg, J. (2011). Diabolic caminos in the desert and cat fights on the Río: A posthumanist political ecology of boundary enforcement in the United States–Mexico borderlands. *Annals of the Association of American Geographers*, 101 (2), 318–336.
Tarcan, B., Pettersen, I., & Edwards, F. (2022). Making-with the environment through more-than-human design. *DRS Biennial Conference Series*.
Tokarczuk, O. (2019). The tender narrator. [Nobel Lecture in Literature]. The Nobel Foundation. Swedish Academy, Stockholm. www.nobelprize.org/uploads/2019/12/tokarczuk-lecture-english-2.pdf
Toland, A., Noller, J. S., & Wessolek, G., Eds (2018). *Field to Palette: Dialogues on Soil and Art in the Anthropocene*, 1st edition. Boca Raton: CRC Press.
Tsing, A., Swanson, H., Gan, E., & Bubandt, N. (2017). *Arts of Living on a Damaged Planet*. Minneapolis: University of Minnesota Press.
Tuuling, I., Bauert, H., Willman, S., & Budd, G. (2011). *The Baltic Sea – Geology and Geotourism Highlights*. Tallinn: NGO Geoguide Baltoscandia.

7

DESIGNING WITH PLANETARY ARTIFICIAL INTELLIGENCE

Marcos Chilet, Martín Tironi, Iohanna Nicenboim and Joseph Lindley

Introduction

Human-nonhuman entanglements are especially vibrant in Artificial Intelligence (AI). Its impact is undeniable, and its potential is limitless. AI's materiality comes from distant and diverse regions, such as lithium salt lakes in Bolivia and mines in the Congo. It is used to navigate drones over Yemen, direct immigration police in the United States of America, and modulate credit scores worldwide to assess human behaviour and risk (Crawford, 2021, p.218). AI operates on a planetary scale, affecting humans and nonhumans on Earth as a 'registry of power' (Crawford, 2021, p.8); for that reason, the challenges posed by AI cannot be underestimated. To address these entanglements, it is crucial to adopt design approaches that are attuned to the complexities of AI. This chapter asserts that more-than-human design (MTHD) perspectives are needed to address significant epistemological and practical challenges that emerge with the broad introduction and application of AI in our societies.

The impact of AI on the Earth's territories and communities raises multiple concerns around extractivism, from the intensive extraction of resources for its operation (Crawford, 2021) to the ubiquitous *data extraction* (Lehuedé, 2024; Couldry & Mejias, 2019). As corporations expand their digital reach and exert control over information and resources, it is crucial to consider regulations and measures to mitigate potential risks. This chapter argues that one of the most significant risks posed by AI is the repackaging of its potential into an agenda for reproducing the forces of anthropocentrism.

According to Braidotti (2022), the anthropocentric socio-political order is challenged by the joint acceleration of posthuman convergence: patterns of climate change, technologies, growth and volatility. This contested order, anthropocentrism, lacks the analytical tools to deal with the convergence between the climate crisis and the expansion of Artificial Intelligence. Against this backdrop, designers must explore and develop practices and perspectives beyond the anthropocentrism that dominates the spheres where AI models and infrastructure are developed.

The ubiquity of AI in all dimensions of life calls for a critical understanding of how it produces knowledge, images, dialogues and all types of interactions, human and nonhuman. Following Haraway (1988), the question of the position where AI claims its legitimacy is crucial.

DOI: 10.4324/9781003467731-9

In the forthcoming sections of this chapter, and in line with Haraway's thought, we will discuss the question: Is it possible to talk about the *situatedness* and *embodiment* (or position) of AI? Due to its distributed composition, defining AI positionality is a challenge; in many ways, AI is an unprecedented technological system constituted by all classes of novel human and nonhuman entanglements acting at a planetary scale. AI requires an expansion of the technological *situatedness* problem.

Looking ahead, this chapter aims to explore how MTHD practices can influence the future of AI. The chapter proceeds as follows: The first section describes the limitations of human-centred design as a paradigm for facing the challenges posed by Artificial Intelligence. Then, the chapter presents the double problem of AI extractivism: the impact on resources in ecosystems and the extractivism of knowledge through "datification". The chapter then argues that AI's positionality problem is critical to its accountability. Building upon Haraway's idea of the 'god trick' (1988), the chapter argues that AI could become a new disembodied objectivity (god trick) if designers and academics fail to position its agency correctly. After that, we present 'the planetary' as a positionality for AI and MTHD practices through two case studies: the exhibition *Hybrid Ecologies* created by the designer Manuela Garretón and the sociologist Martín Tironi dealing with Earth extractivism and the experiment *MoTH,* developed by Iohanna Nicenboim and Joseph Lindley dealing with knowledge abstraction and extraction (Crawford, 2021). Finally, the chapter proposes ideas and questions to advance an MTH AI agenda.

To engage with these ideas from the perspective of design practice, we will address two AI implementation examples: generative images through *Hybrid Ecologies* and conversational agents through *MoTH*. The intention is not to offer closed solutions to the problem of extractivism in AI but rather to show, through the development of two practical cases, how design can overcome critical misunderstandings and help problematise the dominant logic whilst making visible alternative thinking methods about the development of Artificial Intelligence.

The Limitations of the Human-Centred Design Approach in AI

The fields of Human-Computer Interaction (HCI) and design are increasingly recognising that the prevalent Human-Centred Design (HCD) approach might be limited in designing complex technologies (Giaccardi & Redström, 2020; Frauenberger, 2019). These might go from applications based on intricate AI models to generative images or Large Language Models (LLM). In the context of technologies like AI, where there are many connected actors,[1] "the notion of human-centred ceases to be useful" (Coulton & Lindley, 2019, p. 467) because of the inherently entangled animal, mineral and moral relationships encoded within AI itself. Laura Forlano explains that "human-centred design is founded on understanding the human as a discrete, individual subject. However, our new relations to the natural world and socio-technical systems are calling these previous understandings into question" (Forlano, 2017, p. 17). HCD seems limited in positioning AI within broader infrastructures and fails to account for the increased agency that AI-powered tools have in people's everyday lives. The reason is that AI, like any contemporary technology, blurs the traditional boundaries between actors, users and designers (Giaccardi & Redström, 2020), extending issues of responsibility and trust beyond immediate end users and single interactions (Coulton & Lindley, 2019; Forlano, 2017).

Beyond the limitation of HCD to attend to complex relations between humans and nonhumans, other scholars have highlighted its shortcomings in addressing the impact of technologies on the Earth and other species (Hermansen & Tironi, 2018; Hermansen & Guerra, 2024). Whilst understanding human-AI interactions is essential, we should remember that each interaction with AI relies on a vast planetary network of human and nonhuman resources, encompassing the extraction of materials and knowledge (Crawford & Joler, 2018). Considering the limitations and the risks of HCD, Wakkary (2021) formulates a fundamental question: "What if human-centred thinking (and its underlying humanism) is not the answer to [the planetary] problems but rather, in its dominant role, may be part of the problem?" (p. 1).

At a time of climatic and social crisis, it seems risky to project and predict AI's future exclusively from an anthropocentric perspective without considering the complex assemblies and entities that constitute it. Instead of focusing on humans, a relational approach is promising:

> We need new ways of understanding the empires of AI. We need a theory of AI that accounts for the states and corporations that drive and dominate it, the extractive mining that leaves an imprint on the planet, the mass capture of data, and the increasingly exploitative labour practices that sustain it.
>
> (Crawford, 2021, p. 11)

A suitable approach for designing AI should recognise and address the profound impacts these technologies have on our world (Crawford & Joler, 2018). Since "there is nothing about artificial intelligence that is inevitable" (AI Now Institute Report, 2023), the central question is an exploration of what MTHD, as a relational approach, can do to respond to the myriad of challenges posed by the adoption of AI. The field of action in the areas of design highly influenced by HCD perspectives is not wide enough to contain the problems that AI presents to the designer's practice. To avoid significant misunderstandings in the relationship between AI and design, a different comprehension of the scale of design action is required. In the context of AI technology permeating every corner of the planet, human-centred design should move towards a more planetary-oriented approach.

The Challenges of AI Extractivism

Design has played a role in creating products and services that have deepened the eco-social crises by ignoring the fact that there is a political dimension to its impact (Fry, 2010). Artificial intelligence depends on *multiple frontiers of extraction* (Mezzandra & Neilson, 2017), for example, practices of *data colonialism* (Couldry & Mejias, 2019), terrestrial exploitation (Crawford, 2021), or the capture of knowledge and labour (Pasquinelli, 2023). We cannot understand the extractivist dependencies of AI using the same definitions of extractivism for producing consumer goods during the 20th century. For example, AI is built on enormous quantities of social media data – daily extracted from our *platformed sociality* (van Dijck, 2013). Design and designers are fundamental to creating interfaces and devices for data extractivism. The same fundamental relation to design is valid for the physical infrastructure of Artificial Intelligence, from the design of data centres to smartwatches or smartphones; the future of design and AI are deeply interconnected.

The problem is the need to understand that AI intertwines old and new regimes of extractivism. Crawford (2021) describes the current view of AI as depending on the *twin moves of abstraction and extraction*: abstracting away the material conditions of its making while extracting more information and resources from those least able to resist it. Of course, one should not accept without question the discourse of AI as an abstract, unaccountable, and disembodied machine. The first realisation should be that AI is not a disembodied system situated nowhere and without corporeality. Instead, AI requires resources, data, knowledge and labour at an unprecedented scale and currently challenges all sorts of social, cultural and legal frameworks.[2]

If designers and design researchers fail to understand that AI planetary entanglements and its new forms of abstraction and extraction require the formulation of a new theory and practice for planetary design, then design, as a discipline, is at risk. Design will, therefore, have a reduced role in the production of future technologies. To deal with AI, we need to recognise and enable design as a mode of planetary thinking and action. If this is not correctly understood, designers will envision the interfaces, devices and interactions based on AI models; however, they will have a minor role to play in shaping AI, together with a restricted role in the academic and social understanding of this phenomenon. In the face of this technology's accelerated and uncertain advance, design should not be content with continuing to reproduce the "business as usual" logic dictated by human-centrism and an anthropocentric scale of design ethics. In the words of Tony Fry (2010), an a-critical design could become a *defuturing* agent in the development of AI.

One of the most significant differences between HCD and an MTHD approach to AI is the potential of *decentering* the humanist conception of the user (Nicenboim et al., 2023) and enabling design to care for the more-than-human agencies that constitute AI. This critical point has also been made by Coulton and Lindley (2019), for whom HCD, when applied to complex contexts and, in a dogmatic manner, reveals itself to be an incomplete philosophy that lacks a sense of accountability beyond the end-user interactions (p. 466). This incompleteness shapes the main problems of the HCD confronted with a planetary (or MTH) understanding of AI. Anthropocentric design perspectives are usually restricted to business innovation, individual progress and modernisation. The 'human' is typically an individual or an aggregation of individuals who are far from becoming a genuine political agency or have a broader understanding of the complex consequences of design.

The Situatedness *of AI*

The value of design practices derived from MTHD lies in harnessing advances in understanding concrete ways of thinking about interaction with digital interfaces (nonhuman artificial agency) and, at the same time, recognise that the complexity of AI entanglements places the agency of design in a new position: the planetary one. A crucial political first step of MTHD AI practice is to acknowledge the two-headed problem of the political relationship between AI and the planet: firstly, as an infrastructure dependent on terrestrial extractivism and secondly, as models for unprecedented knowledge extractivism.

As it has been shown, the risk of an unaccountable AI is closely related to the problem of not recognising it as a situated and embodied technology. The discourse of AI "enables the illusion of their distanced neutrality with respect to their object of knowledge" (Hakopian, 2024, p. 31); there is no such distance between AI and ecosystems or cultures. We cannot design AI responsibly if we just consider it as a technical system for solving human problems (presented as abstract, objective or transcendental), mainly if the definition of 'human' is narrow and restricted to extractivist agendas and universalising knowledge formation. How can designers avoid the trap of dealing with AI as a non-situated machine at the service of anthropocentrism?

In 1988, in her seminal work *Situated Knowledge: The Science Question in Feminism and the Privilege of Partial Perspective,* Donna Haraway warns us of the "god tricks", where knowledge is categorised as *non-located* and disembodied and – in consequence – "unable to be called into account" (p. 583). Haraway argues for "politics and epistemologies of location, positioning and situating, where partiality and not universality is the condition of being heard to make rational knowledge claims" (p. 589). Following Haraway's argument, positioning AI is a critical starting point for MTHD practices.

The question articulating the problem of AI positionality is *where* can we account for a profoundly local and planetary agency without abstracting or universalising it as an entity of study, intervention and critique? MTHD AI practice should care deeply about AI's positions, and in the same way that it acknowledges that AI expands the modes of extractivism, it also should recognise that AI expands (and invents) novel modes of being situated and embodied, "recognising the positionality of generative AI models seems crucial for comprehending and critically evaluating the potential biases and limitations of AI-generated knowledge" (Nicenboim & Lindley, 2024). A more-than-human approach sees AI not as an extension of the human brain or nervous system but, following Parikka, as an extension of the planet (Crawford, 2021); the challenge is to care deeply about the expanded positionality of AI: the planet, without overlooking the local impacts of AI on communities and ecosystems.

Towards a Planetary AI Design

To deepen our understanding of the MTHD dimensions of AI, we will present two concrete design cases that seek to open up ways of understanding the new challenges in which design can work. Firstly, the exhibition *Hybrid Ecologies* by Manuela Garretón and Martín Tironi, deals with planetary extractivism of natural resources. Secondly, *Conversations with AI*, by researchers Iohanna Nicenboim and Joseph Lindley, explores the complex entanglements of non-situated knowledge formation at a planetary scale. Both examples explore, from their emphasis, alternative visions to HCD, extractivist tensions in the planet and at the level of knowledge, and an attempt to relocate the concept of the planetary from an abstract-universal to a mode of accountable positionality for AI.

1. Hybrid Ecologies *Exhibition*

In 2023, the Chilean government commissioned, through the Ministry of Science, the creation of the Futures of Artificial Intelligence Research group (FAIR). This is an "interdisciplinary Millennium Nucleus of research and creation focused on Artificial Intelligence's (AI) cultural, social, and environmental implications" (Fair, 2024, para.1). The agenda for the research nucleus is understanding AI as a profoundly political and cultural agent with un-tested consequences for the country's future. Chile is far from Silicon Valley but close to the mining operations and ecological implications in the Artificial Intelligence infrastructure supply chain.

The *Hybrid Ecologies* exhibition curators, which was displayed at the "La Moneda Cultural Center" in Chile, are the designer Manuela Garretón and the sociologist Martín Tironi, who work at the School of Design, Pontifical Catholic University, Chile and the FAIR Research Centre (https://www.nucleofair.org). The exhibition focused on revealing the water footprint of image generation with *Stable Diffusion* AI. AI's water impact is present through its operation, from the water used to extract the minerals critical to its function to the construction of the data centres and the different server cooling systems.

The curators calculated the hydric cost of processing an image produced with AI in Chile's data centres to explore an embodied AI design. The visitor to the exhibition entered a dark room

FIGURE 7.1 *Hybrid Ecologies* and *Conversations with AI*. Top left: Overview of *Hybrid Ecologies* installation in Santiago de Chile. A stream of water falls on the floor from a mycelia sphere in the centre of the room when an image is generated with an AI model; these images are projected on the walls. Photography by Verónica Aguirre. Top right: Performance piece in which actors improvise with Alexa and provide alternative responses. Image by Iohanna Nicenboim. Bottom: Conversation Starters was a series of interactive prototypes that explored how conversations with AI can be situated. These CAs are designed to listen and respond to more-than-human human voices, i.e., sounds from the home and other species. Image by Iohanna Nicenboim.

(see Figure 7.1 top left) where they could "request" an image through a prompt on a touch interface. Then, this request was processed, and the resulting generated image was projected onto a wall to provide an immersive experience for the visitor. Simultaneously, a stream of water, falling from a mycelia sphere in the centre of the room, represented the equivalent water footprint the AI process consumed to respond to the viewer's request. This cycle was repeated each time a visitor entered a system prompt.

This work aimed to use the installation to demonstrate how AI works as an agent of the ecological challenges the planet faces, as expressed by Garretón and Tironi during the "Congreso del Futuro", Santiago, Chile. In the words of the curators, "Design can contribute to make visible the geological materiality and relationships of dependence that AI has with the planet and also constitutes a space for materialising alternatives to the relationship between technology and environment in alignment with biodiversity and earthly habitability" (Tironi & Garretón, 2024).

The exhibition's viewers articulated the ethical question about the quantity of water used during the exhibition, whilst it also allowed them to delve into the perception of value between the aesthetic experience of the projected images and the water cost of their production. The recognition of the terrestrial footprint of Artificial Intelligence does not seek the closure of this technology but rather for people to be aware of the ecological footprint the creation generated. The installation prompted visitors to think of AI in terms of water demand, constituting a wider ontological setting for reacting in the same way they would to a typical prompt to ChatGPT. The unpredictable outcome of the visual experience and the materiality of the cascading water gave the installation an organic – even biological – quality. It is biological life, which is liquid, and supposedly, it is not the dry performance of electronic machines.

This exhibition questioned the discourse of an AI detached from ecological costs by making explicit the hydric entanglements between prompts and AI images. The installation was an invitation to bring AI down to Earth, proposing a conceptual and experiential language that allows visitors to recognise the decisive role earthly materiality plays in the constitution of the function of any technological device (Tironi & Garreton, 2024). This highlights the importance of considering how data centres can impact local communities and ecosystems, something which may not have been previously recognised by the exhibition visitors who are more familiar with media coverage of disputes between mining companies and farmers over water use policies. The exhibition sheds light on a previously ignored issue: the significant impact of large data centres on hydric resources. Despite the global scale of the calculation required to produce generative images, cooling individual servers, for example, can have significant regional effects. *Hybrid Ecologies* exemplifies the critical role of water in AI systems. The exhibition showcases water as a powerful force that exceeds the impact of the calculated and projected image. The falling water is not hidden in abstraction; it is present, splashing and creating noise as it spills onto the floor.

This design case study explores the importance of connecting AI processes (image generation) with environmental costs and impact. In the exhibition *Hybrid Ecologies*, there is no such thing as a 'god trick' in image generation; the calculations of the image models extract resources (water, minerals, energy) from the Earth. Garretón and Tironi use the design discipline to exemplify this extractivist entanglement and make it explicit to a broader public audience.

2. *Conversations with AI*

In 2020, designer and researcher Iohanna Nicenboim conducted several experiments to decenter the traditional conception of humans in AI design, focusing on conversational AI applications.

The first experiment consisted of workshops in which designers and researchers across different fields were invited to interview conversational agents (CAs) like Alexa and Siri. The workshop's outcomes, via a series of questionnaires and videos, illustrate how people and CAs were entangled at different scales, from the intimacy of the home to more extensive infrastructures of labour and power. Beyond exposing the infrastructures and biases of CAs, the workshop also highlighted that CAs were not accounting for the user's position in the world nor acknowledging the agent's worldview (Nicenboim et al., 2023). Building on Donna Haraway's feminist epistemology (1988), this issue was articulated as a limitation of anthropocentric design approaches in AI to be situated (Nicenboim et al., 2022).

To explore a situated AI design, Nicenboim conducted other experiments, including designing speculative responses for Alexa (Figure 7.2), developing a performance piece in which actors improvise with Alexa and offer alternative responses (Figure 7.1 top right), and designing CAs that listen and respond to MTH voices (Figure 7.1 bottom). These experiments showed the importance of not only situating the user and the CAs agent's position in the world but also of situating the designer's knowledge and perspective when engaging with AI. They also extended conversations beyond human voices and proposed more plural and inclusive ways of listening-with.

Based on these experiments, the design researchers Iohanna Nicenboim and Joseph Lindley explored how *situatedness* plays a role in designing with OpenAI's GPT-4. They created *MoTH*, an experimental method to explore the question: How can tools created using GPT-4 support more-than-human design with or in practice? *MoTH* is based on a prompt titled Professor Synapse by Synaptic Labs:

> The prompt uses 'chain of reason'. This encourages GPT-4 to break any given task down into small bitesize tasks. The second technique the prompt employs is to ask GPT-4 to imagine several expert agents. Each of these agents should adopt the chain of reason approach to break down the challenge from different perspectives.
>
> (Nicenboim & Lindley, 2024)

One of the examples used when engaging with *MoTH* expands the idea of a "Jane Bennet and Rosi Braidotti-inspired kettle that whistles when it boils" (Nicenboim & Lindley, 2024); this concept came from a previous experiment with a tool they named *Oblique*. Nicenboim and Lindley told *MoTH* they were interested in designing a kettle, using Braidotti's *The Posthuman* as the theoretical framework for the experiment. The result was a conversation with GPT-4 where the different 'experts' assumed roles. For some questions about the kettle, *MoTH* referred to a virtual expert with a background in *science and engineering* to address the topic of the acoustic properties of kettles; for other questions, it referred to an expert on *design research, creativity and art* for a discussion on different ceramic artefacts.

A relevant observation was that the large language model (LLM) of GPT-4 tends to 'flatten' the particularities of different theories from the perspective of situated knowledge. An example is the similar GPT-4 results across different MTH theories when put to the test (Nicenboim & Lindley, 2024). AI represents a challenge to the positionality of the theories and thinking from which the model is trained. Interacting with GPT-4 complicates issues of accountability and transparency. Consequently, questions arise regarding the role of design in situating human-AI co-produced knowledge.

While the first experiments focused on designing conversational agents to reveal their positionality, *MoTH* introduces another approach by employing multiple and differently positioned agents. By embodying agents' perspectives, *MoTH* takes a step towards situating knowledge. Situating AI-co-produced knowledge may involve considering how biases and perspectives embedded in training data and algorithms influence AI-generated knowledge. Perhaps, openly training models and reflecting on data collection and labelling could shed some light on the positionality of the model and the designer/trainer. Ultimately, Nicenboim and Lindley conclude that since metaphors are crucial in AI design, those employed in designing interfaces and applications for generative AI have the potential to either express or obscure positionality.

The critical question arising from these practices is whether researchers and designers should consider AI as a proper *situatedness* or just another form of 'god trick' (Haraway, 1988).

Both of these cases, *Hybrid Ecologies* and *MoTH*, touch on the two-headed problem of AI positionality: the *embodiment* of AI infrastructure and the *situatedness* of AI production of 'knowledge'. Both examples also present the dualism of 'abstraction' and 'extraction' (Crawford, 2021); *Hybrid Ecologies* points out that a vision of AI as an abstract and disembodied machine promotes unaccountable modes of extractivism (like hydric resources); *MoTH* points out that the abstract machine of AI requires extractivist practices of knowledge, potentially making knowledge *non-situated* and ultimately unaccountable. The cases explored in this chapter are essential because they allow us to approach expanded notions of extractivism (Mezzandra, Neilson, 2017). For example, in Chile, water is a critical resource as the country has suffered long periods of drought for more than a decade, with significant social movements fighting to defend and protect their water rights. However, it is essential to acknowledge that the extractivist relationship between AI and water is jointly situated in Chile and many other countries; this awareness could constitute a crucial first step to designing accountable AI applications. Something similar happens with Braidotti's posthuman feminist ways of knowing; in *MoTH*, a firmly situated theory is flattened by GTP-4, which is, in consequence, very difficult to account for. Any form of situated knowledge is at risk of non-accountable forms of machinic extraction. The local and the planetary planes have collapsed into a single domain of understanding and a call to action for designers.

Conclusion and Final Thoughts

When addressing the issues of AI positionality and accountability, an MTHD approach challenges the anthropocentric assumptions regarding the scale and agents of design action and

FIGURE 7.2 Situated conversations with Alexa based on the Anatomy of an AI System map from Crawford & Joler (2018) were collected in a series of videos. Image by Iohanna Nicenboim.

responsibilities, as well as the role of nonhuman intelligence in shaping the future. The contested order of anthropocentrism Braidotti (2022) creates blind spots that are not only unethical but also pose a threat to the survival of living species on the entire planet. The problem arises when, following Crawford (2021), the new power of 'abstraction' requires new powers of 'extraction', which remain unaccountable because designers and researchers fail to address and critique them simultaneously on a local and planetary scale.

When designers achieve advances on the MTHD agenda for AI, it may not translate into radical scepticism but, on the contrary, the recognition that AI is a more-than-human agency capable of questioning anthropocentric biases, therefore opening new radical interdependencies and making AI extractivism accountable. While global media is quick to present the 'non-humanity' of AI as a threat to the future, designers should consider that the 'non-humanity' of AI is also an opportunity to challenge the supremacy of the anthropocentric gaze over the planet, environments, economics, politics, etc. Here lies the potential power of an MTHD approach to AI. There is a need for a shift towards MTHD practice in AI to collaborate in developing a non-anthropocentric AI.

The discussion around the experimentations developed in *Hybrid Ecologies* and *MoTH* suggests the need to reveal latent misconceptions to help advance an agenda for design practices that consider more-than-human perspectives. A critical misconception – following the philosopher Emanuele Coccia (2024) is that "design is a set of human and only human practices and that precisely because of this, the domain of its application is limited to a set of objects and contexts with which human beings have to deal", and another misconception is that design operates at a bounded scale: "In fact, the object of design is always the planet in its totality, and first of all in a very literal sense of the word: that there is no place or area on the planet that cannot or has not been the object of design, conception" (Coccia, 2024). Overcoming this misconception helps MTH designers focus their method on producing critical knowledge and practice about new and expanded forms of AI extractivism and exploring open and symbiotic futures for AI.

Designers working on projects involving AI might start with the following questions: Can I design AI systems that are accountable for their actions? From what position does AI-generated knowledge become accountable? Can I make this accountability explicit to human and nonhuman actors? What worldviews am I extracting or feeding into the AI models or datasets with my design? Is this application of AI essential? Furthermore, what values and responsibilities are part of and justify the design process?

Design possesses speculative capabilities and project-based knowledge that allow us to draw on forms of AI that can be more conscious of the ecological and social crisis we are experiencing. As a practice of generating alternative futures and presents, design can contribute to abandon anthropocentric exceptionalism and explore more generative, local and interspecies ways of thinking about technological development. Design should not be a veil (Simondon, 2014) that hides or obscures AI's power; on the contrary, it should be an actor that keeps open its political agency and the enormous possibilities AI can create with MTH actors in the future.

Questions

- Can I design AI systems that are accountable for their actions?
- What worldviews am I extracting or feeding into the AI models or datasets with my design? Is this application of AI essential?
- What values and responsibilities are part of and justify the design process?

Notes

1 This chapter uses the actor concept, which is in line with the actor-network theory (ANT) understanding of the relations between humans and non-humans.

2 *Critical AI: A Field in Formation,* Rita Raley & Jennifer Rhee (2023) describe the problem of authorship in the current conversational AI landscape: "To use OpenAI's API (application programming interface) to experiment with GPT-3 is to produce text for which there is no proper subject, or for which there can only be a retroactive subject effect produced via an appended claim of authorship that enables the delineation of a difference between deliberative, reflective, expressive writing on the one hand and the real-time, automatic manipulation of symbols on the other. The real lesson of a Turing test in this context is not that language models and conversational AI systems are good enough to deceive but, rather, that actants, training data, input, and output are all now so entangled that the determination of linguistic property and, by extension, responsibility is essentially foreclosed" (p. 190).

References

Amunátegui, C. (2021). *Arcana Technicae, El derecho y la Inteligencia Artificial*. Tirant 4.0.

Braidotti, R. (2022). Feminismo posthumano. Editorial GEDISA.

Bunz, M. (2019). The calculation of meaning: On the misunderstanding of new artificial intelligence as culture. *Culture, Theory and Critique*, 60(3–4), 264–278. https://doi.org/10.1080/14735784.2019.1667255

Coulton, P., & Lindley, J. G. (2019). More-than human centred design: Considering other things. *The Design Journal*, 22(4), 463–481. https://doi.org/10.1080/14606925.2019.1614320

Crawford, K. (2021). *Atlas of AI: Power, Politics, and the Planetary Costs of Artificial Intelligence*. Yale University Press

Crawford, K., & Joler, V. (2018). Anatomy of an AI System. The Amazon Echo as an anatomical map of human labor, data and planetary resources. https://anatomyof.ai

Coccia, M. (2024). Prologue. In Tironi, M., Chilet, M., Ureta Marín, C., & Hermansen, P. (Eds), *Design for More-Than-Human Futures*. Routledge.

Couldry, N., & Mejias, U. A. (2019). *The Costs of Connection: How Data is Colonising Human Life and Appropriating it for Capitalism*. Stanford University Press.

FAIR – Futures of Artificial Intelligence Research (n.d.). FAIR. Retrieved February 14, 2024, from www.nucleofair.org

Florian, J. (2024). Ground truths are human constructions. *Issues in Science and Technology*, 85. https://doi.org/10.58875/LXWC2075

Forlano, L. (2017). Posthumanism and design. *She Ji: The Journal of Design, Economics, and Innovation*, 3(1), 16–29. https://doi.org/10.1016/j.sheji.2017.08.001

Forlano, L. (2021). A new AI lexicon: human. *AI Now Institute*. 2024. https://ainowinstitute.org/publication/a-new-ai-lexicon-human

Frauenberger, C. (2019). Entanglement HCI: The next wave? *ACM Transactions on Computer-Human Interaction*, 27(1), 1–27. https://doi.org/10.1145/3364998

Fry, T. (2010). *Design as Politics*. Berg.

Giaccardi, E., & Redström, J. (2020). Technology and more-than-human design. *Design Issues*, 36(4), 33–44. https://doi.org/10.1162/desi_a_00612

Haraway, D. (1988). Situated knowledges: The science question in feminism and the privilege of partial perspective. *Feminist Studies*, 14(3), 575–599. https://www.jstor.org/stable/3178066

Hakopian, M. F. (2024). Art histories from nowhere: On the coloniality of experiments in art and artificial intelligence. *AI & Soc*, 39, 29–41. https://doi.org/10.1007/s00146-023-01768-0

Hermansen, P., & Guerra, J. (2024). Furrowing the Maraña. Designing to sail out of the Anthropocene. In Tironi, M., Chilet, M., Ureta Marín, C., & Hermansen, P. (Eds), *Design for More-Than-Human Futures*, Routledge.

Hermansen, P., & Tironi, M. (2018). Pedagogical impugnation: Interspecies prototyping and cosmopolitical encounters. *Diseña*, (12), 196–227. https://doi.org/10.7764/disena.12.196-227

Kak, A., & Myers West, S. (2023). AI Now 2023 Landscape: Confronting Tech Power, AI Now Institute, April 11, 2023. https://ainowinstitute.org/2023-landscape

Nicenboim, I., Giaccardi, E., & Redström, J. (2023). *Designing More-Than-Human AI: Experiments on Situated Conversations and Silences*. https://doi.org/10.30682/diid8023c

Lehuedé, S. (2024). The double helix of data extraction: Radicalising reflexivity in critical data studies. *Technology and Regulation*, 2024, 84–92. https://doi.org/10.26116/techreg.2024.009 (Original work published March 18, 2024)

Mezzadra, S., & Neilson, B. (2017). On the multiple frontiers of extraction: excavating contemporary capitalism. *Cultural Studies*, 31(2–3), 185–204. https://doi.org/10.1080/09502386.2017.1303425

Nicenboim, I., & Lindley, J. (2024). Designing generative AI tools to support more-than human design with/in practice. *DRS*, June, Boston, USA. Manuscript approved for publication.

Pasquinelli, M. (2023). *The Eye of the Master*. Verso Books.

Raley, R., & Rhee, J. (2023). Critical AI: A field in formation. *American Literature*, 95(2): 185–204. doi:https://doi.org/10.1215/00029831-10575021

Simondon, G. (2008). *El modo de existencia de los objetos técnicos*. Prometeo.

Simondon, G. (2014). Psychosociologie de la technicité. In *Sur la Technique* (1953–1983). Paris: Presses Universitaries de France, 27–30.

Taylor, L. (2024). AI lacks ethic checks for human experimentation. *Issues in Science and Technology*, pp. 80–81. https://doi.org/10.58875/LCFM3937

Tironi, M. (2024). Introduction: Design for more-than-human futures: towards post-anthropocentric and decolonial perspectives. In M. Tironi, M. Chilet, C. Ureta, & P. Hermansen (Eds), *Design for More-Than-Human Futures*. Routledge.

Tironi, M., & Chilet, M. (2023). Planet-oriented design: A proposal for new ethical transitions in Design Education. In De Sainz Molestina, D., Galluzzo, L., Rizzo, F., Spallazzo, D. (Eds), *IASDR 2023: Life-Changing Design*, 9–13 October, Milan, Italy. https://doi.org/10.21606/iasdr.2023.506

Tironi, M., & Garretón, M. (2024). Hybrid ecologies of artificial intelligence: Prototyping terrestrial practices through a design installation. *DRS* 2024, June, Boston, USA. https://doi.org/10.21606/drs.2024.1104

van Dijck, J. (2013). *The Culture of Connectivity: A Critical History of Social Media*. Oxford University Press.

Wakkary, R. (2021). *Things We Could Design: For more than Human-Centered Worlds*. The MIT Press.

8

CREATIVE AI AS MORE-THAN-HUMAN

Design Practices, Aesthetics and Cultural Imaginaries

Petra Jääskeläinen

Introduction

While technology has often been perceived and understood through the modernist epistemological separation of man and nature (Latour, 1993), there is a need to question and critique (both in theory and in practice) these perspectives and their related socio-material arrangements. Often these arrangements re-produce and maintain power structures that place humans at the forefront of ethical consideration while deprioritizing and neglecting care over non-humans – animals, the environment and technological artefacts – in design practices, and generally epistemically highlight the separation between humans and non-humans (*e.g.* Wakkary, 2021). Several scholars have asked questions regarding *who gets to matter* (*e.g.* Puig de la Bellacasa, 2017), a question that also extends to non-humans, and this applies to the case of AI technology used in creative practices (Creative AI).

To shed more light on the *case of Creative AI technology*, the design stances that drive and fuel these emerging Creative AI technologies often rely on human-centred notions leaving non-human subjectivities and entanglements outside of the scope. We can, for example, see this in how a large majority of the AI ethics and AI design guidelines are focusing on 'human-centred AI' (Shneiderman, 2022; Auernhammer, 2020), which consequently leads to efforts in "human-centred design of generative AI" (Chen et al., 2023). There is a lack of (design) research in more-than-human-centred Creative AI, with only a handful of cases exploring it (see Jääskeläinen et al., 2022b), and so far, only a few people aimed to explore such directions in the case of AI *generally* (Nicenboim et al., 2020).

One concrete example of non-human concern in the context of Creative AI technologies is how visual AI aesthetics represent non-humans in certain limited ways, which also emergently configures our relation to non-humans. It is also problematic how such practices of producing visual aesthetics in an increasingly homogenous manner can contribute to the hegemony of certain worldviews, such as Euro-Western cultural depictions of human and non-human relations. Extending this list, yet another critical concern is found in how the construction of Creative AI technologies (*e.g.* hardware) has also significant direct impacts on the non-human

DOI: 10.4324/9781003467731-10

stakeholders, who are often treated in an extractivist manner and materially exploited (Jääskeläinen et al., 2022b). These potential critical impacts largely relate to the imaginaries and culture that surround the design of such technologies.

In this chapter, I start critically unpacking Creative AI from the more-than-human perspective, focusing on values and worldviews, which are carried on or absent in the material and cultural aspects of Creative AI technology. I discuss how framing technology as more-than-human and understanding it as a part of the more-than-human entanglements simultaneously configures our environmental ethical stance and has implications for its material design in practice. I begin by discussing what Creative AI is, and how it relates to valuing or disregarding non-humans: primarily the environment and animals, while also briefly discussing our relation to AI as non-human. Secondly, I will discuss what implications designing Creative AI technologies would have from a more-than-human perspective, and what specific issues are associated with it from a more-than-human perspective – with a particular focus on aesthetics that are produced by visual Creative AI tools (including examples from a specific technology: Craiyon). Thirdly, I will present critical design and speculation case studies of ongoing work that are focused on critically examining and re-imagining Creative AI technologies (such as VQGAN-Clip) and their insights by taking a more-than-human design perspective.

I hope for this chapter firstly to lay the ground for more-than-human design of Creative AI technologies by highlighting the need to reflect on our ethical commitments as designers of these technologies, while also providing a few light beacons for future directions in more-than-human design research in Creative AI.

What is More-Than-Human Creative AI?

Creative AI technologies (often also framed as Generative AI) have become prominent in recent years in various domains; film (Totlani, 2023), illustration (Adobe, 2023), music (Civit et al., 2022), performance (Wallace et al., 2023), contemporary art (Jääskeläinen et al., 2022a), literature and writing (Doshi et al., 2023), design (Van der Maden et al., 2023; Van der Maden et al., 2024), and many others. This newly emerging technology is about to transform the practices that involve production of culture by changing the predominant cultural-aesthetic landscape, and the creative labour practices behind them. First, this means that an increasing amount of labour is going to be automated in creative practices. This leads to the transforming nature of work when increased automatization of cultural production reduces manual work in certain parts of the processes. It arguably also re-distributes agency and power in creative (AI) work, by changing, for example, the power of control over the aesthetics so that it is dependent on the training of the AI models, rather than the design aesthetical decisions being primarily those of the creatives. While any digital or analogue creative tools arguably influence the potential of aesthetics outputs in creative work, AI introduces a new development in which the aesthetics are deeply integrated into the tool design (as data largely constitutes the AI tool), further limiting the agency and potential of creatives at certain stages of creative work. And, last, it clearly leads to re-configuring materiality, for example when the increased computational load of AI requires establishing new server farms. Thus, we can see that this ongoing AI adoption in creative practices is significant and expected on many levels from aesthetics to labour, and power. This raises the question of how these socio-material transformations are supporting and undermining the relational more-than-human worldviews, or non-human stakeholders?

Designing Creative AI Technologies with a More-Than-Human Perspective

Arguably, a more-than-human approach in design can be understood as an orientation or stance for (design) research, among other modes of knowledge-making, such as human-centred design that aims to centralise humans in design practices. There is an increasing number of researchers who are not only interested in more-than-human design, but furthermore actively promoting and driving this design stance both generally and within AI (Wakkary et al., 2021; Nicenboim et al., 2020), and it can be thought that more-than-human designers (at least to a certain extent) share the belief and assumption that non-humans should be included as stakeholders in the design processes, and also that we are part of relational, entangled ecologies. In taking the 'more-than-human stance', as I see it, we as designers are *de facto* taking an environmental ethical stance in design practice. This environmental ethical stance is also a design political (Fry, 2010) stance, actively arguing and advocating for the subjectivities of non-humans while simultaneously acknowledging the relational nature of our being (Latour, 1991; Haraway, 1991). While the more-than-human design stance is less often discussed with traditional environmental ethics (see environmental ethics by Thompson, 2019) in contemporary design research, related themes are generally common in the posthumanities discourse (e.g. Puig de la Bellacasa, 2017). Regardless, turning into traditional theories of environmental ethics can help us to unpack the questions of who gets to have value in design practice – and particularly in the design practice of Creative AI technologies.

Essentially, research on environmental ethics outlines the central ethics perspectives of assigning value to 'things': biocentrism, ecocentrism and human-centrism (Thompson, 2019). In short, this means whether we assign high value and prioritise *biological things*, *ecologies*, or *humans*. We can see that human-centred design is closely related to human-centric values and ethics. Biocentrism and ecocentrism, instead, require a bit more unpacking. Biocentrism implies that 'anything that lives' ought to have value, while ecocentrism suggests that 'ecologies matter'. The main difference between these perspectives is that biocentrism focuses the intrinsic value on the individual biological beings, whereas the ecocentric perspective puts ecologies and entanglements at the forefront, thus taking a more relational orientation. This relational orientation is something that has been argued for in contemporary sustainability research (West et al., 2020, but it is also a central perspective in more-than-human design philosophy (Wakkary, 2021) that emphasises entanglement and relations. Arguably, the more-than-human perspective that acknowledges and values 'entanglement' can thus be seen as related to an ecocentric perspective, but furthermore, a more-than-human perspective urges to value all that lives despite their humanness-ness, which relates it also to biocentric values and ethics. This demonstrates how the ethical stances in some ways can co-exist with each other in more-than-human design practice, and designers who engage in designing Creative AI should critically acknowledge their environmental ethical positionality – e.g. when relational eco-centric perspectives are present, and when the focus is on the biocentric perspectives – to understand how assigning value is configured in the design practice.

I discussed earlier in this chapter the human-centred worldviews that are predominant in the design of Creative AI. Extending this, their design and development also often embody technological enthusiasm and/or determinism. Techno-deterministic worldviews position technology as something that is integrally part of human development, thus deprioritizing critical lenses on technology and de-prioritizing critical design approaches, such as slow design (Fuad-Luke, 2008), de-technologization, or computing within limits (Penzenstadler, 2018).

The techno-enthusiasm discourse, in turn, considers technology and technological advancement as something positive to be celebrated. Thus, these two worldviews (techno-enthusiasm and techno-determinism) share the commonality of driving technological development forward in society. Many Creative AI technologies are emerging in the Global North and are fuelled by developments in Silicon Valley (particularly companies such as OpenAI, Microsoft, Google, etc.) and stem from these techno-positivist, if not borderline techno-religious attitudes (McElroy, 2024) and worldviews that see technology as a saviour or solution to our multiple planetary crises. However, it has been acknowledged that the very technology that we develop to solve our problems is constantly creating new problems, and many prominent thinkers have, thus, criticised this techno-optimist ideology and worldview (Nardi et al., 2018; Kalfagianni et al., 2019). This frames a situation, in which Creative AI technologies are "argued to be justified by their capacity of pushing the humanity forward … an argument that is strongly based on unproblematized human-centred value sets" (Jääskeläinen et al., 2022b). As discussed throughout this chapter, the central tension that we can identify between the emerging Creative AI technologies and more-than-human design lies in how the prevalent design approaches are – at best – human-centred, rather than expanding the scope of sensitivity and ethical consideration to a wide range of non-humans. There is no need to look further than at any common Creative AI technology, and we can see that they have not been designed with prioritising a more-than-human approach and sensitivities. Thus, there is a need to deeply question these assumptions that are being carried on to the design of Creative AI technologies from the wider trends in AI development.

So then, if we take on the assumption that more-than-human perspectives (particularly the perspectives of natural non-humans such as the environment and animals) and relational more-than-human perspectives are overlooked in the design of Creative AI technologies, critical questions emerge regarding *what consequences their absence leads to for these non-human stakeholders*, and *what would it mean for the design practices of these technologies to prioritise more-than-human worldviews?* In the subsequent parts of this chapter, I will start unpacking these questions by discussing the empirical design work that I have conducted in this area.

Addressing the Lack of Care for More-Than-Human Actors and Relations in Creative AI

My work as a critical design scholar largely focuses on the questions of Creative AI, tying together environmental ethics and more-than-human designerly practices. For example, in an earlier work, I highlighted the insensitive more-than-human practices in the context of Creative AI technologies (Jääskeläinen et al., 2022b) that seemed to indicate the lack of care and sensitivities towards the more-than-human constellations and stakeholders that relate to, and are part of, Creative AI technologies. Various stakeholders are connected to Creative AI: the users, designers, non-humans (for example 'materials' and 'nature') that are impacted in the processes of constructing the technology, and so forth. There is an apparent lack of care for more-than-human actors and relations in the configurations and cultures around Creative AI, from how hardware is constructed to the interactions enabled by these technologies, both regarding human and non-human. For example, in Jääskeläinen et al. (2022b) I describe the apparent lack of more-than-human care in extracting nature as a resource to construct AI hardware that Creative AI is run on, while simultaneously coinciding with disregard for humans in the form of low-wage labour. Thus, the more-than-human perspective opens a perspective on Creative AI as a site that needs to be reconfigured toward more-than-human care.

In other cases of my on-going research, I also witnessed non-humans being disregarded in the aesthetic and cultural imaginaries that surround Creative AI practices. I will provide here a few examples of this aesthetic dimension of Creative AI. Earlier in this chapter, I discussed how human perspectives are centralised in the training of the AI models. This reflects, for example, what data the AI trained on and how the data is annotated – which directly influences the visual AI imagery and aesthetics. In Figure 8.1, we can see an example of what the prompt "*how a worm sees the world*" produces in the visual Creative AI tool Craiyon.[1] While on the one hand, it is an immediate choice of the person who labelled the data used in training the AI model, what images they chose and what textual labels they put on those images, on the other hand, the images that the Creative AI is trained on are also simultaneously a continuum of a long practices of portrayals and visual and textual culture. Thus, while there is an immediate agency, the cultural embeddedness of the training data is also a key consideration.

Furthermore, the visual AI imagery itself is highly human-centric in how we humans largely rely on and prioritise our visual sense, unlike some other species (see the work of Hayward, 2008). In the example illustration in Figure 8.1, the system does not understand how the worm sees the world, but the produced imagery is rather a depiction of a worm from a strongly human-centred perspective. While it is difficult for us to place ourselves into the perspective of other species (if not impossible), it is fairly absurd that a prompt "how a worm sees the world" should produce a depiction of how we as humans see the worm (see foundational work on species' points of view by Nagel, 1974). This paradoxical depiction is an example of how human-centred perspectives are deeply embedded in technology. To unpack this further, the image is completely devoid of context, such as "Where does a worm live? What does it do? How is it part of a habitat?", representing a god-eye view (as described by Haraway, 1991) of worm split separate from its context and relations to the world. It almost resembles a view from a microscope, being split from its environment to be an object for modernist human scientific inspection. The critical questions come down to whether these visual aesthetics and portrayals contribute to disregard and devaluing of worms' subjectivity in presenting them as such objects of scientific examination. The implications could be significant if the cultural and aesthetic values that are being enforced are radically devaluing the subjectivities of non-humans. I will next discuss another demonstrative (and crude) example of a case like this.

In Figure 8.2, we have the imagery resulting from another prompt set "a good animal" and "a bad animal". Effectively, the imagery that these tools are (designed to) produce is evaluating certain species (or rather aesthetics tied to them) higher and certain lower. The "good animal" has a certain set of visual features that are deemed desirable to humans, featuring pleasant species such as dogs and deer in pastel background portraits. The "bad animal", in turn, features scary-looking predators with fangs, perhaps drawing from representations of antagonistic

FIGURE 8.1 What the prompt "how worm sees the world" produced in Creative AI tool Craiyon.

Source: Generated on www.craiyon.com

non-human characters in visual culture. Regardless, these aesthetics are valuating some animals as better (or something different) than others based on their relationship to humans. The AI seems to have an inbuilt assessment of which species are friendly (good) and which are not (bad). Furthermore, the diversity of the animals displayed in this imagery is very homogenous: where are the birds? Where are the worms? Where are the fishes? Where are all the other types of diverse creatures that exist in our world? This very concretely displays how what is included or left out is actively configuring our relation to other species, either privileging or marginalizing them. These remarks are nothing new but have been discussed in human-animal studies (HAS) regarding how mammals are seen as the 'animal par excellence', while in actuality they occupy only a small proportion of the world's creatures and critters (e.g. Cesaresco, 2010). The case of Creative AI is a particular type of visual medium, but similar representations have long genealogies in other visual mediums (film, photography, illustration) – which in turn, have been used as training material for the current day Creative AI. When we reflect on the impact that such tools can have on our relation to biological non-humans, it is clear that they are actively configuring our cultural and aesthetic landscape – and through that, our sensitivities and relations to various non-humans. While these examples are a scratch on the surface of this phenomenon, we need to carefully consider the more-than-human cultural politics (how non-humans and humans are portrayed) within Creative AI technology. In my prior work, I proposed the term "cultural politics of portrayal" to refer to such design political negotiations of visual culture (Jääskeläinen, 2024), while focusing more on human representations that establish social categories and discrimination. It is of utmost importance to expand such efforts into examining cultural portrayals and valuation of non-humans to ensure that Creative AI aligns with more-than-human design commitments.

While many of these legacies are tied to our visual culture and the imaging conventions within it, as well as the legacies of our language, it is nevertheless clear that the visual culture and imagery produced by Creative AI technology is most often de-prioritizing the non-humans at the expense

FIGURE 8.2 AI-generated imagery with prompts "a good animal" (first image) and "a bad animal" (second image).

Source: Generated on www.craiyon.com

of humans, and valuing non-human species in different ways (see speciesism by Singer, 1975). This could have significant implications for our imaginaries in terms of how we can imagine and relate to non-humans' perspectives', and furthermore in how we 'objectify' and disregard the subjectivity of non-humans. Referring to Haraway's critique (1991) of the god-eye perspective, there is a danger that the visual depictions slice non-humans from their natural environment and portray them as objects for human scientific (or creative?) examination and manipulation, effectively positioning them as objects for human use. Thus, there is a need to rethink how these Creative AI technologies could be designed with increased more-than-human sensitivities, in terms of aesthetics and materiality. Fundamentally, a critical question that concerns more-than-human design of Creative AI technology involves maintaining and producing the current value arrangements between various stakeholders; the human, the Creative AI technology, and other non-humans. How these stakeholders get to matter within the scope of ethical consideration is a critical concern from a more-than-human perspective, particularly when AI enables a large increase in the magnitude of production for the cultural artifacts that embed those value judgements. For example, how could the aesthetics be configured in a way that they would support care and sensitivities toward biological non-humans, rather than extracting and examining them?

Challenging the Modes of Knowing and Making in the Design of/through Creative AI Technologies

In the following section, I explore strategies of how challenging the modes of knowing and making could be done in design (research) practice, in the context of Creative AI technologies. I discuss these themes that resulted in a meta-analysis of my own exploratory research work; (1) framing Creative AI as more-than-human, (2) expanding imaginaries of more-than-human, as well as (3) cultivating reflectivity and sensitivity among the stakeholders of Creative AI technology.

Framing Creative AI as More-Than-Human

The first challenge is to understand and frame Creative AI as more-than-human, and to reconsider the boundaries that are drawn in knowledge-making practices to separate the human from the non-human. One prominent example of doing so in the context of AI technology is Crawford's mapping of AI systems which includes the expanded materiality from minerals to data, and other non-human actors that are part of the entanglement (see Crawford, 2021). Crawford essentially provides a visual map of non-human and human stakeholders that take part in the design of a certain AI technology, from the extraction of raw minerals for the hardware to the labelling work conducted by humans, and the resulting AI technology. So, first, we can shift the framing of Creative AI by attempting to frame the technology as part of more-than-human entanglements, rather than as designers and design researchers 'making cuts' that separate and abstract the technology from its entangled relations. These approaches can be taken in knowledge-making through writing, visualisation, and other forms of knowledge communication in research and beyond. In (Jääskeläinen et al., 2022b), I have engaged in the first attempt to frame Creative AI and Creative AI technologies as more-than-human, by providing an image of the de-valuing processes that underlie this technology – from non-human towards the human actors that partake in these entanglements.

Expanding Imaginaries of the More-Than-Human

In that same work (see Jääskeläinen et al., 2022b), I also performed a speculation in which I imagined alternative ways for how Creative AI could look like, using care ethics as a starting point for the speculation. In this case, in speculative interaction design, I used clay, a fictile material, standing in as a medium for exploring the cultural assumptions of Creative AI. The design process included, for example, workshops in which 'violence' and 'care' towards non-humans' were explored in interactions with clay to surface features of violent and caring embodied interaction, which later resulted in conceptualizing a speculative design. To describe this speculation, in Figure 8.3, there is a mapping of an imaginary Creative AI technology (a critical re-design of VQGAN+Clip) with interactions that couple the environmental harm of Creative AI technology with how the 'user' interacts with the Creative AI interface. For instance, if the user would practise embodied sensitivity towards non-human (e.g. caring touch), the visual Creative AI would tweak its parameters so that the generated outcome would require less energy consumption in the generative process. Furthermore, if the user uses violent embodied interaction strategies (such as forcefully handling the technology, pulling, or ripping it apart), the environmental impact of the tool would ramp up. This speculative interaction design work was primarily aimed to raise critical questions and discussion about how we are (or are not) prioritising non-human and more-than-human care in the design of Creative AI technologies, but also through speculative fabulation to expand imaginaries of what these technologies could look like.

Cultivating Reflectivity and Sensitivity among the Stakeholders

Furthermore, to be able to engage in more-than-human design practices in the context of Creative AI, we need to first question our own perspectives and framings of things – and reflect on

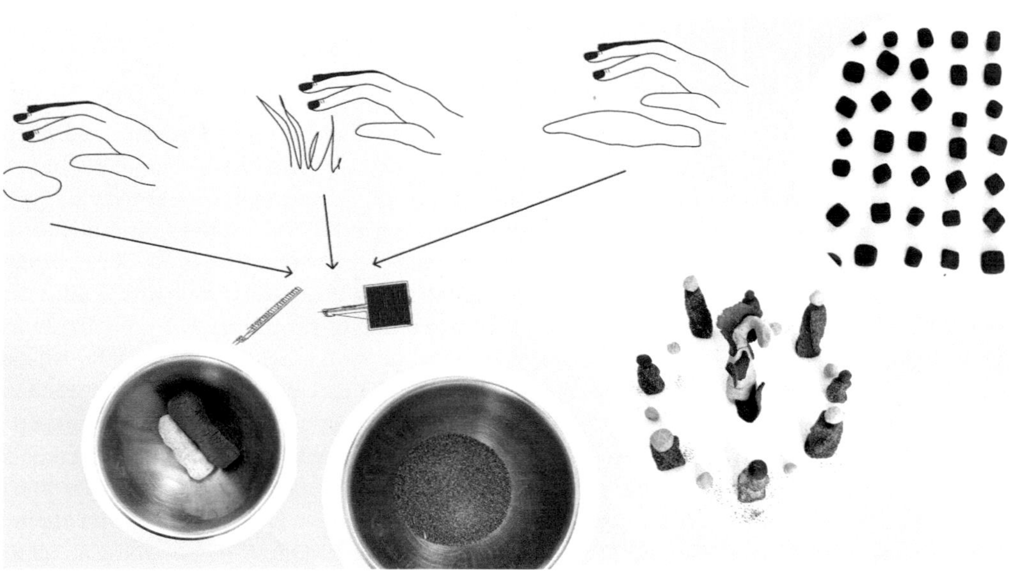

FIGURE 8.3 Speculation on what Creative AI technology VQGAN could look like with interactions that couple the environmental harm of Creative AI technology with the interaction.

them from the perspective of the more-than-human as design researchers who engage in work with Creative AI. This calls for developing a capacity for reflective practices about our own positionalities of environmental ethics and values, as well as critically examining our imaginaries of Creative AI.

This reflexivity work is something that I have begun in the context of Creative AI, by organising workshops for Creative AI artists and technology engineers to reflect on their own imaginaries and ethical stances in materialisation and knowledge-making, as well as their implications. These reflections include questions, such as, is the research and design of Creative AI that I do and build my work on using more-than-human perspectives or some other framings? Am I committing to an ecocentric, biocentric or human-centric ethical framing in this practice? What kind of values are reflected in and through these ethical commitments? And what are the epistemological, cultural and material genealogies behind my 'way of making design'? These questions can serve as starting points for design inquiries that aim to frame Creative AI as more-than-human, through developing sensitivity to our own positionality in the more-than-human entanglements, as well as how the Creative AI is framed in those design processes and knowledge-making practices. In practice, these questions can be brought also to examining the design artefacts – including the aesthetics that the Creative AI systems are putting out – to uncover how our own worldviews are impacting the design work.

Speculation has been a large part of critically examining and challenging the human-centred outlook on Creative AI towards a more-than-human perspective with the stakeholders that partake in their design. For example, I engage in speculation with the various stakeholders that develop and design Creative AI technologies through writing *speculative news articles* and *speculative research abstracts* and imagining *speculative Creative AI technologies*. For example, in the last one of these workshops, artists engaged in imagining Creative AI technologies/artefacts for their own creative practice. Afterwards, the workshop participants engaged in ethical reflection around the perspectives that became implicitly embedded in the speculative Creative AI artefacts. An outcome of all these workshop studies is that non-human perspectives were de-prioritised and absent from the imaginaries, regardless of the mode of speculation, if not specifically prompted for. This calls for the importance of 'expanding imaginaries' of Creative AI, which we discussed earlier in this chapter. The perspectives of biological non-humans were entirely absent in these imaginaries, but the stakeholders were more concerned with Creative AI taking over certain creative tasks, and other dystopian scenarios from a human-centred perspective.

Conclusion

In this chapter, I have discussed Creative AI from a more-than-human perspective, with the central themes being the environmental, ethical, and aesthetic stances that designers are taking in the context of Creative AI. Furthermore, I have discussed the strategies to be explored further to strive towards the more-than-human design of Creative AI technologies. These included framing AI as more-than-human in the practices of knowledge-making, cultivating more-than-human sensitivities among the designers of Creative AI technologies, expanding the imaginaries of Creative AI towards more-than-human worldviews and sensitivities, and re-configuring the Creative AI aesthetics to avert extractivist orientation toward biological non-humans.

The empirical cases covered in this chapter (aesthetics examples, the caring re-designs of interactions, and stakeholder speculations to open up alternative imaginaries) have shown how Creative AI technologies are oftentimes designed with disregard towards both human and non-human stakeholders in more-than-human entanglements (Jääskeläinen et al., 2022b). At times, due to the historical practices of deprioritising non-humans, it might be challenging to understand how Creative AI technologies could be approached from a more-than-human perspective. Thus, there is a need to explore alternative imaginaries and extend the current technological imaginaries that we have of Creative AI. Likewise, a more-than-human perspective urges us to question 'how should we value things?' drawing from the perspectives of biocentrism, ecocentrism and human-centrism. These reflections can aid us in shifting the future of Creative AI technologies toward a more-than-human design stance. As discussed earlier, the efforts to push AI development toward a human-centred ethos actively undermine more-than-human design approaches. Going towards the future, we need to keep asking critical questions from a more-than-human perspective, regarding "What would more-than-human Creative AI mean?" and "How can we re-configure the Creative AI imaginaries and aesthetic materiality to be more inclusive toward non-humans?"

I hope for this chapter to lay ground for more-than-human-centred design of Creative AI technologies, highlighting both the need for reflecting our ethical commitments as designers of these Creative AI technologies, and providing a few guide signs for future directions in more-than-human design research within Creative AI.

Questions

- Who gets to matter in the design of Creative AI technologies?
- What critical consequences can the absence of consideration and sensitivities toward non-human stakeholders lead to in the design of Creative AI technology?
- What would it mean for Creative AI technologies to embody more-than-human worldviews?
- How can we re-configure the imaginaries and aesthetic materiality of Creative AI to be more inclusive toward non-humans?

Note

1 www.craiyon.com.

References

Adobe (2023). Retrieved November 17, 2023, from www.adobe.com/products/illustrator/generative-recolor.html

Auernhammer, J. (2020) Human-centered AI: The role of human-centered design research in the development of AI. In Boess, S., Cheung, M. and Cain, R. (Eds), *Synergy – DRS International Conference 2020*, 11–14 August, Held online. https://doi.org/10.21606/drs.2020.282

Cesaresco, E. M. (2010). *The Place of Animals in Human Thought*. Kessinger Publishing.

Chen, X. (2023). Anthony. In Burke, J., Du, R., Hong, M. K., Jacobs, J., Laban, P., Li, D., Peng, N., Willis, K. D. D., Wu, C.-S., & Zhou, B., *Next Steps for Human-Centered Generative AI: A Technical Perspective* (Version 1). arXiv. https://doi.org/10.48550/ARXIV.2306.15774

Civit, M., Civit-Masot, J., Cuadrado, F., & Escalona, M. J. (2022). A systematic review of artificial intelligence-based music generation: Scope, applications, and future trends. In *Expert Systems with Applications*, 209, 118190. Elsevier BV. https://doi.org/10.1016/j.eswa.2022.118190

Crawford, K. (2021, April 6). *The Atlas of AI*. Yale University Press.

Doshi, A. R., & Hauser, O. (2023). Generative artificial intelligence enhances creativity. In *SSRN Electronic Journal*. Elsevier BV. https://doi.org/10.2139/ssrn.4535536

Fry, T. (2010, November 1). *Design as Politics*. Berg.

Fuad-Luke, A. (2008). Slow design. In Erlhoff, M., & Marshall, T. (Eds), *Design Dictionary*. Board of International Research in Design. Birkhäuser Basel. https://doi.org/10.1007/978-3-7643-8140-0_251

Haraway, D. (1991). *Simians, Cyborgs, and Women*. London: Routledge.

Hayward, E. (2008). More lessons from a starfish: Prefixial flesh and transspeciated selves. *Women's Studies Quarterly*, 36(3/4), 64–85. http://www.jstor.org/stable/27649785

Jääskeläinen, P., Pargman, D., & Holzapfel, A. (2022a). On the environmental sustainability of Creative AI(s). In *Eighth Workshop On Computing Within Limits 2022*. https://doi.org/10.21428/bf6fb269.c46375fa

Jääskeläinen, P., Holzapfel, A., & Åsberg, C. (2022b). Exploring more-than-human caring in creative-AI interactions. In *Nordic Human-Computer Interaction Conference*, NordiCHI '22. ACM. https://doi.org/10.1145/3546155.3547278

Jääskeläinen, P. and Åsberg, C. *What's the Look of "Negative Gender" and "Max Ethnicity" in AI-Generated Images? A Critical Visual Analysis of the Intersectional Politics of Portrayal.* CHI EA '24, May 11–16, 2024, Honolulu, HI, USA. https://doi.org/10.1145/3613905.3644057

Kalfagianni, A., Fuchs, D., & Hayden, A. (2019). *Routledge Handbook of Global Sustainability Governance*. Routledge.

Latour, B. (1993). *We Have Never Been Modern* (C. Porter, Trans.). London, England: Harvard University Press.

McElroy, E. (2024). *Silicon Valley Imperialism. Techno Fantasies and Frictions in Postsocialist Times*. Durham, USA: Duke University Press.

Nagel, T. (1974). What is it like to be a bat? *The Philosophical Review*, 83(4), 435–450. https://doi.org/10.2307/2183914

Nardi, B., Tomlinson, B., Patterson, D. J., Chen, J., Pargman, D., Raghavan, B., Nicenboim, I., Giaccardi, E., Søndergaard, M. L. J., Reddy, A. V., Strengers, Y., Pierce, J., & Redström, J. (2020). More-Than-Human Design and AI. In *Companion Publication of the 2020 ACM Designing Interactive Systems Conference*. DIS '20: Designing Interactive Systems Conference 2020. ACM. https://doi.org/10.1145/3393914.3395912

Penzenstadler, B. (2018). Computing within limits. In *Communications of the ACM*, 61(10), 86–93. https://doi.org/10.1145/3183582

Puig de la Bellacasa, M. (2017). *Matters of Care: Speculative Ethics in More Than Human Worlds*. University of Minnesota Press.

Shneiderman, B. (2022). *Human-Centered AI*. Oxford University Press.

Singer, P. (1975). *Animal Liberation: A New Ethics for Our Treatment of Animals*. Random House.

Thompson, A. (2019). *The Oxford Handbook of Environmental Ethics* (S. M. Gardiner, Ed.). New York, NY: Oxford University Press.

Totlani, K. (2023). The evolution of generative AI: Implications for the media and film industry. *International Journal for Research in Applied Science and Engineering Technology*, 11 (10), 973–980. https://doi.org/10.22214/ijraset.2023.56140

Van Der Maden, W., Van Beek, E., Nicenboim, I., Van Der Burg, V., Kun, P., Lomas, J. D., & Kang, E. (2023). Towards a design (research) framework with generative AI. In *ACM Designing Interactive Systems Conference*, DIS '23: Designing Interactive Systems Conference. Association for Computing Machinery, New York, NY, USA. https://doi.org/10.1145/3563703.3591453

Van Der Maden, W., Van Beek, E., Halperin, B. A., Jääskeläinen, P., Kang, E., Kun, P., Lomas, J. D., Merritt, T., Lindley, J., Muller, M, Nicenboim, I., Zhu, J., & Van Der Burg, V. (2024). Death of the design researcher? Creating knowledge resources for designers using generative AI. In *Companion Publication of the 2024 ACM Designing Interactive Systems Conference.* Association for Computing Machinery, New York, NY, USA. https://doi.org/10.1145/3656156.3658398

Wakkary, R. (2021). *Things We Could Design*. London: MIT Press.

Wallace, B., Hilton, C., Nymoen, K., Torresen, J., Martin, C. P., & Fiebrink, R. (2023). Embodying an interactive AI for dance through movement ideation. In *Creativity and Cognition*, C&C '23: Creativity and Cognition. ACM. https://doi.org/10.1145/3591196.3593336

West, S., Jamila Haider, L., Stålhammar, S., & Woroniecki, S. (2020) A relational turn for sustainability science? Relational thinking, leverage points and transformations. *Ecosystems and People*, 16 (1), 304–325. doi:10.1080/26395916.2020.1814417

PART 2

Methods and Pedagogy

9

MULTISPECIES ETHNOGRAPHY IN DESIGN RESEARCH AND PRACTICE

Heidi R. Biggs, Anton Poikolainen Rosén, Emīlija Veselova, David Sánchez Ruano and Katerina Cerna

Introduction

In this chapter, we introduce the approach of multispecies ethnography and provide examples of methods that can be used by designers seeking to include non-humans in their design thinking and practices. Ethnography is a research practice that stems from anthropology, where the researcher immerses oneself in a particular community to closely observe their behaviour and interactions to make sense of their culture, and to understand how its members make sense of the world. Ethnography has a history as a method for design research, used to extend the scope of the design and its impact by investigating the lived experience and holistic environment of the stakeholders (Salvador, Bell, and Anderson, 1999; Dourish, 2006; Dourish and Button, 1998; Salvador and Mateas, 1997; Khovanskaya et al., 2017). In addition, ethnography is often used to foreground power imbalances by richly describing the workings of the world as a phenomenon consisting of complex processes – not fixed or naturalised.

Multispecies ethnography extends the framing of ethnography to include non-human others. Multi-species ethnographies extend the framing of "community" and "culture" to include the whole more-than-human world (e.g. animals, plants, microbes, materials, technologies) (Hamilton and Taylor, 2017; Kirksey and Helmreich, 2010). Hence, who and what the researcher observes and views as significant in a culture and community, and the related sense-making, expands. The politics of expanding ethnography beyond the purview of the human asks how human life and culture include and are shaped by non-human others, as well as how human goings-on impact non-human others. As a methodology, it asks how we might shift thinking around design to include and care for non-human impacts and influence. In some ways, design is already intuitively a more-than-human practice as it often grapples with materials and how they mediate relations to the world for humans. The material push-back at the centre of craft and making praxis places designers into latent more-than-human relations. However, multi-species ethnography extends this sensibility toward non-human animals and plants, using techniques from ethnography to foreground the human/non-human entanglements at the heart of emerging issues like climate change, AI, and beyond (Liu, Bardzell, and Bardzell, 2019; Liu, Byrne, and

DOI: 10.4324/9781003467731-12

Devendorf, 2018; Giaccardi and Redström, 2020; Akama, Light, and Kamihira, 2020; Janicki et al., 2024; Nicenboim et al., 2020).

In this chapter, we will introduce and critically reflect on a variety of methods that we and others have used as part of multispecies ethnographies. We break the chapter into an overview of the foundations of ethnography and multi-species ethnography in design research then discuss foundational principles in multispecies design, and finally discuss concrete methods we and others have used to implement a multi-species ethnographic approach to design research. These methods and approaches are by no means a comprehensive list but should rather be seen as examples and starting points for how to work creatively and rigorously through methods based on *observing one's immersion in more-than-human relationships*. We discuss a variety of post-human noticing practices through listening, recording, and making to engage with non-humans. Specific methods addressed include perception walks; embodiment probes; the application of naturalist practices the shifting of perspective of devices for observation; and biological empathy maps. As design is an inherently interventionist practice, not simply describing the world, but also actively building it in material ways, we suggest that employing multi-species ethnographic methods sensitise designers and leads to emergent orientations to design that reframe human-beingness considering our entanglements with other organisms, species, and beings.

Toward a Multi-Species Ethnographic Praxis for Design

Ethnography and Design

To understand the impact and practice of a multi-species ethnography and its place in design, we first briefly review the basic tenets of ethnography and the onto-epistemological shifts that are implied by shifting the practice to more-than-human design. As this is a practice by and for designers and design researchers, it is necessary to remember the subtle shift in goals from an ethnographic practice employed by social scientists or anthropologists, and that employed by designers. In design's adoption, research becomes the grounds for intervention, materialization, and as some have argued, ontological orientations and world-building (Fry, 2009; Willis, 2006; Blevis, 2007; Escobar, 2018). This section can be read as a framework or methodology overview, whereas the following sections will offer various methods or approaches to multi-species ethnographies.

Ethnographic practice has become a major way of understanding the context, impact, and points for intervention for designed artefacts, systems, and services. Ethnography is a qualitative research method that stems from anthropological traditions. The research practice originated in the early 20th century to study cultures typically far removed from the immediate cultural context of the researcher. The ethnographic researcher explores the culture and practices of communities by the researcher's immersion in 'the field' – whereby the field is the place where the subjects of study conduct their daily lives, work, or other activities in question. The immersion engages the senses and requires the researcher to undergo an experience as an 'outsider' and it is through the researcher's subjective experience that knowledge is produced (Pink, 2015; O'Reilly, 2005). The practice is usually done by undergoing an experience for a length of time (often years in traditional anthropological practices, but much shorter in design settings), taking detailed notes informed by sociological questions, which are synthesised into themes and findings (Emerson, Fretz, and Shaw, 2011).

Ethnography is thus the study of social and cultural practices and habits, through careful attending, embodied experience, and exposure and consideration of others. It is also quite critical, according to Hamilton and Taylor (Hamilton and Taylor, 2017), that ethnographic focus on thick description, experiential/phenomenological ways of knowing, and poetic writing, allows for critical, nuanced reflection on the world that highlights how things *become* not just how they are. While ethnography has been critiqued for its role in and connection to colonialism and imperialism, contemporary practices have considered the need for researcher reflexivity and positionality as critical parts of the process of ethnographic knowledge production (Doucet, 2008).

Design practice and research have come to embrace ethnographic methods over time in ways that remain under development. Some see design as being driven by ethnographic studies, others see design guiding ethnographic practice, and still others frame practices as a hybrid "design ethnography" practices that combine elements of both into interdisciplinary, interventionist approaches (Dourish, 2006; Pink et al., 2022). But, ultimately, to borrow tools from design and ethnography is to engage in the ways that design is ever entangled in the social and cultural worlds of people. Designed artefacts and systems build worlds (Dunne and Raby, 2013), and have ontological orientations (Willis, 2006). and are entangled in the cultures and social lives of the people who use the technologies. Pink et al. discuss how ethnography has an everyday, phenomenological sensibility to it, and that design, crossed with ethnography, engages knowledge production in embodied processes, collaborative practices, and interventionist intentions (Pink et al. 2022).

Multispecies Ethnography and Design

We see another complex space emerge as designers examine multi-species ethnography through ethnographic practice, design of research methods, and borrowing from scientific practice to inform new sensitization toward non-human others. All these things, as part of the design discipline, are taken into material, embodied, or digital interventions, which might be the subtle distinction between design and social science ethnographic practice. We showcase these 'doings' and some of their impacts in the section "Methods: Exemplifying multispecies design ethnography".

To include notions of a multi-species ethnography already extends and complicates ethnographic practices by moving beyond the purview of the human as the goal of research and asking how humans and non-humans "co-constitute the world" (Hamilton and Taylor, 2017, 2). Multi-species approaches can be used anthropologically, not just by the sciences like ecology or biology, toward new orientations (Kirksey and Helmreich, 2010). We see, as is typical in much more-than-human or posthuman research, the value in playful interdisciplinarity and the erosion of boundaries between the scientific, the social, and the new stories that emerge in points of breakdown between nature and culture, "where encounters between Homo sapiens and other beings generate mutual ecologies and coproduced niches" (Kirksey and Helmreich, 2010, p. 4). Human exceptionalism is reified through, "a dominant and hegemonic belief that animals do not have *selves* or identities, and by extension, that they do not matter or at least do not matter as much as humans" (p. 3). To include non-humans in the ethnographies, the study of human life and culture, or to attempt to grapple with extending the sphere of concern and care is the active project of many posthuman and more-than-human scholars, which, in fact, often requires an un-learning of these human-centric and human-exceptional values. Some of what happens in multispecies ethnography is learning, and some of what happens is unlearning.

The field of design has picked up on these multi-species starting points as well. While in fields like the humanities and social sciences, a multispecies approach reflects a prior constant foregrounding of human concerns in research, in design, a multispecies approach stands in contrast to human-centred design paradigms and their associated methodologies of tracking things like ease of use, cultural and social contexts of use, and product adoption. This emerging way of working beyond the human in design is grounded in theories of technological mediation and feminist science and technology studies (STS), new materialism, and object-oriented ontology – which begin to notice the ways that human actions are entangled with networks of non-human agencies, challenge binaries and hierarchies, and push the boundaries of scientific and humanistic ways of knowing. Early discussions by Disalvo and Lukens leveraged these theoretical texts to argue for a non-anthropocentric approach to design, which was followed by approaches that explore the intersection of nature and cultures through design of smart cities or coexistence (Disalvo and Lukens, 2011). Light et al. called for a design practice that acknowledged the existential entanglements humans were currently facing such as climate change, and asked how we might find other narratives which shape the design of technologies that latently assume human exceptionalism and control over nature (Light, Shklovski, and Powell, 2017). She asks how we might instead, "design to unseat humans from the centre of the universe and support a more equitable gaze?" (Light, Shklovski, and Powell, 2017, p. 728). In his book on designing for more-than-human-centred worlds, Wakkary also discusses how human-centred thinking and approaches might ironically be the root of many human problems (Wakkary, 2021). Much discussion has ensued with one of the major questions being, how do we do the work of decentring, post-anthropocentric, and more-than-human-centred design, and design research? Multi-species ethnography seems to be one of the major strategies.

Multispecies ethnography sensitizes the designer to non-human others, who learns in-situ in the field by undergoing experiences. The resulting research offers thick descriptions and critical proclivities of ethnographic practice, and draws from and expands other naturalistic forms of noticing toward multi-species encounters and design.

Concepts in Practice

We see multiple examples of ethnographies used toward a better understanding of these concepts such as Liu et al '.s ethnographic exploration of symbiotic human/non-human encounters in small farming operations in Taiwan (Liu, Bardzell, and Bardzell, 2019). In practice, the ideas of hybrids and naturecultures can sensitise the multispecies design ethnographer to become aware of the inseparability of culture and nature in the context they study and design for. Others have used ethnographic practices to unpack the multispecies aspects of design practices like weaving (Oogjes and Wakkary, 2022), or to study skilled craftspeople, like carpenters, who hold rich understandings and relationships to lively materials like wood (Dew and Rosner, 2018). On the other hand, some designers have used design practice to get closer to non-human others, like silk worms (Ikeya, Wakkary, and Barati 2023), kombucha SCOBY (Ofer and Alistar, 2023) and mushrooms (Liu, Byrne, and Devendorf, 2018) or future ecologies with climate change (Biggs and Desjardins, 2020). We trace and exemplify these different ways 'in' to multi-species ethnographies, and trace how they showcase different strategies for multi-species ethnographies, below.

Strategies of Multispecies Ethnographies

We now move to introducing some theoretical concepts that are often touched on in multi-species ethnographies. The concepts, well served by multispecies ethnography, are 'noticing', 'decentring', 'hybrids' and 'natureculture". We touch on each briefly in the following and then illustrate them with more concrete examples of methods in the subsequent section.

Noticing and Decentring

To notice is to become aware of, and to treat something as worthy of recognition. More specifically, multi-species ethnographer Anna Tsing describes noticing as a culturally and politically sensitive skill that recognises the interconnectedness of ecological, economic, and cultural systems – and how these systems function from more-than-human perspectives (Tsing, 2015). Tsing states that in noticing, "we must reorient our attention … look around rather than ahead … [and] notice the divergent, layered, and conjoined projects that make up worlds" (Tsing, 2015, p. 22). In design practice and research, noticing can be used as a method that enacts the process of becoming aware of, experiencing, and understanding the world from more-than-human perspectives (Liu, Byrne, and Devendorf, 2018; Liu, Bardzell, and Bardzell, 2019; Poikolainen Rosén, Normark, and Wiberg, 2020; Biggs, Bardzell, and Bardzell, 2021). Here, the situated experience of noticing shifts attention from the experience of "self" to the experience of oneself as part of the environment. This experience includes practical benefits, such as identifying the needs and interdependencies of other organisms, but also more affective dimensions, such as feelings of connectedness. The insights from noticing a phenomenon over time can be used to inform more-than-human-centred design. The idea of "noticing" can thus be understood as the steps taken before isolating a particular matter to design for. In addition, the practice of noticing does something fundamental by bringing forward the non-human other. In Sara Ahmed's work on queer phenomenology, bringing something that is normally in the background to the foreground can have critical or queering effects (Ahmed, 2006). We see the act of noticing as an embodied way of bringing non-humans to the foreground of our attention and noticing what shifts take place as we decentre human-centred approaches. Noticing, in multispecies ethnography, builds from the ethnographic tradition that is in and of itself a process of phenomenological undergoing. In noticing, one can trace a growing understanding and a personal subjective experience that leads to growth and new perspectives. We see this type of close-attention to the shifting of the human away from human-centric perspectives in some of the work we share below.

Hybrids and Naturecultures

Multi-species ethnography engages with posthuman concepts like "hybrids" and naturecultures, concepts developed by Donna Haraway (Haraway, 2003; 2010). Ethnographic work, which showcases the processes of actors, traces, "entanglements and boundaries that are helpfully blurred through method" (Hamilton and Taylor, 2017, p. 12). Hybridity and natureculture are concepts which both push back on the delineation of the human/non-human separation. In her work on species encounters, Haraway suggests that the separation of humans and nature is part of the oppressive logic that led to the objectification of non-human others, and instead, challenges her audiences to see nature and culture as intimately entangled and co-constitutive (Haraway, 2003; 2016).

> As Haraway points out, nature cannot stand outside of culture just as culture cannot stand outside of nature. This is because the meaning of nature – what we identify as natural – is not just determined by culture but is also the result of specific historical, material, and political conditions of possibility.
>
> (Latimer and Miele, 2013, p. 11)

Hybridity is another concept by Haraway which relies on the erosion of boundaries between human and machine, or human and nonhuman towards politically liberatory feminist agendas (Haraway, 2010). Ultimately, to see ourselves as hybrids (not purely human) and to see culture as entangled in nature do seem to be at the heart of what is trying to be achieved through noticing practices and through multispecies ethnographies.

Methods: Exemplifying Multispecies Design Ethnography

In what follows, we illustrate how the concepts addressed in the above section are carried out in specific methods for multispecies ethnography. In these examples, we draw both from our practice and research published by others. These examples may begin answering the question of how multispecies design ethnographies can be carried out, the type of results they achieve, and how they are used toward designed outcomes. The following examples are not exhaustive, simply illustrative – offered in hopes of inspiring more participation, more method explorations, and a growing cannon of design practitioners working toward multispecies ethnographic practices and sensitivities in design.

Noticing as a Method and Decentring

We now turn to the more specific principle of noticing within a multispecies ethnography, noticing practices are, as mentioned above, methods that bring focus to the entanglements of humans and non-humans through repeated attention that also, often, blurs the boundaries between scientific and naturalistic observation, and human-centred sociological analysis.

In one example, Biggs et al. use autoethnographic bird-watching practices to notice how birdwatching reoriented or decentred them in relation to their neighbourhood bird companions (Biggs, Bardzell, and Bardzell, 2021). In this work, they used making explorations (short video sketches, sound compositions, and sketching) to reflect on their experiences with birds, ultimately finding a new, more diffuse, and multi-species way of attuning to the world. The experience scaffolded birdwatching, naturalistic, techniques found through reading the recommendations for bird-watching practices in books, and then used the fundamentals of bird-watching practice to explore concepts of noticing and decentring from posthuman theory. Interestingly, in Tsing's (2015) work on more-than-human noticing, she suggests noticing is the hybrid approach of natural history and ethnographic observational practices, and this is what Biggs accomplishes in this paper.

As Biggs et al. (2021) notice, they also undergo what they claim is a kind of decentring of themselves as birds take on a larger and larger part of their sensory lifeworld. The process of becoming decentred is uncomfortable, and they describe it as ecological 'abjection' – or the Ecological Abjection: the "feeling of being surrounded and penetrated by [ecological] entities that I can't peel off" (Morton, 2016). Abjection is a concept from feminist theorist Julia Kristeva that describes, "the most archaic form of negativity, an exclusion or expulsion which functions by securing the borders of self, carving a space, marking a divide, out of which the ego can emerge". Noticing this abject feeling suddenly makes one aware that the discomfort of decentring has to do with the discomfort of having one's boundaries with non-human others blurred, one's ego diminished. Noticing non-humans, and attending to them, shifted Biggs et al'.s (2021) relationship to the birds and decentred them through practice.

The work offered some thoughts for design such as the use of making to catalogue and process interactions with non-human others like birds where there is no shared language or lifeworld, as well as some reflections on the ways that technologies orient attention and might embrace non-optimal, non-forward orientations that include non-human voices more openly and easily. Again, one of the main contributions was also an account of changing and the documentation of one's shifting relations and attunements to non-human others and what this could offer to design practitioners.

In another example, Poikolainen Rosén (Poikolainen Rosén, Normark, and Wiberg, 2020) approached noticing both to guide their ethnographic participant observation studies – how they noticed as design researchers – and to understand how users such as urban farmers notice the environment, sometimes with the support of sensing technologies. This dual understanding of noticing practices blurs who or what a designer is, including researchers, urban farmers, animals, plants, microbes, and autonomous technologies and what is designed, such as relations in a garden. The work suggests that we can design technologies that support more situated and nuanced noticing practices. This includes sensing technologies, but also more experimental designs as discussed in the following section, "Embodiment probes".

Embodiment Probes

As discussed above, noticing can be supported by more specific methods and tools. Embodiment probes are physical objects intended to remind their wearer to notice differently. The material qualities of these probes are intended to emphasise phenomena one is interested in noticing by for example limiting and enforcing certain movements or senses. Such probes can be ready-made or made by their wearer. Embodiment probes are not end-goals in themselves but rather tools to support more over-arching methodological approaches such as "noticing differently", and the ethnographic sensitivity of "making strange". Embodiment probes are a design-oriented method for multispecies ethnography in development. The method has also been signified as '*Umwelt* Apparatus'.[1]

One example of embodiment probes is found in a workshop facilitated by Poikolainen Rosén and Cerna. In the workshop, they asked: Could we design probes that help us perceive our being in the world from the perspective of other species? If so, can this experience help foster more nuanced more-than-human design skills? In other words, the method does not claim that we can ever exactly know what it is like to be another being, however, we can wonder, imagine, and experience in ways that might be constructive for a multispecies design ethnography. In a way these probes provide a designerly method for achieving hybridity – as one takes on (physically) and morphs into a new shape and mindset.

For example, an ant embodiment probe (see Figure 9.1, right) created an experience of having ant-like jaws and antennas. Sounds from encountering materials transferred through the sticks onto the cheekbones, amplifying the experience of contact-based sound. This prompted design reflection on alternative modes for mediating sound, and how different organisms may have different thresholds for when sound pollution becomes problematic. The probe additionally simulated a more constricted experience of having an exoskeleton and a body divided into three parts. This prompted design reflection for example on how the same space provides different affordances depending on an organism's specific capabilities. In sum, embodiment probes can be used as part of a multispecies ethnography as a creative way to help the wearer perceive their context from novel, unexpected and multiple perspectives. In addition, the artefacts and the enactment of new perspectives bring forward a new quality of hybridity, the performance of hybrid bodies that help to extend the human imagination into more-than-human points of view.

FIGURE 9.1 Embodiment probes. Examples of embodiment probes that are used to respond differently as part of a multispecies ethnography. Left: A probe that restricts the vision of the wearer. Right: A probe made of branches intended to embody ascertain capacities of an ant.

Perception Walks and Other Movement-Based Ethnographies

Another way to notice differently is walking. Walking is a method used in many forms of research to break free from static thinking and conventions, and direct attention and cognitive effort towards one's environment The core idea of perception walks is to move in an environment with a particular intentionality. This intentionality can be guided by specific tasks, the use of devices that alter perception, roleplaying, or elements from focusing and meditation. By noticing differently in this way, the designer may gain a new sense of the environment/design space which makes them more sensitised when designing for a multispecies context. The walk can be completely open-ended using elements from meditation to help maintain a heightened state of attention and awareness. The walk may also use instructions to guide attention to specific phenomena. More specific examples include letting an animal, such as a dog, take the lead, following its move in a drifting way, and emphasizing what the animal's movements tell us is important in a space (Cabrera et al. 2022). In another example, participants walk in an urban area wearing masks that represent animals such as rats and doves. The purpose is to roleplay this animal and view the city from the perspective of a particular species (Clarke et al., 2019). Both these examples help the multispecies ethnography view the context from the perspective of multiple species.

Shifting Perspectives through Multi-Species Observation

The following example experiments with the idea of from whose perspective one is noticing. The core idea behind shifting perspectives is to reconsider the placement of our observational devices to foster human decentring in design. Practically, it can, for example, mean placing cameras into plants and conducting participatory observation in front of them. Rather than aiming to bring us closer to plant vision, it aims to bring us closer to a situation when more than one species' perspective is considered. This exercise goes beyond a simple shift in placement of the observation device and prompts us to re-think: what can we learn when we let another species "hold" the observational perspective? A concrete example of such observation is a small study focused on the exploration of relationships between humans and plants in the homes of participants, in an ongoing study of Cerna. By discussing where to place the camera with participants, sensorially engaging with the plants during interviews or considering leaving the camera in a plant when the researcher leaves the field, valuable insights are gained into the lived experience of cohabiting with plants, not just from the human observer's standpoint but also from the unique perspective of a plant-"held" camera.

Biological Empathy Mapping

Many of the above examples have focused on noticing, observing, or gathering of data. However, this is not enough, in the next step one must make sense of the knowledge that is available of a species or multispecies setting. Here the biological empathy mapping method developed by Sánchez Ruano can be beneficial. This map was developed as part of the tools for the Symbiotic Design Practice framework which includes aspects of biophilic thinking and biomimetic design. This tool has been used constantly in undergraduate and postgraduate courses specifically in teaching about Nature Based Solutions, bio-design, and regenerative design. This method is used for mapping the key personhood of a species. By researching the key features, needs, emotions, and behaviour using questions and gathering biological data. It involves drawing and looking for stories. Through the empathy map, the values we perceive in a particular context support designers to have a closer understanding of a particular organism or ecosystem. The biological empathy map is a fieldwork activity that helps to begin the exploration or direct conversation with a living organism. Through using a template (see Figure 9.2) the observer incorporates features of the Goethean method of observation (Wahl, 2005) that can be used on-site looking directly at the organism, parts of the organism (e.g. feathers, bones, branches, or flower) or pictures of the organism we want to interact with. Beyond observation, the Goethean method deepens the perception of a species, as we focus on the drawing, the dynamic of observing forms, textures, and colours, and also the expression of gestures that the organism may have seen as a person itself.

The template of the method integrates a space for drawing and answering questions such as: What is your common name? Where do you live? How do you travel? What senses do you use more? What do you eat usually? How do you socialise? How do you learn? and an open question. These questions require going deep into biological research in databases or scientific publications to understand habits, behaviours, and physical features. Some extra questions are directed to the local people or context where the species lives, such as "What stories have you heard about this species?" and "What are the values you found in interacting with this species?" which can help us to find insights about ancient practices or current traditions, which represent

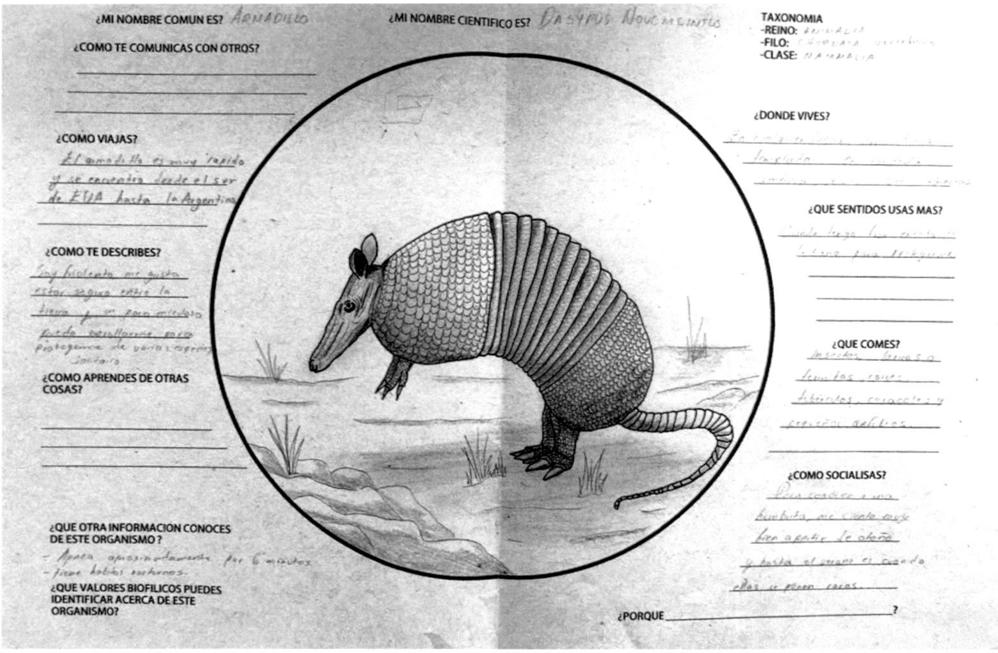

FIGURE 9.2 Sample of a completed Biological Empathy Mapping template with a drawing and text in Spanish.

multispecies ethnographic aspects. In this way the map will help the practitioner in going deeper and straight to the biological features, to analyze further. At the same time, it opens a biophilic and biomimetic understanding that manifests deep empathy with any kind of species.

Concluding Discussion

In this chapter, we have discussed the basic tenets of more than human ethnographies and how they are being transposed into design research which offers interventions at the level of epistemology, ontology, and design goals and considerations which extend beyond the human. We also offered several concrete examples of multispecies ethnography in practice and particular methods that can be used within these.

The hallmark of multispecies ethnography is sensitization to the more-than-human organisms, species, and beings in a setting, pulling them into focus through undergoing some embodied experience, which leads to empathy, care, and attentiveness. This attention and phenomenological undergoing, as well as a narrativization of the process of 'becoming' are key qualities of this research as we slowly unlearn the focus on the human and teach ourselves through various methods, how to consider and include the more-than-human world in the design process.

In this work, we saw the various ways that designers have taken multispecies ethnography – the understanding of the lifeworld of a non-human other – into consideration. Two concepts guide these ethnographies, one is *noticing* which is often enacted by an ethnography-first approach that can decentre, build empathy, or find other ways of being with non-human others that shift perspectives. We saw this in methods that use forest walks, bird watching, and empathy mapping. Another way to work uses design-first approaches to imagine hybrid or more-than-human entanglements through artefacts or systems. In some of the related works, we saw others use design to engage more deeply with silk worms, kombucha scoby, and mushrooms (Ikeya, Wakkary, and Barati, 2023; Ofer and Alistar, 2023; Liu, Byrne, and Devendorf, 2018). In the example of embodiment probes above, we saw how design can open spaces to build hybrid bodies and more-than-human imaginaries. This is a design-first approach to more-than-human ethnographies that explores how artefacts can extend, awaken, and connect our awareness and interaction with non-human others to see further into their worldviews.

As always, the intersections of ethnography and design are complex. Sometimes ethnography leads, other times design practice is turned into an ethnography of its own, and sometimes art and design-based ways of knowing are used to open onto or engage with non-human species. However, these practices, which use embodied, phenomenological experiences, help decentre the ethnographer and bring up questions about the life-worlds of non-humans, their needs, their ways of being in the world, and how we might attend more to our intersection toward collective, careful, and sustainable worldbuilding. We see this as epistemological shifts (inclusion of non-humans in research methodologies) toward ontological shifts toward human-beingness that are reframed to critically consider the inclusion of non-human species in design processes. As Anne Marie Willis (2006) argues, design is world-building, ontological, and builds or forecloses futures on Earth. Fry goes as far as to argue we must re-design the practice of design (2009). Multispecies ethnography seems to offer one path toward doing just that.

Note

1 www.danimetcalfe.com/index.php/project/umwelt-apparatus-/.

References

Ahmed, Sara. 2006. *Queer Phenomenology*. Duke University Press.

Akama, Yoko, Light, Ann, & Kamihira, Takahito. 2020. Expanding participation to design with more-than-human concerns. In *Proceedings of the 16th Participatory Design Conference 2020-Participation (s)*, 1, 1–11. https://doi.org/10.1145/3385010.3385016.

Biggs, Heidi R., & Desjardins, Audrey. 2020. High water pants: Designing embodied environmental speculation. In *Proceedings of the 2020 CHI Conference on Human Factors in Computing Systems*, 1–13. New York, NY, USA: ACM. https://doi.org/10.1145/3313831.3376429.

Biggs, Heidi, Bardzell, Jeffrey, & Bardzell, Shaowen. 2021. Watching myself watching birds: Abjection, ecological thinking and posthuman design. In *Proceedings of the 2021 CHI Conference on Human Factors in Computing Systems (CHI '21)*.

Blevis, Eli. 2007. Sustainable interaction design: Invention & disposal, renewal & reuse. In *Proceedings of the SIGCHI Conference on Human Factors in Computing Systems (CHI '07)*, 503–512. https://doi.org/10.1145/1240624.1240705.

Cabrera, Andrea Botero, Dolejsova, Marketa, Choi, Jaz Hee-jeong, & Ampatzidou, Cristina. 2022. Open forest: Walking with forests, stories, data, and other creatures. *Interactions*.

Clarke, R., Heitlinger, S., Light, A., Forlano, L., Foth, M., & DiSalvo, C. (2019). More-than-human participation: Design for sustainable smart city futures. *Interactions*, *26*(3), 60–63. https://doi.org/10.1145/3319075

Dew, K. N., & Rosner, D. K. 2018, April. Lessons from the woodshop: Cultivating design with living materials. *Conference on Human Factors in Computing Systems – Proceedings*, 1–12. https://doi.org/10.1145/3173574.3174159.

Disalvo, Carl, & Lukens, J. 2011. Nonanthropocentrism and the nonhuman in design: Possibilities for designing new forms of engagement with and through technology. In *From Social Butterfly to Engaged Citizen*. MIT Press. https://doi.org/10.7551/mitpress/8744.003.0034.

Doucet, A. 2008. 'From her side of the gossamer wall(s)': Reflexivity and relational knowing. *Qualitative Sociology*, 31, 73–87.

Dourish, Paul. 2006. Implication for design. *Proceedings of the 2006 Conference on Human Factors in Computing Systems*.

Dourish, Paul, & Button, Graham. 1998. On 'technomethodology': Foundational relationships between ethnomethodology and system design. *Human-Computer Interaction* 13 (4): 395–432. https://doi.org/10.1207/s15327051hci1304_2.

Dunne, Anthony, & Raby, Fiona. 2013. *Speculative Everything: Design, Fiction, and Social Dreaming*. Cambridge, MA: MIT Press.

Emerson, Robert M., Fretz, Rachel I., & Shaw, Linda L. 2011. *Writing Ethnographic Fieldnotes*. University of Chicago Press.

Escobar, Arturo. 2018. *Designs for the Pluriverse: Radical Interdependence, Autonomy, and the Making of Worlds*. Duke University Press.

Fry, Tony. 2009. *Design Futuring*. University of New South Wales Press.

Giaccardi, Elisa, & Redström, Johan. 2020. Technology and More-Than-Human Design. *Design Issues* 36 (4).

Hamilton, Lindsay, & Taylor, Nik. 2017. *Ethnography after Humanism: Power, Politics and Method in Multi-Species Research*. Palgrave Macmillan. https://doi.org/10.1057/978-1-137-53933-5.

Haraway, Donna J. 2003. *The Companion Species Manifesto: Dogs, People, and Significant Otherness. Manifestly Haraway*. Chicago: Prickly Paradigm Press. https://doi.org/10.5749/minnesota/9780816650477.003.0002.

Haraway, Donna. 2010. 'A Cyborg Manifesto' (1985). *Cultural Theory: An Anthology*, 454.

Haraway, Donna J. 2016. *Staying with the Trouble: Making Kin in the Chthulucene*. Duke University Press.

Leshed, Gilly, Rosca, Masha, Huang, Michael, Mansbach, Liza, Zhu, Yicheng, & Hernandez-Aguilera, Juan Nicolas. 2018. "Calcucafé: Designing for Collaboration among Coffee Farmers to Calculate Costs of Production." *Proceedings of the ACM on Human-Computer Interaction* 2 (CSCW). https://doi.org/10.1145/3274418.

Janicki, Sylvia, Teixeira Riggs, Alexandra, Howell, Noura, Sullivan, Anne, & Parvin, Nassim. 2024. Sensing bodies: Engaging postcolonial histories through more-than-human interactions. In *Proceedings of the Eighteenth International Conference on Tangible, Embedded, and Embodied Interaction (TEI '24)*. New York, New York, USA: Association for Computing Machinery.

Khovanskaya, Vera, Sengers, Phoebe, Mazmanian, Melissa, & Derrah, Charles. 2017. Reworking the gaps between design and ethnography. *Conference on Human Factors in Computing Systems – Proceedings* 2017 – May: 5373–5385. https://doi.org/10.1145/3025453.3026051.

Kirksey, S. Eben, & Helmreich, S. 2010. The emergence of multispecies ethnography. *Cultural Anthropology* 25 (4), 545–576. https://doi.org/10.1111/j.1548-1360.2010.01069.x.

Latimer, Joanna, & Miele, Mara. 2013. Naturecultures? Science, affect and the non-human. *Theory, Culture & Society* 30 (8): 5–31. https://doi.org/10.1177/0263276413502088.

Light, Ann, Shklovski, Irina, & Powell, Alison. 2017. Design for existential crisis. *Proceedings of the 2017 CHI Conference Extended Abstracts on Human Factors in Computing Systems (CHI EA '17)*, 722–734. https://doi.org/10.1145/3027063.3052760.

Liu, Jen, Byrne, Daragh, & Devendorf, Laura. 2018. Design for collaborative survival: An inquiry into human-fungi relationships. In *Proceedings of the 2018 CHI Conference on Human Factors in Computing Systems (CHI '18)*, 1–13. https://doi.org/10.1145/3173574.3173614.

Liu, Szu Yu, Bardzell, Shaowen, & Bardzell, Jeffrey. 2019. Symbiotic encounters: HCI and sustainable agriculture. *Conference on Human Factors in Computing Systems – Proceedings*, 1–13. https://doi.org/10.1145/3290605.3300547.

Morton, Timothy. 2016. *Dark Ecology: For a Logic of Future Coexistence.* Columbia University Press.

Nicenboim, Iohanna, Giaccardi, Elisa, Juul Søndergaard, Marie Louise, Venugopal Reddy, Anuradha, Strengers, Yolande, Pierce, James, & Redström, Johan. 2020. More-than-human design and AI: In conversation with agents. *DIS 2020 Companion – Companion Publication of the 2020 ACM Designing Interactive Systems Conference*, 397–400. https://doi.org/10.1145/3393914.3395912.

Ofer, Netta, & Alistar, Mirela. 2023. Felt experiences with kombucha scoby: Exploring first-person perspectives with living matter. In *Proceedings of the 2023 CHI Conference on Human Factors in Computing Systems*, 1–18. New York, NY, USA: ACM. https://doi.org/10.1145/3544548.3581276.

Oogjes, Doenja, & Wakkary, Ron. 2022. Weaving stories: Toward repertoires for designing things. In *In CHI Conference on Human Factors in Computing Systems (CHI '22)*.

O'Reilly, Karen. 2005. *Ethnographic Methods*, 2nd edition. Routledge. http://ebookcentral.proquest.com/lib/gatech/detail.action?docID=958470.

Pink, Sarah. 2015. *Doing Sensory Ethnography*. Sage.

Pink, Sarah, Fors, Vaike, Lanzeni, Debora, Duque, Melisa, Sumartojo, Shanti, & Strengers, Yolande. 2022. *Design Ethnography: Research, Responsibilities, and Futures*. Routledge.

Poikolainen Rosén, Anton, Normark, Maria, & Wiberg, Mikael. 2020. Relating to the environment through photography: The smartphone camera as a tool in urban farming. In *32nd Australian Conference on Human-Computer Interaction (OzCHI '20)*, 506–19. https://doi.org/10.1145/3441000.3441026.

Salvador, Tony, & Mateas, Michael. 1997. Introduction to design ethnography. *Conference on Human Factors in Computing Systems – Proceedings* 22–27-Marc (March): 166–67. https://doi.org/10.1145/1120212.1120325.

Salvador, Tony, Bell, Genevieve, & Anderson, Ken. 1999. Design ethnography. *Design Management Journal (Former Series)* 10 (4), 35–41.

Tsing, Anna Lowenhaupt. 2015. *The Mushroom at the End of the World: On the Possibility of Life in Capitalist Ruins*. Princeton University Press.

Wakkary, Ron. 2021. *Things We Could Design: For More than Human-Centered Worlds*. MIT press.

Willis, Anne-Marie. 2006. Ontological designing. *Design Philosophy Papers* 4 (2): 69–92. https://doi.org/10.2752/144871306x13966268131514.

10

FOUR QUESTIONS FOR SYSTEMIC MORE-THAN-HUMAN DESIGN IN PRACTICE

Anton Poikolainen Rosén, Yuxi Chen, Suzanna Törnroth, İdil Gaziulusoy and Tatu Marttila

Introduction

More-than-human issues are inherently systemic and complex in their character. More-than-human design in practice thus requires systemic thinking that is concrete and action-oriented while simultaneously remaining ontologically open, critical and questioning. In this chapter, we present a set of questions and related considerations that highlight our thinking when we have approached more-than-human design as a systemic issue from diverse epistemological positions.

Systems thinking as an academic discourse was developed in the 1920s by biologists, Gestalt psychologists, ecologists and quantum physicists (Hansson et al., 2021). Several shifts of perspective characterise systems theory: from the parts to the whole, from disciplines to multidisciplinary, from objects to relationships, from measuring to mapping, from quantities to qualities, from structures to processes, from objective to epistemic science, and from the 'universal' truths of Cartesian certainty to approximate and situated knowledge (Hansson et al., 2021). In particular, the posthuman perspective on systems does not think of a system as fixed, finite and entirely visible. Instead, systems are understood as fluid and transitional, haphazard, and spontaneous, organic and unforeseen. In the words of Anna Tsing (2015) many systems are "an open-ended assemblage, not a logical machine" (p. viii).

Designers need to take such organic and open-ended systemic aspects into account. However, given the long history of systemic thinking, it is impossible to give a complete and comprehensive overview of systemic and relational thinking for more-than-human design. Instead, we offer four questions in this chapter 1) How are systemic relationships established? 2) How do we select who or what to consider in a system? 3) How can more-than-human systems be mapped? 4) How can designers intervene in systems? We identified these questions by reflecting on the design process.

First one needs to be aware of the wider ideological setting in which you are designing. Then you need to define and concretely illustrate what your systemic design space is. Only thereafter can you begin to consider how to potentially intervene. These questions are also reflected in recent critiques towards systemic thinking within the field of design research where Buchanan (2019) argues that it is important to critique a) whether the system exists b) what is systemised

DOI: 10.4324/9781003467731-13

c) how the system operates and d) why the system exists. The contribution of our questions is thus twofold. First, the questions may act as concrete and practical guidance in design processes, prompting consideration of various systems thinking approaches in multiple stages of the design process. Second, the questions act as a scaffold for structuring a more nuanced discussion in this chapter of tension in systemic thinking, such as the risk of oversimplification, rendering phenomena invisible or utilizing systemic patterns for oppression. In addition, we point to concrete exercises, methods and probes that can be used to address the four questions. Our questions and their related considerations are intended to help others to problematise more-than-human design spaces as relational and systemic, and encourage efforts to intervene in them, despite their complexities.

How are Systemic Relationships Established?

A first question to consider in systemic more-than-human design is how systemic relationships are established. As illustrated in Figure 10.1 and Table 10.1 there are many parallel discourses on how sustainable and more-than-human systems are established. In this section, we provide an overview of some of these perspectives.

Greek philosophers discussed a 'system' as comprising of smaller components that in some ways form a larger whole. More recent attention to the term started to emerge in connection with ecosystem science and ecology, and later also industrial management discourse. With contemporary discourse connecting design with systems, the concept has in many cases been taken over by casual business talk, while it is inevitably also present in connection with interdisciplinary research, innovation and sustainability transitions. In studies of contemporary design, a systems lens has been an especially salient concept. The field of science and technology studies (STS) emerged between the 50s to 70s to consider large scale energy transitions and their societal governance (e.g., coal mining and labour issues in the UK, nuclear energy) (Ropohl, 1999). Here, actor-network theory (ANT) (Latour, 1999) connected the governance of technology systems with social inquiries connecting both humans and non-humans. A few decades later, transition management (TM) building on 'multi-level perspective' (Geels, 2002) emerged as a more positivist lens to assess dynamics and challenges in managing technology transitions and innovation.

These diverse lenses and discourses illustrate how our understanding of system dynamics is affected by different interpretations of systems. According to Ropohl (1999), some approaches apply a "structural concept" of a system, in which the system is perceived to consist of a set of elements and relations between them. By contrast, approaches aligning with the more contemporary "functional concept" suggest that a system is an entity – sometimes a "black box" – that transforms inputs to outputs according to "specific internal states" (Ropohl, 1999) The interplay of approaches may eventually suggest also a "hierarchical concept," in which sub-system elements connect to a larger "supersystem". Lastly, these three concepts connect in a "general systems theory" that outlines general "laws" for thinking about systems (Ropohl, 1999).

In contemporary sociology, in contrast with the formal rationalist tradition of modernity, one main view is that the world around us becomes structured through communication (Habermas, 1992). In this view, the social reality is created of networked elements emerging through public and expert discourse and interaction, and the system as the realm of formal rationality requires a critical approach. The philosophical concept of "autopoiesis" ('self-creation' (Maturana & Varela, 2012)), originally taken to sociological discourse from biological studies, is suggesting that anything which functions as unity in a system – element, operation, structure, or boundary – results from the systemic production processes, consequently emerging as a subcomponent of the system itself. With a similar notion of the communicative rationale, systems theorist Niklas Luhmann (2012) proposes that a modern society is based on emergent differentiation between innumerable interaction systems and more long-term organizations, however with deep and interconnected hierarchical structures. Consequently, particularly under the condition of 'functional differentiation' that dominates the postmodern era an increasing amount of synchronization is required (Albert, 2016). This latter perception distinguishes between society and functionally differentiated social subsystems, suggesting the existence of both structured hierarchy and emergence between systemic levels of order, while the first viewpoint emphasises the power of meanings in things and communication. In sum, comprehension on various systemic lenses is of need in setting up collaborative design and action with a more-than-human approach.

The underlying philosophical standing points affect how systems are understood. Ropohl (1999) links this with how we understand unity and diversity, holism versus atomism, and the modelling of reality in general. Different perspectives connect with understanding on how systems emergence, equilibriums and transitions are taking place. For example, according to ANT each part of a system should be viewed as equally important and as belonging together in an interactive relationship, while in contrast, TM is more hierarchically structured in projecting a "multi-level perspective" into technological change (Geels, 2002). Figure 10.1 illustrates that while different systemic perspectives perceive the environment and socioeconomic processes as more or less connected, the hierarchies between them can have fundamental variations. The triple-bottom-line model (Figure 10.1, left; Elkington, 1994) separates the environment from human economic and sociocultural activities, thus enforcing an anthropocentric principle that seems integral through emphasis on projecting systems separately from humans despite already facing the Anthropocene (Steffen et al., 2011). By contrast, the nested approach to sustainability (Munro, 1991) (Figure 10.1, middle), suggests a hierarchy that is based on physical reality, however with human activity at the centre. Here sociocultural emerges from the environment, and the economy is an expression of the sociocultural (as in Stockholm Resilience Institute's wedding cake model (*The SDGs Wedding Cake*, 2016)). The nested approach to sustainability progresses logically also further towards more non-anthropocentric models, with a focus on interactions between the different 'realms' of earth–atmosphere, hydrosphere, lithosphere and biosphere – that have offered a main framework for the major international sustainability programmes for over half a century –with a "human sphere" emerging as one important new realm amongst them (Himiyama, 2020). A possible non-anthropocentric model for sustainability (Figure 10.1, right) would perhaps consequently frame human systems only as a minor part of the whole biosphere that in turn develops within the geosphere, atmosphere and hydrosphere.

The examples of different perceptions to sustainability exemplify how the various hierarchical settings and different disciplinary and philosophical emphases play a crucial role in studying and managing interactions within a system. Consequently, to properly set up the stage for an inquiry into systems involving both human and non-human actors, at least the following considerations are necessary to be reflected on. First, the epistemic and ontological standing points need to be openly discussed (see Table 10.1). What is expected and why? What is the focus of possible interactions? This derives into understanding various positions in relation to systems and their dynamics. Second, the above standing points trickle down into questions on how agency and the dynamic settings of a system are conceived.

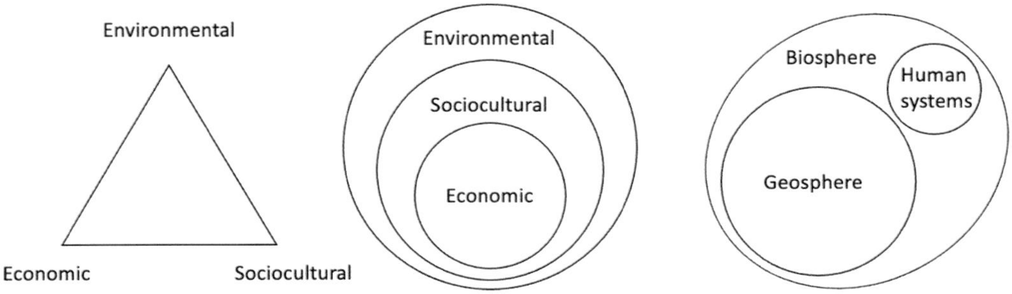

FIGURE 10.1 Different systemic approaches to sustainability issues: triple-bottom line perspective (Elkington, 1994), nested model (IUCN), and a non-anthropocentric model Source: the authors

TABLE 10.1 Elements of discourse, practice and emphasis in sustainable design action today

	Epistemological and ontological approach	*Focus on interaction and development*	*Emphasis for sustainable design*
Singular design approach:	Modernism; techno positivism	Focus on non-human aspects, materials, production networks	Eco-design
Systemic design approach:	Socio-technical systems theory; multi-level perspective	Focus on optimization of human-environment systems	Product-service system design
Pluralistic, transdisciplinary design approach:	Deep ecology; emergent systems theory	Focus on supporting interaction in actor networks, natural systems	Design for sufficiency; critical design

To conclude, an open setting for discussions on a variety of systemic perspectives leaves more room to consider more-than-human design. Perspectives differ on which extent the world can be governed by understanding how to affect systems. Thus, the way we interpret the discourse on systems connects with how we understand relations and agency in a system. Here, a focus on more-than-human design renders the question of agency both more crucial and more complicated. This implies that more-than-human components in a system need to be brought under critical scrutiny, as addressed in what follows.

How Do We Select Who or What to Consider in a System?

A second question to consider in systemic more-than-human design is how we select who or what to consider in a system. In other words, it matters what we assign inside and outside the system – and thus the focus of our concerns. However, it is not always easy to define or name entities of a system. As touched upon in the above section, one of the key characteristics of systemic perspectives, regardless of disciplinary origins, is the attention paid to the relationships between elements. It is through these interrelations that system behaviour emerges. A significant challenge in studying or intervening in systems is to make decisions on how and where to set systemic boundaries, given that each system is part of larger systems in a nested way. In this section, we provide two approaches that may help more-than-human designers frame and scope their systemic focus. The first concept – agential cuts – is a philosophical position, while the second concept is a more practically oriented typology of more-than-human entities in systems. By juxtaposing these quite different approaches to answering the question "How do we select who or what to consider in a system?" we hope to illustrate the breadth of what systemic thinking can be in more-than-human design.

First, it may be beneficial to be aware of the process itself of defining what is inside and outside a system. Helpful for this is the theory of agential realism as proposed by the feminist theorist and physicist Karen Barad (2014). In agential realism, the universe consists of phenomena which are "the ontological inseparability of intra-acting agencies". While these entities are inseparable – in other words systemic – we may make momentarily analytical separations to better make sense of phenomena. For this, Barad uses the term agential cuts. Here "agential" stands for a focus on the inseparable flow of agency in a system, while "cuts" stands for the momentarily analytical separations of such agency when we choose to focus on a particular phenomenon. Agential cuts thus momentarily stabilise specific qualities of a phenomenon of interest. This is phrased by Barad as the "cutting together/apart within phenomena in a two-folded movement that produces the boundaries through which something is made" (Barad, 2014). Agential cuts can thus be a way of determining what actors are relevant in the system through enacting what is "inside" and "outside". We introduce agential cuts in this chapter, as a sensitizing concept that may help more-than-human designers navigate analytical spaces where they simultaneously need to recognise "everything as connected" while being able to recognise the uniqueness of situated phenomena. For more-than-human design in practice, a useful and generative agential cut creates a design space that is manageable while recognizing potential systemic consequences. Another beneficial practice could be shifting the agential cuts you are making, taking different perspectives on the design space by framing the space itself in new ways, thus creating conditions for understanding more systemic aspects by re-framing the design brief.

To concretise this thinking Poikolainen Rosén developed "Agential Cutters" cards (see Figure 10.2) that are intended to 1) make designers aware of the agential cuts they are making in their design process and 2) prompt designers to make agential cuts that would not necessarily be made automatically. These cards should not be viewed as fully encompassing or explaining the theory of agential cuts, but rather as probes for exploring its applicability to understand and inspire design processes. Each Agential Cutter includes guiding questions that can help to both open and narrow down the scope of the space related to a particular topic (see Figure 10.2). References to additional readings are provided on the backside of each card to offer paths for a more in-depth engagement. While the Agential Cutters are purposefully ambiguous and open-ended they are intended to be used to stabilise the focus and consider the opportunities for design in this space. For example, one card (see Figure 10.2) prompts designers to define their design space as extending into both past and present, thus requiring considerations of historic injustices (Sheikh et al., 2023) and future consequences. While another card prompts questions on the economic viability of the system, and who economically gains from it. By contrasting different cards with each other, different aspects become important – or inside, with agency – in the design space, helping the designer to be more aware of what is left outside or rendered invisible, seeing how parts are connected and which parts seem incompatible.

Our second example focuses on a more concrete level, on how frameworks and typologies may facilitate a more holistic consideration of diverse systemic entities. In Table 10.2, we offer one example of a typology of more-than-human entities in nature as identified by Veselova and Gaziulusoy (2021), to illustrate how this approach can make more-than-human stakeholders

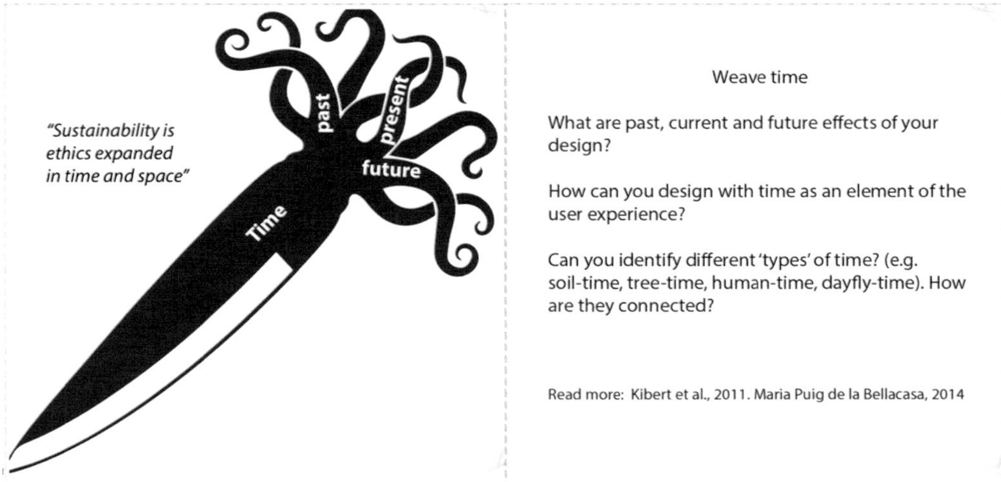

FIGURE 10.2 Example of an Agential Cutter card, front and back. Available at www.poiros.com

TABLE 10.2 A systemic typology of naturally occurring more-than-human entities (Veselova & Gaziulusoy, 2021)

Type	*Example*
Individual organism an organism is typically seen as an independent living entity (but is also a system in itself).	Plants, animals, incl. mammals, birds, reptiles, insects, amphibians, crustaceans and molluscs
Single species collective A collective of organisms from a single species that live together and might have a special organization of their life.	Social insect colonies bryophytes, incl. mosses and hornworts, algae, fungi.
Multispecies collective A collective of living organisms (such as microorganisms, insects and worms) that jointly partake in life processes.	Bacterial collectives, lichens, soil, compost animal manure.
Life Process Flows of elements between living and non-living parts of the biosphere.	Photosynthesis, decomposition of organic matter, respiration, nitrogen fixation.
Living system A location-tied system of living organisms, collectives and the organic and inorganic matter and gasses that jointly partake in life processes.	Garden, lawn, greenhouse, forest and river.
Biogeochemical cycle A cyclical flow of an element between the living and non-living parts of the biosphere).	Carbon cycle, nitrogen cycle, phosphorus cycle, water cycle.
Processes of the atmosphere A short-, mid-, or long-term process in the atmosphere that determines the presence of elements and energetic resources for life processes.	Weather, season and climate.

more present and pronounced in design processes. While the typology was not created with the theory of agential cuts in mind, it can be seen as an example of the making of agential cuts in practice, of conducting analytical framings of "the ontological inseparability of intra-acting agencies" (Barad, 2014) that make sense when we want to practically understand and intervene in and ecology.

This typology was developed empirically using multispecies ethnography (Kirksey & Helmreich, 2010). Veselova spent five immersive weeks in her family's garden in Latvia collecting rich, qualitative data using multiple methods including participant observations, interviews with the garden owners, recording ad hoc conversations of people working in the garden, photos, short and long audios/videos of the garden, its elements and inhabitants. Then Veselova and Gaziulusoy iteratively and collaboratively analysed the data using the DSRP (distinctions, systems, relationships, patterns) model of systems thinking (Cabrera & Cabrera, 2022). This resulted in the typology presented in Table 10.2 of seven key types of natural more-than-human stakeholders. However, typologies are guilty of simplifying, sometimes quite important, complexities of reality. So, Veselova and Gaziulusoy underlined that despite the seven types presented as distinct, a single natural entity represents several stakeholder types simultaneously because of the nestedness of complex systems. There is no easy way to map an observed more-than-human stakeholder onto a single type – an apple tree is an individual organism, a multispecies collective, a part of life processes and biogeochemical cycles all at the same time. Depending on what agential cut you make the apple tree can appear as any of these types.

The implication of this for design is that designers need to understand this complexity of interlinked interdependencies, needs, causes and effects to practice more-than-human design at any level of sophistication that complex systems call for. No stakeholder in a more-than-human world is an isolated entity; it is a part of nested living systems, an entangled part of a collection of several systemic stakeholders. Therefore, in more-than-human design practice, a shift is necessary towards a systemic mental model about who and what is considered a stakeholder. In practice, this might for example imply a shift from viewing compost as a material usable to produce lettuce, to viewing compost as a multispecies collective that sustains the individual organism of lettuce. This shift may help designers better see the interdependencies or needs of all in the system, not just the end outcome of producing food for humans. Similarly, there is a need to rethink what participation means in design, given that almost none of the stakeholder types in Table 10.2 can participate in design processes through direct and deliberate communication. What would design look like were we try to invite the multispecies collective of a compost into our design process? While there is still a significant need for further research and tool/method development to cope with such questions of participation in more-than-human design practice, the typology can assist designers in analysing the stakeholders and interrelationships between them as a starting point. The typology can also assist as a reflective and heuristic tool to make system boundary judgements in more-than-human design projects and enable transparent documentation of the implications of such boundary judgements.

How can More-Than-Human Systems be Mapped?

Once designers have a better understanding of who and what they should focus on they may begin to map them in a system. By representing the connections between entities and elements, designers may gain insights into the interdependencies within a system, enabling a nuanced understanding of how different entities – such as those outlined in Table 10.2 – influence one another. One aim of system mapping is to *understand complexity and dynamics* by revealing intricate relationships between human and non-human entities (Hansson et al., 2021). Another aim is to *identify root causes and effects* by identifying the dynamics of relationships (Hansson et al., 2021). These understandings may lay a foundation for future interventions such as: identifying strategic leverage points, promoting holistic problem-solving approaches, understanding – often unintended – cause and consequence, and facilitating communication and stakeholder alignments.

There are different methods of mapping a system, depending on the purpose, scope and perspective of the analysis. Table 10.3 provides a selection of some mapping methods that can be potentially useful for more-than-human design. Our intention in including this table is to emphasise how different methods make visible different aspects of a system – and that no systems map will ever be complete or fully comprehensive. Another way to phrase it, is that different

TABLE 10.3. Systems mapping methods and approaches that can benefit more-than-human design.

Mapping methods	*Typical application scenarios – what does the approach emphasise in the system?*
Value chains and networks focus on the sequential activities and processes that add value to a product or service within a specific industry or organization. Value networks expand this to encompass the complex web of relationships and collaborations among various stakeholders in a network (Büchel & Raub, 2002).	Business operations, economic systems, industrial systems and production-consumption systems.
Food chains and ecosystem mapping illustrate the hierarchical feeding relationships in an ecosystem, showcasing the flow of energy as organisms consume one another. While ecosystem mapping captures the complex interactions between living organisms and their environment, showcasing the ecological relationships within an ecosystem i.e., it recognises more kinds of relationships than feeding. (Martínez-Harms & Balvanera, 2012)	Ecological research and conservation.
Stakeholder networks map out the interconnected relationships between different individuals, groups, or organizations that have an interest or stake in a particular system or project (Roloff, 2008).	Social systems, project planning and management.
Circular mapping visualises information in a circular format, often used to represent cyclical processes or interconnected elements (Richardson, 1999). Relates to the concept of *feedback loops* described in the section "How can designers intervene in systems"	Process visualization and management.
Spatial mapping depicts the physical layout, arrangement, or distribution of elements within a space, offering insights into spatial relationships (Ramachandra & Shruthi, 2007). This technique harnesses various forms of geographic data, such as soil composition, land use patterns, hydrological features, elevation, or solar exposure. Sometimes 3D scanning is used to transfer the real-world environment into virtual representations.	Urban planning, geography, environmental studies.
Gigamapping is super-extensive mapping across multiple layers and scales to construct a rich picture of real-life complexity (Sevaldson, 2011).	Systemic design and urban planning.

mapping methods will result in different agential cuts. Notably, most of the approaches and methods are not mutually exclusive, meaning that you can shift between these perspectives when mapping a system. We have found that creative design processes often benefit from more informal systems mapping that includes elements from one or several of the methods and approaches in Table 10.3. Next, we provide two examples of how this can unfold in practice in the projects of two of the authors of this chapter "Being with Wetland" by Chen and "Urban Permaculture Farm" by Poikolainen Rosén.

In the project 'Being with Wetland' Chen approached a wetland in China from a socio-environmental systems perspective. The work used visual communication tools – such as system maps (see Figure 10.3), introduction brochures and animations – to stimulate and structure systemic thinking and thus allow the complexities of the wetland system to be communicated The project was thus a mix of methods of spatial mapping, ecological mapping and stakeholder mapping (see Table 10.3), leveraging their strength in capturing the complexity of the ecosystem and socioeconomic system. Through these methods, a comprehensive image of the wetland was revealed by combining existing knowledge from natural science and local ethnographic observations. The wetland was considered a living system (see Table 10.2) that supports the individual organisms' living. The individual organisms living in the wetland, such as fish, migrant birds and plants were considered stakeholders. The stakeholder network mapping methods (see Table 10.3) were applied to critically analyse the potential stakeholders and possibilities to intervene in the social-environmental conflicts. Statistical information related to the ecology system (wetlands, birds) in government reports, journal articles, monographs and press releases was used to make these judgments. The human stakeholders' relationship was mapped via interviews and participatory observation. However, it is important to acknowledge that this method also has its limitations, such as potentially overcomplicated information, capturing a relationship wrong or incompletely, or illustrating a false cohesiveness. In terms of supporting concrete design ideation and development, the mapping for example revealed a communication gap between the decision-makers and the on-site operators. Therefore, an introduction brochure and a related animation video were designed to build a shared understanding of the wetland, the surrounding environments and the human and non-human activities. These two introductory materials attempted to communicate the wetland's geological formation process, ecological condition, original flora and fauna and the modern history of social-ecology development to the public in a compassionate, understandable way. The local youngsters, visitors and tourists were considered as the main audience group of these materials.

The study further intervened in the complex social-ecological system by starting with birds' habitat conservation practices and then exploring the possibilities of restoring sustainable ecological resources within the local context. Here, the composition of stakeholder mapping and ecosystem mapping emphasised the wetland as a living system where humans and nonhumans are actively living with the system and others – and became a tool for communicating these insights to others. This example illustrates how more-than-human systems mapping is beneficial as a *communication engagement tool* for supporting communication and engagement in more-than-human design, fostering collaboration and shared understanding. It is further an *insight generator* in design practice that can spark creative ideas and innovative solutions.

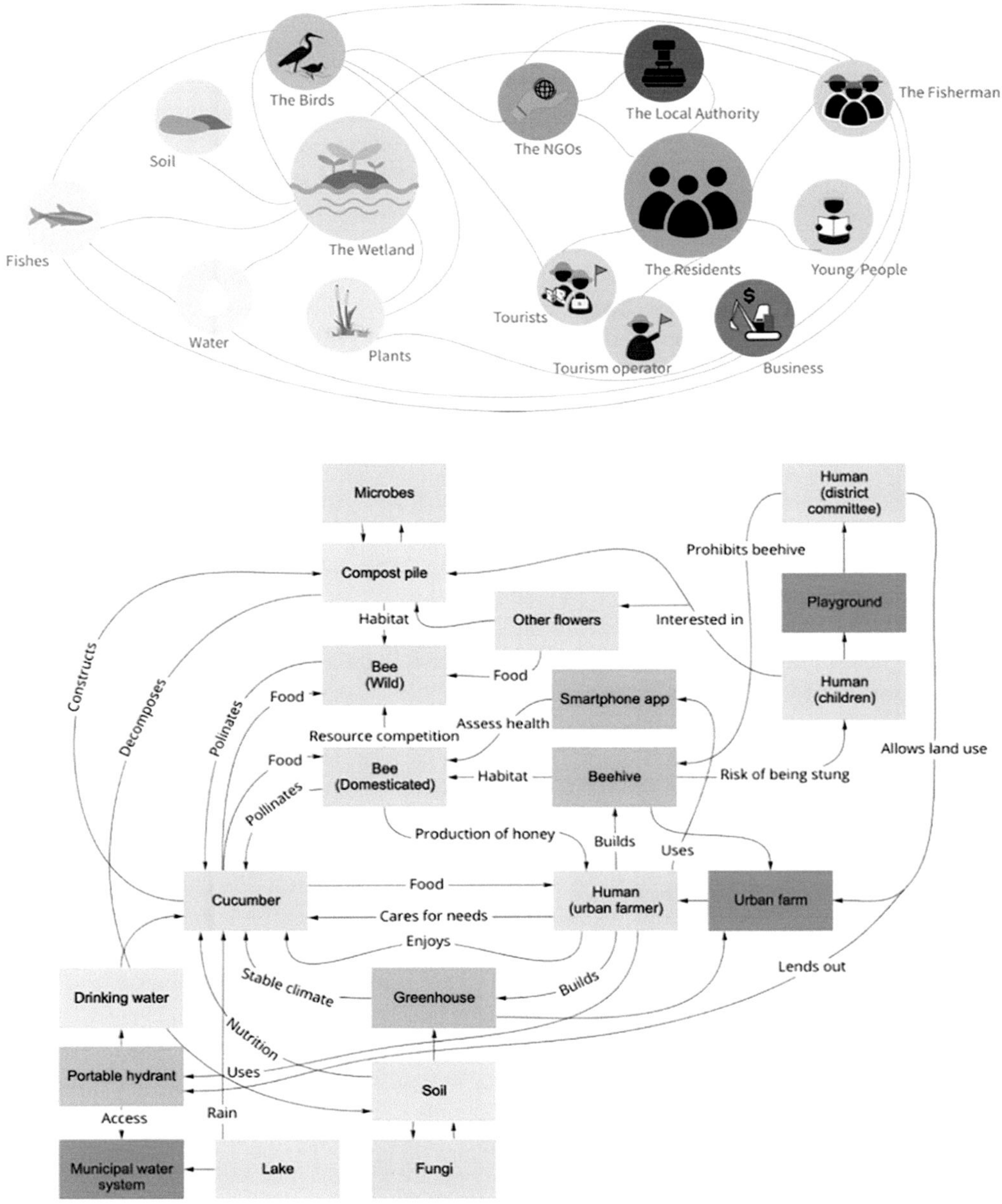

FIGURE 10.3 System maps. Top: Example of a design-oriented system map of a wetland in Tianjin, China. The mapping emphasises both ecological, social and economic relationships. Bottom: Example of a design-oriented system map of an urban farm in Stockholm, Sweden. Naming the relationships and adding a directionality emphasises interactions in the system.

Lastly, it can be used for the *alignment of stakeholders*, as in the example above, including residents, NGOs, policymakers and others, in the pursuit of a common vision and strategy in more-than-human design.

The process of creating the system map involved creatively and freely combining elements of food chains, ecosystem mapping and stakeholder networks (see Table 10.3). At the centre of this system map is the relationship between the individual organisms' bees, cucumbers and humans. Based on insights from a multispecies ethnography, relationships to additional related entities were sketched out such as beehives (single species collective), municipal water systems (an infrastructure for life-processes) and compost piles (multispecies collective). Elements of food chains involve identifying "who eats who" while a focus on broader ecosystem mapping helped to reveal more mutual relationships such as pollination. Elements from stakeholder mapping helped to nuance how the same species (bees, microbes, humans) could take on different systemic roles depending on their relational placement in the system – such as the three ways of being human "urban farmer", "child" and "district committee member", or the two ways of being a bee "domesticated" and "feral". The map also emphasised biochemical cycles such as the recycling of carbon and nitrogen through compositing. This illustrates how the selection of what entities to include in the map is a reflexive process of deciding what matters to design for and who or what belongs in the design space, where inclusion and exclusion happen, thus requiring extra efforts of care and being aware of own biases and potential lack of empirical understanding. Here, naming the relationships and adding a directionality – i.e., identifying and indicating system dynamics – emphasises interactions in the system, thus offering opportunities to identify mutual reciprocity such as how bees feeding on cucumber nectar also act as pollinators. Notably, relationships can be practical, political, economic, emotional etc. Therefore, this design-oriented mapping of a more-than-human system includes for example both that cucumbers provide food for humans, but also that urban farmers enjoy looking at the plants and caring for them. Lastly, the map supported concrete design ideation. For example, an application where the systemic needs could be communicated to the district administration, as the map revealed the administration as a central stakeholder in anything from lending out portable hydrants, banning beehives and contributing with organic matter to the compost.

We want to end this section with some practical considerations on technological and material choices of how to map systems since such considerations are directly relevant to design practice. The two above examples (see Figure 10.3) show a digitally fleshed-out map. However, it is often sufficient and more time-efficient to iteratively sketch a system with pen and paper – as this already offers a great first step of opening the design space as more systemic (see Figure 10.4). The purpose of such "quick and dirty" systems maps is to support processes of ideation and knowledge compiling of the design space for the designer, rather than presenting it as an outcome of the design process to be communicated externally. However, it may sometimes be beneficial to capture and visualise the dynamic changes, expansiveness and interactions of systems. In such cases digital and interactive maps are beneficial – if there is time, knowledge and resources to develop these within the scope of the design project. One prominent example of a creative digital systems map is the Feral Atlas (www.feralatlas.org) which uses the affordances of links to make visible unexpected systemic connections on multiple levels.

In sum, mapping supports working with design spaces as more systemic and relational and offers an entry point into thinking of how to affect the system positively. We will address the intervention in systems in what follows.

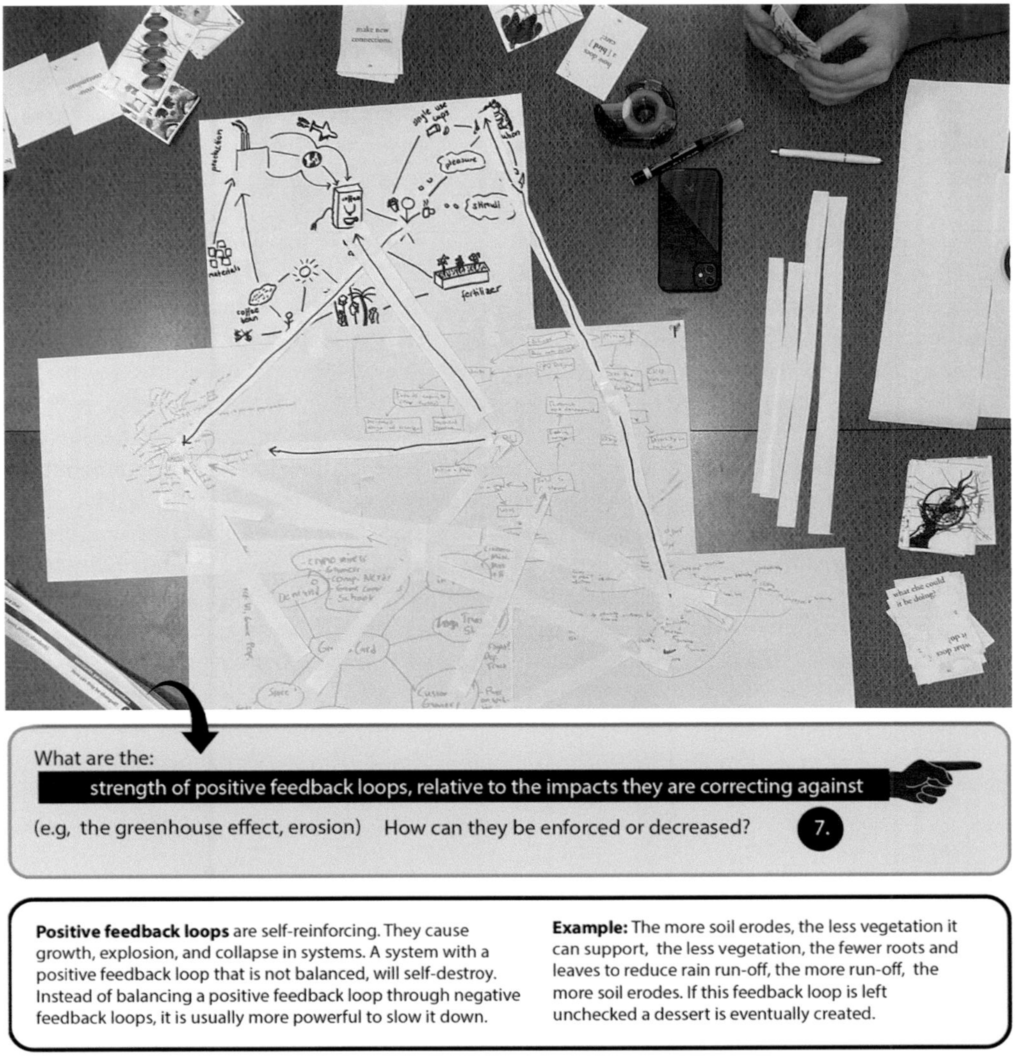

FIGURE 10.4 Systems sketched out with pen and paper by students in a design course. In this task, the students start from an IT device they own. They then trace economic/emotional/political etc. relations to at least one other person, non-human being, mode of production and ecology. All individual systems are then connected. This exercise reveals how a design never acts in isolation and how also very human-centred technologies have implications for the more-than-human world. Below is an example of a used Leverage Pointer, front and back. Available at www.poiros.com

How can Designers Intervene in Systems?

When we understand how systemic relationships are established; we have identified who and what to focus on, and have mapped these entities and relationships, a particularly salient question emerges. How can designers intervene in systems?

Several principles for systemic intervention have been identified in more-than-human design research. One way to intervene in systems is by *providing a scaffold* for natural processes to unfold on their own such as in fermentation, or drop glazing (Liu et al., 2019; Ståhl et al., 2017). Others have suggested that it is sometimes most beneficial to *refrain from intervention* (Poikolainen Rosén et al., 2022) in these cases, the system benefits from quitting a particular practice, such as stopping mowing grass to allow flower meadows to grow, thereby providing food for insects.

Another promising direction is to identify *leverage points* in the system (see Figure 10.5). Leverage points are locations within a system where a slight change in one aspect can result in significant system-wide change (Meadows, 1999). High-level points are hard to change but have massive systemic effects. Lower lever points are easy to change but have a limited effect on the system. This typically works by enforcing positive or negative feedback loops. A positive feedback loop is a self-reinforcing mechanism where a change leads to further changes in the same direction. In other words, it amplifies the original change, creating a feedback loop that can either enhance or destabilise a system. A negative feedback loop is a self-regulating mechanism in a system where a change in a particular direction leads to responses that work to counteract or diminish the original change. In contrast to a positive feedback loop that amplifies the change, a negative feedback loop tends to stabilise the system and maintain a relatively steady state. If designers are aware of and understand such feedback loops their design interventions may submerge skilfully in systems, creating the desired change. Likewise, ignorance of how an intervention may enforce a positive feedback loop may cause systemic havoc.

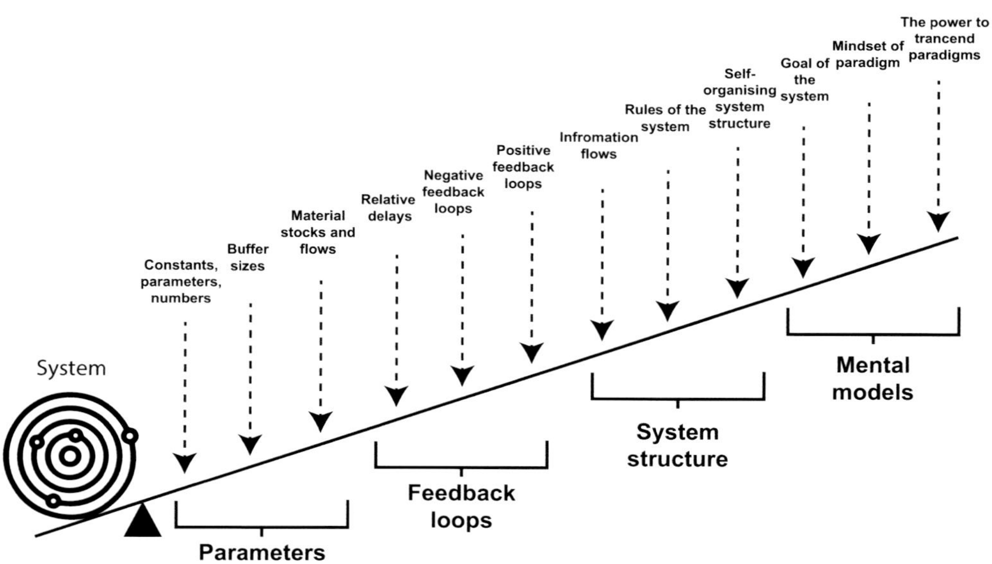

FIGURE 10.5 Leverage points: places to intervene in a system. Our illustration is based on Meadows (1999) and Angheloiu & Tennant (2020).

While the benefits of identifying leverage points are clear, it is not always easy to communicate these to others. Therefore, Poikolainen Rosén developed *Leverage Pointers* cards (see Figure 10.4). Like the Agential Cutters, these cards should not be viewed as fully encompassing or explaining the theory of leverage points but should rather be viewed as probes for exploring the applicability of Leverage Point theory in design processes. The theory is made more accessible as, on each card, a question focuses on how the leverage point can be affected while the back side contains text that explains the leverage point further. Concretely, the Leverage Pointers are intended to be placed out on a mapped system to point at leverage points i.e., places for potential design intervention.

To end this section with an illustration of the concrete applicability leverage points, in the project "Being with Wetland" depicted in Figure 10.3, a low-level (easy to change) leverage point could be to increase the water flow of a dam, i.e., changing a parameter that affects the overall landscape. An example of more mid-level leverage was changing the information flows in ways that increased the public's recognition of wetlands' value and the understanding of wetland reserves, operating systems and policies. This approach of communication contributed to a positive feedback loop as increased community engagement and stewardship for the wetland, in turn, led to more information being spread and discussed, thus resulting in a larger socioeconomic impact for the local authority – and benefitting the reservation area's ecological condition. An example of a high-level leverage point would be to change the people in power in the municipal government to a party with another ideology that would change the policies regulating the wetland, thus changing the foundational rules of interaction in the system or changing the mindset of the community from a human-centred to a more-than-human centred. Here, identifying the leverage points can help designers better understand the systemic impact of their potential interventions.

Concluding Discussion: Human Being and Becoming in More-Than-Human Systems

We have outlined considerations for systemic more-than–human design through four questions: 1) How are systemic relationships established? 2) How do we select who or what to consider in a system? 3) How can systems be mapped? 4) How can designers intervene in systems? The sections focusing on each question relay philosophical grounding balanced with practical possibilities to explore what more-than-human design might mean in practice. What becomes clear in these discussions is the paradoxical position of being a human trying to understand and design for a complex more-than-human world – i.e., the impossibility of capturing all details and complexities of reality yet striving to partly do so – and the messiness and entanglement of designer researchers being beings that inevitably affect systems both physically and socially through our mere presence.

In dealing with this, we have arrived at a reflection regarding the epistemology and performance of more-than-human design (MtHD). The MtH designer researcher learns and performs MtHD, by themself or with others, often with the responsibility of taking decisions that influence research outputs within MtHD. In our examples, this involved utilising the approach of agential cuts to make sense of complex systems by mapping stakeholders in a system in Chinese wetlands and Swedish urban farms and categorizing typologies of stakeholders as in the work of multispecies ethnography in Latvia.

In such processes, making educated decisions is intrinsic to the role of a designer researcher. However, these decisions are limited by the perceived involvement or embeddedness within systems. As MtH Designers we are often periodically and sporadically entering systems of study, and it is that particular entrance that might affect the system in unintended or intrinsic ways. For example, the introduction of a designer researcher in an existing complex ecological environment might result in rippled effects; on a very physical level, this might involve increased local temperature generated by bodily heat, additional carbon dioxide concentration from respiration, or disturbance of wild animals and insects. Accounting for such implicit change is thus necessary. In the words of Donna Haraway (2016, p. 101), "[it] matters which stories tell stories, which concepts think concepts. Mathematically, visually and narratively, it matters which figures figure figures, which systems systematise systems". Right now, we – as humans, designers and researchers – are the ones telling the stories. Thus, a meaningful starting point for human designer researchers when approaching design as a systemic multispecies challenge is an in-depth reflection on assumed hierarchies and agency of more-than-human phenomena within processes of design. It is a matter of ethics, a "looking around rather than ahead" (Tsing, 2015, p. 22) – a certain noticing of who or what gets to tell their stories, which histories allow agency and where voices are (un)heard. This also means recognising that a political dimension is embedded into both human and nonhuman living. Who or what can afford privileged and unreserved storytelling by highlighting their place in systems (Tsing, 2015)? Here, multispecies storytelling through mapping– as shown in the case of the "Being with Wetland" project where local youth was engaged in mapping the role of birds in their wetlands – is a way of focusing on more-than-human narratives in a systemic understanding of a whole.

In the introduction to the chapter, we discussed Buchanan's critiques (2019) arguing that it is important to critique whether the system exists, what is systemised, how the system operates, and why the system exists. Tied particularly to the discussion here on the human physicality of being in systems, *whether the system exists* is directly delimited by the human perception and sensorial ways of experiencing the world around us since we cannot escape our human bodies. The positioning that this entitles us to is the ability to perceive what is visible to us and make sense according to logical reasoning. However, as mentioned when discussing the development of the more-than-human typology in this chapter, frameworks formed by us as designers and researchers *might be guilty of oversimplifying rather important complexities in reality.* This leads to an important yet slightly ironic conclusion: performing MtHD meaningfully links to our understanding and acceptance as human stakeholders – physiologically being a human being, and existentially, of our *humanness.* This ties in with an emphasis that was placed at the beginning of this chapter, that *more-than-human design in practice requires systemic thinking that is concrete and action-oriented while simultaneously remaining ontologically open, critical and questioning.* Rosi Braidotti and Simone Bignall (2019) similarly discuss becoming-human. In their words:

> The 'posthuman turn' – defined as the convergence of posthumanism with postanthropocentrism – is a complex and multidirectional discursive and material event. It encourages us to build on the generative potential of the critiques of humanism developed by radical epistemologies that aim at a more inclusive practice of becoming-human. And it

> also supports an opening out of our conceptual imagination, the power (potentia) of thinking beyond the established anthropocentric frame, towards becoming-world.
>
> (p. 1)

Becoming-human is thus a process that happens alongside MtHD thinking and performance, in both time and even by being entangled in linear and temporal immersion, growth and development of the person doing MtHD thinking and performance (with its subsequent interactions) within a complex world. *We ourselves are shifting, and thus we shift the systems around us consequentially*. This is neither a positive nor negative suggestion, but simply the way our relational world works. Thus, performing MtHD has its inbuilt assumption – that we are only able to observe and explain what we sense and know at a particular time of how the world is, and who we are. Relating to more posthuman thinking, Donna Haraway in *When Species Meet*, also suggests this similarly in saying that a part of being – or *becoming* – human thus means, "to become worldly and to respond" (Haraway, 2013, p. 41). Therefore, "to respond", is to be reflexive and iterative in exploring the four questions within this chapter. As these questions are not final destinations in themselves, but rather, points of departure for further exploration.

In conclusion, it is both a challenge and an opportunity to creatively structure the inherently emergent and complex more-than-human world through designerly systems thinking methods. There are myriads of ways to go about systemic thinking in more-than-human design, of which this chapter has only briefly touched upon a few. Beyond our four central questions for guiding design processes as more systemic we thus ask the reader:

- How can we make more space for more-than-human systemic thinking in the design process?
- What are the hegemonic ways of systemic thinking in contemporary design and what is made 'possible' or 'impossible' by these approaches?

References

Angheloiu, C., & Tennant, M. (2020). Urban futures: Systemic or system changing interventions? A literature review using Meadows' leverage points as analytical framework. *Cities*, *104*, 102808. https://doi.org/10.1016/j.cities.2020.102808

Barad, K. (2014). Diffracting diffraction: Cutting together-apart. *Parallax*, *20*(3). www.tandfonline.com/doi/abs/10.1080/13534645.2014.927623

Braidotti, R., & Bignall, S. (2019). *Chapter 1: Posthuman Systems in Posthuman Ecologies: Complexity and Process after Deleuze*. Rowman and Littlefield International. https://rowman.com/ISBN/9781786608239/Posthuman-Ecologies-Complexity-and-Process-after-Deleuze

Buchanan, R. (2019). Systems thinking and design thinking: the search for principles in the world we are making. *She Ji: The Journal of Design, Economics, and Innovation*, *5*(2), 85–104. https://doi.org/10.1016/j.sheji.2019.04.001

Büchel, B., & Raub, S. (2002). Building knowledge-creating value networks. *European Management Journal*, *20*(6), 587–596. https://doi.org/10.1016/S0263-2373(02)00110-X

Cabrera, D., & Cabrera, L. (2022). DSRP theory: A primer. *Systems*, *10*(2), Article 2. https://doi.org/10.3390/systems10020026

Elkington, J. (1994). Towards the sustainable corporation: Win-win-win business strategies for sustainable development. *California Management Review*, *36*(2), 90–100. https://doi.org/10.2307/41165746

Geels, F. W. (2002). Technological transitions as evolutionary reconfiguration processes: A multi-level perspective and a case-study. *Research Policy*, *31*(8), 1257–1274. https://doi.org/10.1016/S0048-7333(02)00062-8

Habermas, J. (1992). Themes in postmetaphysical thinking. In *Postmetaphysical Thinking: Philosophical Essays* (pp. 28–57). MIT Press.

Hansson, L. Å. E. J., Cerratto Pargman, T., & Pargman, D. S. (2021). A decade of sustainable hci: connecting SHCI to the sustainable development goals. In *Proceedings of the 2021 CHI Conference on Human Factors in Computing Systems* (pp. 1–19). Association for Computing Machinery. http://doi.org/10.1145/3411764.3445069

Haraway, D. (2016). *Staying With The Trouble*. Duke University Press.

Haraway, D. J. (2013). *When Species Meet*. University of Minnesota Press.

Himiyama, Y. (2020). Human sphere: The earth surface realm created by human. In Y. Himiyama, K. Satake, & T. Oki (Eds), *Human Geoscience* (pp. 47–57). Springer. https://doi.org/10.1007/978-981-32-9224-6_5

Kirksey, S. E., & Helmreich, S. (2010). The emergence of multispecies ethnography. *Cultural Anthropology*, *25*(4), 545–576. https://doi.org/10.1111/j.1548-1360.2010.01069.x

Latour, B. (1999). *Pandora's Hope: Essays on the Reality of Science Studies* Harvard University Press.

Liu, S.-Y., Bardzell, J., & Bardzell, S. (2019). Decomposition as design: Co-creating (with) natureculture. *Proceedings of the Thirteenth International Conference on Tangible, Embedded, and Embodied Interaction*, 605–614. https://doi.org/10.1145/3294109.3295653

Luhmann, N. (2012). *Introduction to Systems Theory.* Wiley.

Martínez-Harms, M. J., & Balvanera, P. (2012). Methods for mapping ecosystem service supply: A review. *International Journal of Biodiversity Science, Ecosystem Services & Management*, *8*(1–2), 17–25. https://doi.org/10.1080/21513732.2012.663792

Maturana, H. R., & Varela, F. J. (2012). *Autopoiesis and Cognition: The Realization of the Living*. Springer Science & Business Media.

Meadows, D. (1999). Leverage points: Places to intervene in a system. *The Solutions Journal*. www.thesolutionsjournal.com/article/leverage-points-places-to-intervene-in-a-system/

Munro, D. A. (1991). *Caring for the Earth: A Strategy for Sustainable Living*. IUCN. https://portals.iucn.org/library/node/6439

Poikolainen Rosén, A., Normark, M., & Wiberg, M. (2022). Towards more-than-human-centred design: Learning from gardening. *International Journal of Design*, *16*(3), 21–36.

Ramachandra, T. V., & Shruthi, B. V. (2007). Spatial mapping of renewable energy potential. *Renewable and Sustainable Energy Reviews*, *11*(7), 1460–1480. https://doi.org/10.1016/j.rser.2005.12.002

Richardson, G. (1999). *Feedback Thought in Social Science and Systems Theory*. Pegasus Communications.

Roloff, J. (2008). Learning from multi-stakeholder networks: Issue-focussed stakeholder management. *Journal of Business Ethics*, *82*(1), 233–250. https://doi.org/10.1007/s10551-007-9573-3

Ropohl, G. (1999). Philosophy of socio-technical systems. *Society for Philosophy and Technology Quarterly Electronic Journal*, *4*(3), 186–194. https://doi.org/10.5840/techne19994311

Sevaldson, B. (2011, May 29). *GIGA-Mapping: Visualisation For Complexity And Systems Thinking In Design*. Nordes 2011: Making Design Matter. https://doi.org/10.21606/nordes.2011.015

Sheikh, H., Foth, M., & Mitchell, P. (2023). Reparative futures of smart urban governance: A speculative design approach for multispecies justice. *Futures*, 103266. https://doi.org/10.1016/j.futures.2023.103266

Ståhl, A., Tholander, J., Laaksolahti, J., & Kosmack-Vaara, E. (2017). *Being, Bringing and Bridging: Three Aspects of Sketching with Nature*. 1309–1320. http://urn.kb.se/resolve?urn=urn:nbn:se:su:diva-149546

Steffen, W., Grinevald, J., Crutzen, P., & McNeill, J. (2011). The anthropocene: Conceptual and historical perspectives. *Philosophical Transactions of the Royal Society A: Mathematical, Physical and Engineering Sciences*, *369*(1938), 842–867. https://doi.org/10.1098/rsta.2010.0327

The SDGs Wedding Cake. (2016, June 14). [Text]. www.stockholmresilience.org/research/research-news/2016-06-14-the-sdgs-wedding-cake.html

Tsing, A. (2015). *The Mushroom at the End of the World*. Princeton University Press.

Veselova, E., & Gaziulusoy, I. (2021). When a tree is also a multispecies collective, a photosynthesis process and a carbon cycle: A systemic typology of natural nonhuman stakeholders when designing for sustainability. *Proceedings of Relating Systems Thinking and Design (RSD10) 2021 Symposium*, 25–35. https://research.aalto.fi/en/publications/when-a-tree-is-also-a-multispecies-collective-a-photosynthesis-pr

11

STAYING WITH COMPLEXITY THROUGH MULTISPECIES COMPANIONSHIP AT THE I.N.S.E.C.T. SUMMER CAMP 2023

Svenja Keune, Colleen Ludwig, Katerina Cerna, Anneke ter Schure and Julia Tabet

Introduction

One of the major challenges in more-than-human design is that the inclusion of other life cycles quickly increases the complexity of the design space (Canepa et al., 2022), as just one being can interact with many different other beings. Like a drop of ink in a glass of water or a growing web of mycelium, this is not a linear chain, but a multidimensional growth of connectivity that quickly becomes too much to take into account, too different to comprehend, too entangled to visualise and too specific to know. Most design projects that address the more-than-human simplify this complexity by e.g., limiting the selection of species (Parker et al., 2022), programming their behaviour to let e.g., a yeast produce silk or vanilla (Collet, 2017), or creating a controlled environment (Elsacker et al., 2020). However, relationships between species are integral to the purpose of more-than-human design and cannot be ignored just because of their complexity. To address these challenges the I.N.S.E.C.T. Summer Camp has been established to create a network, strengthen the field and engage in experimental methods. Summer camps are a popular format to engage with the more-than-human and are often used to experiment with unconventional methods (Fentz, 2021; Matt, 2023).

The I.N.S.E.C.T Summer Camp Series

The I.N.S.E.C.T. Summer Camps were launched in the Summer of 2022 by a small group of researchers to explore and develop the field of more-than-human design and contribute with approaches and methods that the members would co-create together. The camps mostly attract designers and researchers who are already invested or interested in multispecies perspectives to design and who long for a group of like-minded people to connect with and engage in meaningful exchange, to learn from and with others, and to deepen their own practice. In 2022, the Summer Camp was facilitated in two parts. Part 1 focused on developing and building a nesting aid for local insects at the HBBE in Newcastle, UK (Parker et al., 2023). Part 2 was our first co-created format, a week of co-living in a Scout camp outside of Copenhagen, in a group of about 30 people.

DOI: 10.4324/9781003467731-14

Afterward, we determined that, while Part 2 of the 2022 Summer Camp successfully used co-creation to generate meaningful activities and experiences and establish an ongoing research community (Keune et al., 2023), we wanted to deepen the impact of our investigations. We set out in 2023 to plan a camp that would generate experiences and develop knowledge to support more-than-human design that would be relevant not only for designers but also attract scholars from biology, ecology and entomology.

We therefore identified three objectives to integrate into developing the methodology for the I.N.S.E.C.T. Summer Camp 2023:

- Find more ways to focus on and maintain sensitivity and awareness to the more-than-human. When a group of humans meet for the first time and are sharing space day and night, they tend to focus on establishing human relations and communication, especially when co-creating.
- Deepen our more-than-human relationships and knowledge generation as non-experts in biology, ecology, or entomology in order to generate meaningful experiences and tangible design outcomes.
- Ground any activities involving speculation or design fiction, so that design outcomes remain plausible and viable.

To address these objectives, we co-created the 2023 Summer Camp by implementing a methodological framework and organizing focus groups around three core themes: Critical Making, Mapping Ecologies and Earth Wisdom Practices. These themes emerged from our pre-camp online co-creation through community building, brainstorming, discussion and mapping exercises to determine what activities the camp members wanted to explore.

Methodological and Philosophical Grounding: Co-creation and Embodiment

Co-creation has been a core value to the I.N.S.E.C.T. community from the beginning because it is a type of collaboration that is grounded in mutual respect, the willingness to work together, curiosity/openness to the perspectives of everyone involved, and a sharing of power (Newton & Geissel, 2011). We used it as a strategy for organising the camp and most community activities because co-creation strengthens ownership of solutions and trust among participants (Fledderus, 2016; Torfing et al., 2016). Sanders and Stappers predicted that the co-creation paradigm would lead to a significant shift in facilitation towards qualities/acts such as guiding, leading, offering and providing nurturing conditions for the individual's levels of creativity (Sanders & Stappers, 2008).

Besides co-creation, we paid special attention to another guiding principle: *embodiment*. Within the camp community, we define embodiment as an understanding of the world that we build through our cognitive abilities in concert with all our senses. Embodiment practices help us create knowledge situated in and impacted by the context of our bodies. The lived experience is then shaped by explicit and implicit knowledge and takes into account intuitions or feelings that are not always easily explained or described by the rational mind. The I.N.S.E.C.T. members have varied training in somatic practices and agree that the human body's sensory system is essential to having direct experiences with the more-than-human world. Through s*ensitization*, *listening*, *noticing* and *reflecting*, we can deepen our engagement with and understanding of the non-human world. During our pre-camp online planning, we established that co-creation and embodiment practices are the basis for everything we aim to do.

Summer Camp 2023: Mapping Embodied Ecologies in Practice

Beyond the two guiding principles *Co-creation* and *Embodiment*, we implemented the *Companion Species and The Ecosystem of the Elder* as foundational methods to engage with the more-than-human as a way to anchor our camp experience and knowledge generation.

The framework for the 2023 Summer Camp therefore consisted of *Co-creation* and *Embodiment* as guiding principles, *Companion Species and The Ecosystem of the Elder* as foundational methods guiding our engagements with the more-than-human, and three themes that we process through the focus groups *Mapping Ecologies, Critical Making* and *Earth Wisdom Practices*.

Companion Species

One of our facilitators who co-founded the community usually hosts the summer camps, facilitates the online co-creation and drives the conceptual and methodological development of the camp. She developed a method for engaging more deeply with complexity in ecosystems by utilising Donna Haraway's *The Camille Stories*, which appears in her book, *Staying with the Trouble* (Haraway, 2016). The story is a short piece of speculative fiction that describes a future society which intentionally raises children to be animal symbionts through gene-splicing, so that the resulting humans embody characteristics of other animals. It is a text that brings up questions and is not meant to be a roadmap, but asks us to imagine how to "live collectively in intimate and worldly caretaking symbiosis with another animal as a practice of repairing damaged places and making flourishing multispecies futures" (Haraway, 2016). We imagined that choosing one organism to gather information about, to sympathise with, and to represent, would help us as individuals with establishing a deep relationship with the local ecosystem. This specific information and hybrid awareness could then help us engage in relational practice and activities, e.g., mapping relationships.

The Ecosystem of the Elder

From the *Camille Stories* and considerations, we came up with the idea of mapping a very small ecosystem to make sure that the 15 members and companions were connected through direct relations, e.g., because they live on the same plant. Our host facilitator lives adjacent to the scout camp and chose a young Elderberry tree as our launch point. We called it *The Ecosystem of the Elder*, referring to the tree species, but also to the wisdom that trees grant us through their deeply entwined roles with ecosystems. Using methods borrowed from entomology, she gently shook the tree and documented the insects that fell out of it and into her umbrella, adding other insects found in the vicinity. Then, each camp member chose a species to be their companion and began doing online research on their biological traits, such as life-cycle, habitat requirements and relationships with others in the ecosystem. Preparing these *Species Profiles* took place in our online meetings and online platforms e.g., Miro. This pre-camp activity supported the members in connecting to one another and remotely to the scout camp environment. As we developed the activities within the three focus groups – Critical Making, Mapping Ecologies and Earth Wisdom Practices – remotely pre-camp and physically on-site, we could continually return to the Species Profiles as our touchstone for co-creating activities that were relevant to the complexity we were trying to understand.

Mapping Ecologies

The intention of the theme "Mapping Ecologies" was to create a series of representations illustrating ecological relationships among the more-than-human, e.g., our companion species. We defined mapping as an approach to developing an understanding of connections and relationships between what is "in the world" and what kind of meanings can be assigned to it. These connections are not something that is pre-defined, but rather something we find through our explorations and by noticing meanings and relationships. Mapping Ecologies entailed a range of experimental "maps" and mapping activities that represented our own approach to finding the type of information that seemed relevant to us. By creating a *representation,* we re-present or bring again to the present what we are interested in; and highlight our own relationship to the thing at hand (Kohn, 2013). Besides making species profiles and mapping their relations, we speculated about our species perceptions of their environment through engaging in imaginary cartographies. We also explored edible landscapes and mapped the plants we foraged and used as food, and listened to the soundscapes of the landscape. These methods built on our experimental and playful approach to getting closer to our companion species in creative ways. We drew upon our professional practices connected to ecology and design, personal experiences and memories as well as relevant digital resources. In what follows, we provide a brief overview of three of the above-mentioned mapping activities.

Image 1 in Figure 11.1 The image shows the young Elderberry tree in the centre of the collage, representing it as the focal ecosystem of the 2023 Summer Camp.

Species Profiles

As part of our online preparation, each camp member chose a species to be their companion and created a digital species profile in our Miro board. Besides the information usually provided by databases and knowledge platforms such as their taxonomy, distribution, morphology, life cycle, behaviour and ecology, we also included personal memories and experiences, basically all the information we found relevant to establish a connection with the species we committed to. After weeks of online preparation we collaboratively developed a paper-based profile template during the camp. The template included a space for representing the species via drawing and a range of information such as name, habitat, seasonality, life stages, sensory perception, physiology, main role in the ecosystem, primary predator, pollination/dispersion, particularity, superpower and food. Each person made a copy of the template and then we visited the species natural habitat and, if possible, observed the species themselves. Our subjective observations, paired with scientific information about the companion complemented the species profile. To round up this activity we gathered in a circle and introduced the companion and our findings to one another.

Image 2 in Figure 11.1 The screenshot shows the digital species profile of the stinging nettle in our Miro board.

Image 3 in Figure 11.1 The image shows the paper-based species profile of a spider, with a drawing in the centre and factual information organised around it.

FIGURE 11.1 A collage of photographs and screenshots arranged around a central image of the Elderberry tree, illustrating the different methods and activities experienced at the 2023 I.N.S.E.C.T. Summer Camp and how they build on one another.

Relational Mapping

As a follow up activity to the species profiles, we engaged in relational mapping. We pinned our profiles to a wall, considering the companions position in relation to the Elder. We created a physical depiction of the different types of relationships our companion species are connected with. These were represented by differently coloured threads, connecting the species profiles e.g., through mapping a predator-prey or a common habitat relationship. The relational map of the Ecosystem of the Elder became a centre piece for our conversations and remained on the wall of the scout camp's living room for the rest of the camp. This map supported our understanding of the species not as discrete points but in the context of our own relationships, situated in the given space of the camp and the particular season. As such, the map was not an objective representation of the existing species but rather highlighted our point of view on the given ecosystem, through our own capabilities and expertise (we chose what we could notice), as well as the available digital expert tools (iNaturalist showed the most common species for the area). This gathered knowledge was later reflected in a talent show at the end of the camp in which we used costumes and performance to express the species characteristics that inspired and impressed us.

> Image 4 in Figure 11.1 The screenshot shows a series of profiles organised around the species profile of the Elder.
>
> Image 5 in Figure 11.1 The pictures shows a camp member connecting two profiles with a coloured thread to illustrate a relationship between them.

Imaginary Cartographies

The workshop "Imaginary Cartographies" invited a speculation of our companion species perspectives of their territory, the Ecosystem of the Elder. Accompanied by a simple hand-drawn template of the environment from a human perspective and guidelines such as to imagine and draw a map in which territorial relations regarding our companions' habitat and food, and environmental factors are represented in different colours, we gathered around the Elder and started drawing (Figure 11.1, Image 6). This workshop built on the knowledge gathered through the species profiles and connected this knowledge to an imagined scenario from the perspective of the companion species, for example moving from one leaf of stinging nettle to another to find a potential mate. A shift from human perspective to companion perspective was necessary and being in the actual environment and observing the potential path of e.g., a nettle weevil combined theoretical knowledge with the embodied experience of e.g., lying on the ground at eye level with the horizontal platforms that connecting leaves of stinging nettles build for weevil mobility. Past the individual speculations of perceptions and cartographies, the illustrations were pinned up for a collective reflection. Different from the expected, a comparison between species was difficult, since it reflected subjective interpretations and different graphic approaches and skills.

> Image 6 in Figure 11.1 The image shows a camp member lying under the Elder tree to approach their companion species "at eye level", imagine the territory from their perspective and draw a map as a representation.
>
> Image 7 in Figure 11.1 The image shows two maps that represent the territory and perspective of the companion species through the speculation of the human camp member.

Critical Making

We see critical making as one methodology that helps us move towards more collaborative relationships with other species. Our objective is to engage in ongoing dialogues with living and non-living material contexts to identify extractive and unsustainable issues with our current ways of making. How can we find material alternatives that regenerate earth's ecosystems and sustain human and more-than-human life?

Critical making is a means of conceptually exploring technological and societal issues through the constructive and shared process of making tangible things, while simultaneously conversing and reflecting on the process (Ratto, 2011; Remko, 2011). Combining the acts of 'critical thinking ', which is abstract, with 'making ', which is physical and embodied, critical making focuses on how material and conceptual exploration within making processes can create innovation in material methods (Ratto, 2011).

Our critical making activities at the Summer Camp included foraging for materials and edibles, isolation and cultivation of wild fungal specimens, making seed yarns, crocheting and making costumes of our companion species. On the last evening of the camp, we created performances to express the characteristics and relationships of our companion species in a playful and celebratory way through a talent show.

The camp lab and kitchen were central work spaces for critical making activities. We set up guidelines for safe community use with the acknowledgment that when working in the field, rigorous sterile methods are difficult or impossible without being extremely wasteful. There were productive discussions about what was possible with our limited space, budget and supplies. Two wall posters outlined the instructions for documenting lab experiments, and workshop leaders reviewed the minimal lab equipment that was available, such as a microscope, supplies for making agar plates, sound-recording equipment and specialised lenses for macro photography. After foraging, we would identify and label them in the lab, make field notes or map drawings, or take foraged items into the kitchen to clean and prepare for eating. Part of the dialogue became analysing these protocols, such as asking, "Is there a need for a separate lab workspace?" "How can we minimise contamination? When is that important?"

Foraging Materials and Edibles

The critical making focus group wanted to explore the local resources within the Ecosystem of the Elder as alternatives for store-bought materials and to connect with the local environment. We started this activity by stating our intentions for the foraging session and discussing how to go about the collection of organisms and materials. This resulted in several concrete guidelines:

- Foraging to sensitise: The main purpose of foraging is to attune to the environment and to collect experiences and knowledges.
- Liminal intimacy: Being as respectful and careful as possible. Every little piece of bark is the home of hundreds of organisms.
- Don't over-forage: Only take what you need, leave enough to the species to survive and reproduce and for other foragers to collect.
- Label the collected species with a possible identification of the organism, name of the collector, date of collection and location.

To forage, we went for walks through the scout camp's neighbourhood, informed by local guidance and members with knowledge about edible plants and mushrooms. The foraging for materials included the gathering of plant material, the fruiting bodies of fungi, or other materials and organisms due to their aesthetics and potential to serve a design idea. We foraged with the intent of gathering resources for the critical making activities (Figure 11.1, Image 8). After the walk, we gathered the edibles in the camp kitchen where they were prepared to accompany the next meal (Figure 11.1, Image 10). The gathered materials were arranged on the tables of the camp lab, where they were annotated with information about their species, and the location of where they were harvested (Figure 11.1, Image 9). A sharing circle rounded up this first foraging experience. This first trip set the frame for future ones and generated a material collection for the next activities, e.g., spore printing.

> Image 8 in Figure 11.1 The image shows a basket with foraged plants and mushrooms.
>
> Image 9 in Figure 11.1 The image shows some camp members in the camp lab, discussing the foraged digital and physical information and materials.
>
> Image 10 in Figure 11.1 The image shows a collective meal co-creation in the kitchen.

Isolation of Wild Fungal Specimens (Mycelium, Mushroom Spores, Mushroom Patches)

Mycelium (thread-like networks of hyphae that fruit into mushrooms) has been seen as a promising bio-material in design and architecture. Mycelium projects are generally made with some combination of human-made and grown materials. The growth process involves a professional or semi-professional lab where the development of a commercially-sourced mycelium can be controlled by providing optimal conditions. To test a more DIY approach that could produce locally-sourced material and thus be more sustainable, we brought equipment necessary for a semi-sterile environment to grow mycelium from wild foraged samples (Figure 11.1, Image 11). As we located rich piles of dead matter that was once a fallen tree, we found the delicate white threads of fungi just below the surface, on leaves, and small chunks of soil and rotting wood. We successfully grew mycelium from one sample, even though we lost it to contamination on the last day of the camp. Still, it was an enriching experience and we believe that with more time and practice, making mycelium material from local samples has promise for the future of the practice.

> Image 11 in Figure 11.1 The image shows a session with mycelium under semi-sterile conditions on the kitchen counter.

Making Costumes for the Talent Show

The culminating camp activity was to celebrate the final night with a talent show, by making costumes to help us to embody our companion species and to co-create a short skit with our focus groups. This activity allowed for a light-hearted moment to bring together all the human and more-than-human relational experiences of the week as well as the breadth of new knowledges we developed together. Costume-making stretched over one entire day and was meant not only to generate a wearable outfit, but also give us space for introspection and integration of the week's events before celebrating the camp's end. One group performed a rap song with a verse for each companion species. Another group performed a skit about the delicate moments when insects and humans meet during multispecies cohabitation. This playful approach played an important role in wrapping up our camp experience by embodying our companions, representing our learning and celebrating our human connections. It represented both the human and species entanglements that we had managed to weave over the short span of the camp. We had a last opportunity to think about the physiology of our companion, how to represent it on a human body and how to perform for each other and our local audience of neighbours.

Image 12 in Figure 11.1 The image shows a group picture of the camp members who participated in the talent show in their companion species inspired costumes.

Earth Wisdom Practices

The Earth Wisdom Practices focus group was formed by group members' desire to make use of embodiment practices throughout the camp experience for deepening our engagement with the human and more-than-human worlds. Meditations, somatic exercises and reflection sessions helped to deepen and ground our mapping, critical making, foraging and other activities through sensitization, listening, noticing and reflecting. We called this the *Earth Wisdom Practices* group to align ourselves with original peoples, who have stewarded lands and ecosystems for thousands of years in helpful, healthful and regenerative ways. The tacit knowledge of natural cycles that indigenous communities have collected over deep time have been all but obscured by the relentless extraction economies of capitalism. We seek to rebuild our original relationships to the land, so that all our activities are rooted in lived experience, and sensed through a combination of body and mind, or *soma*, and integrated with earth's rhythms, as disrupted as they may be.

In order to come to the camp activities with greater awareness, less distraction and a slower pace, we developed a few longer meditation workshops, as well as several short rituals that we weaved into daily routines. We led an opening circle to set the stage and atmosphere of the camp, created a ceremony to greet the Elder tree when we met her in real life for the first time, and organised daily movement breaks to loosen up our bodies, shake out tension and relax our minds.

Shapeshifting Meditation

Umwelt or *Umwelten* is a term coined in 1909 by zoologist Jakob von Uexküll to describe any sentient being's unique sensory experience. To better understand the *umwelten* of our companion species, we wanted to attune our bodies to an expanded sensorium that included the special sensory capabilities of their bodies. Such a practice can be found in many cultures; our ancestors shapeshifted into different animals for purposes of hunting, survival, spiritual practice and healing. Our researched species profiles and speculated relational maps served as a preparation for this shapeshifting workshop, so that we could remember our knowledge of our companion's physiology, behaviour and sensory organs in imagining how they might perceive their individual world. We began the shapeshifting journey like a typical meditation, lying down with closed eyes, bringing awareness to our own breath, the air around us, our skin and heaviness of our bodies. We slowly began to loosen up our human form through the prompts of our guide for this meditation. We started to imagine the particulars of our companion's physiology, invited our companions into our bodies by imagining their feelers, compound eyes, multiple legs or hairy protrusions. We allowed our bodies to respond to this imaginary transformation and to move intuitively, checking out new appendages, imagining what it would be like to move in this body, perhaps interacting with each other or the space around us. After a time of exploration, we let go of our transformation with a metaphorical release of our companions, and returned to our human selves. Some individual and group reflection time helped us process and integrate the experience, to keep these visions and feelings alive to inform later activities.

The shapeshifting meditation served as a body-mind experience that connected us to our companions, let us "slip into their skin," and helped create an embodied experience we could draw from in other activities. Attempting to inhabit the *umwelten* of another species, through a felt practice and paired with the mind's inventiveness, became a hybrid perspective distributed through muscles, sinew and nervous systems. Even if our imaginations were speculative, we engaged in an intentional, slow, informed, guided and holistic method. Our experience with this meditation is that it increases our capacities in empathising with our creature kin, and strengthens our worldview when integrating more perspectives into design practice.

Truth Mandala

The focus group wanted to provide a space for eco-grieving as another method toward deepening members' connection to the local ecosystem. Rather than submitting to despair, eco-griefing in collective experiences can provide healing and a strengthened sense of love and commitment to places, thus allowing people to form deeper relationships with ecosystems and species (Cunsolo & Ellis, 2018). To hold space for this, we chose The Truth Mandala, a ritual developed by Joanna Macy (Macy et al., 2014) which we applied and modified to fit our setting. At the campfire circle we set up a small table with the required elements: a stone, some dead leaves, a thick stick and an empty bowl. The facilitator picked up each one and explained their meaning. The stone represents fear, tightness, hardness. The dry leaves represent our sorrow. This stick is for our anger and outrage. This empty bowl stands for our sense of deprivation and need. Macy reminds us that the truth mandala is not for lectures or reports, but for direct and simple expression of our pain for the world. She also reminds us that we speak not only for ourselves, but, as for all rituals, we speak archetypally, on behalf of all people, and the Earth. It was near the end of the week and by now the group had cultivated a high level of intimacy. Some of the group who had specialised training in somatics allowed them deeper access to feelings more easily and genuinely, in front of other people. One member was able to purge some anger through loud screaming, while another expressed deep grief through sobbing. Others who had never done such an exercise before spoke for the first time about their personal ecological distress. It was probably the most serious moment of the camp, but most people agreed that it was a powerful and useful experience.

Staying with Complexity through Multispecies Companionship

By implementing the strategies discussed in this chapter, such as creating The Ecosystem of the Elder and adopting a representative from this ecosystem as a personal companion during the camp, the I.N.S.E.C.T. community aimed to deepen relationships with a web of more-than-human beings. While it limited each member to concentrating on one species, it created a clear framework that opened up creative relational spaces that we could explore through multiple modalities. These opportunities were plentiful and we did not come close to exhausting our energy for our companions during or after the duration of the camp.

Embodiment practices deepened our creative engagement with The Ecosystem of the Elder. In addition to noticing the outer world by observing a nettle weevil or smelling mycelium in the forest floor, we also noticed inwards, where and how we could sense those experiences inside our bodies, what memories came up when smelling fungi or being stung by a nettle. Only through taking the time to do this embodiment work can we become as entwined as possible with our non-human kin and develop design methodologies that can encompass the complex ecosystems we hope to support.

The Ecosystem of the Elder became the organising framework that helped us continually connect our co-created activities to the quest of deepening our knowledge and our relations. We formed new bonds through modalities such as cataloguing factual information on paper and through digital apps such as iNaturalist. We visited and spent time observing the Elder and our companions, drawing, writing and mapping connections between species with coloured pencils and strings. We embodied our companions through meditations that helped us speculate, transforming our bodies, extending our sensory perceptions and imagining the alternative realities we could live in as part of the Ecosystem of the Elder. We foraged, ate and digested edible plants to experience processing them through our bodies and we made and performed in costumes to imagine looking and acting like our companions.

While many of these exercises require us to use speculation and cannot be verified by science, they helped us maintain an embodied perspective and a more holistic view of ecosystem complexity. Our perceptions went beyond the initial group of animals that were identified as part of the Ecosystem of the Elder and helped us to look at all the forest, lake and field ecosystems around us with new, more complex outlooks as we foraged for plants and fungi to augment our meals, or as raw ingredients to use as lab materials for critical making. This shift in perspective toward our food, our design materials and our relationships were the most significant outcomes of this embodiment work.

Beyond the sense of deeper connectivity to local ecosystems through our companions, the Summer Camp experience also created a profound feeling of human connectivity. While one might argue that our human relationships are less relevant for more-than-human design perspectives and distract us from the more subtle expressions of other species, we found our overall heightened sensitivity to be useful all around. The collective effort of writing this book chapter is one testament of the lasting effects of the camp, as is the innovation of other initiatives that grew out of the camp. The 2023 Summer Camp has so far initiated a more-than-human design exhibition, a new research network, a food-zine and a new focus for the 2024 I.N.S.E.C.T. Summer Camp.

For those interested in designing a similar experience in more-than-human design, consider the following questions:

- How do you balance facilitation with co-creation, so that members feel inspired, open and a sense of ownership in the content of the experience, but there is enough structure and cohesiveness to the activities to create a successful workshop?
- How do you balance speculation with knowledge building so that the outcome remains plausible, viable and contributes to the research community?

References

Canepa, M., Mosca, F., Barath, S., Changenet, A., & Hauck, T. E. (2022). Ecolopes, beyond greening. A multi-species approach for urban design. *Agathòn*, *11*, 238. https://doi.org/10.19229/2464-9309/11212022

Collet, C. (2017). 'Grow-made' textiles. *Alive. Active. Adaptive: International Conference on Experiential Knowledge and Emerging Materials, EKSIG 2017*.

Elsacker, E., Vandelook, S., Van Wylick, A., Ruytinx, J., De Laet, L., & Peeters, E. (2020). A comprehensive framework for the production of mycelium-based lignocellulosic composites. *Science of The Total Environment*, *725*, 138431. https://doi.org/10.1016/j.scitotenv.2020.138431

Fentz, C. (2021). *EARTHBOUND SYMPOSIUM – Earthwise Residency*. Earthwise. https://earthwise.dk/artist/earthbound-symposium/

Fledderus, J. (2016). *User Co-Production of Public Service Delivery: Effects on Trust*. https://doi.org/10.13140/RG.2.1.1315.9440

Haraway, D. J. (2016). *Staying with the Trouble: Making Kin in the Chthulucene*. Duke University Press.

Keune, S., Ilgun, A., & Ludwig, C. (2023). I.N.S.E.C.T Summercamp: Developing multispecies design perspectives, practices, and discourse through co-creating (in) community. In M. Mostafa, R. Baumeister, M. R. Thomsen, & M. Tamke (Eds), *Design for Inclusivity* (pp. 701–715). Springer International Publishing. https://doi.org/10.1007/978-3-031-36302-3_52

Kohn, E. (2013). *How Forests Think: Toward An Anthropology Beyond The Human*. University of California Press.

Matt, C. (2023). *Nature Design Camp 2023*. Studio Matt. https://christophmatt.com/ndc23

Newton, K., & Geissel, B. (Eds) (2011). *Curing the Democratic Malaise?* Routledge. https://doi.org/10.4324/9780203155196

Parker, D., Ilgün, A., Lim, A. C. S., Vašatko, H., Vu, D. V., Piórecka, N., & Keune, S. (2023). I.N.S.E.C.T. wall twin: Designing for and with insects, fungi, and humans. *Temes de Disseny*, *39*, 228–247. https://doi.org/10.46467/TdD39.2023.228-247

Parker, D., Roudavski, S., Jones, T. M., Bradsworth, N., Isaac, B., Lockett, M. T., & Soanes, K. (2022). A framework for computer-aided design and manufacturing of habitat structures for cavity-dependent animals. *Methods in Ecology and Evolution*, *13*(4), 826–841. https://doi.org/10.1111/2041-210X.13806

Ratto, M. (2011). Critical making: Conceptual and material studies in technology and social life. *The Information Society*, *27*(4), 252–260. https://doi.org/10.1080/01972243.2011.583819

Remko. (2011). *Critical making / Matt Ratto | Open Design Now*. Open Design Now. http://opendesignnow.org/index.php/article/critical-making-matt-ratto/

Sanders, E. B.-N., & Stappers, P. J. (2008). Co-creation and the new landscapes of design. *CoDesign*, *4*(1), 5–18. https://doi.org/10.1080/15710880701875068

Torfing, J., Røiseland, A., & Sørensen, E. (2016). *Transforming the Public Sector into an Arena for Co-creation: Barriers, Drivers, Benefits and Ways Forward*. Sage Journals.

12

HOW CAN WE DESIGN WITH A MULTI-SPECIES MINDSET TOWARDS REGENERATIVE PRACTICES?

Julia Lohmann

Introduction

In this chapter, I am trying to distil the insights I gained in 20+ years of critical more-than-human design practice, grown out of my background in communication design, product design and material exploration, into sections that can be read as a methodological approach to practicing more-than-human design. Through autoethnographic case studies involving fauna-based materials, microorganisms and algae, the challenges and frictions, ethics and value systems involved in our interactions with them, as well as their origins, agency and potential futures, I hope to inspire you to immerse yourself in a caring, knowing and acting entanglement of life and weave your actions and reflections into a meaningful multi-species practice that understands humans as one of many creatures in the damaged webs of life we depend on: A strange creature with an immense impact that urgently needs to shift from extraction to regeneration.

I have sought experiences outside my anthropocentric comfort zone, challenging myself to view the world from the perspective of non-human others, from microorganisms and mice to fly larvae and farm animals. These immersive experiences helped form the multi-species mindset that underpins my design practice. To discuss it, I will draw on:

- Early encounters: Navigating the natural and the human-made world, based on my childhood; Process-led immersion: Proximity, physical interaction, and the body as a medium of connection, by example of formative experiences in Iceland and London;
- Reframing: Developing context sensitivity, by example of my *Ruminant Bloom* sheep stomach lamps;
- Multi-sensorial access to material ethics, illustrated by my *Cowbenches* reconnecting leather with its origins;
- Crossing thresholds and decoding experience for discourse, represented by *The Lasting Void*, a cast of the inside of a calf's body;
- Rethinking ourselves as multi-species ecosystems: From human beings to creatures collectively being human, through *Co-existence*, an installation on the non-human organisms inside us;

DOI: 10.4324/9781003467731-15

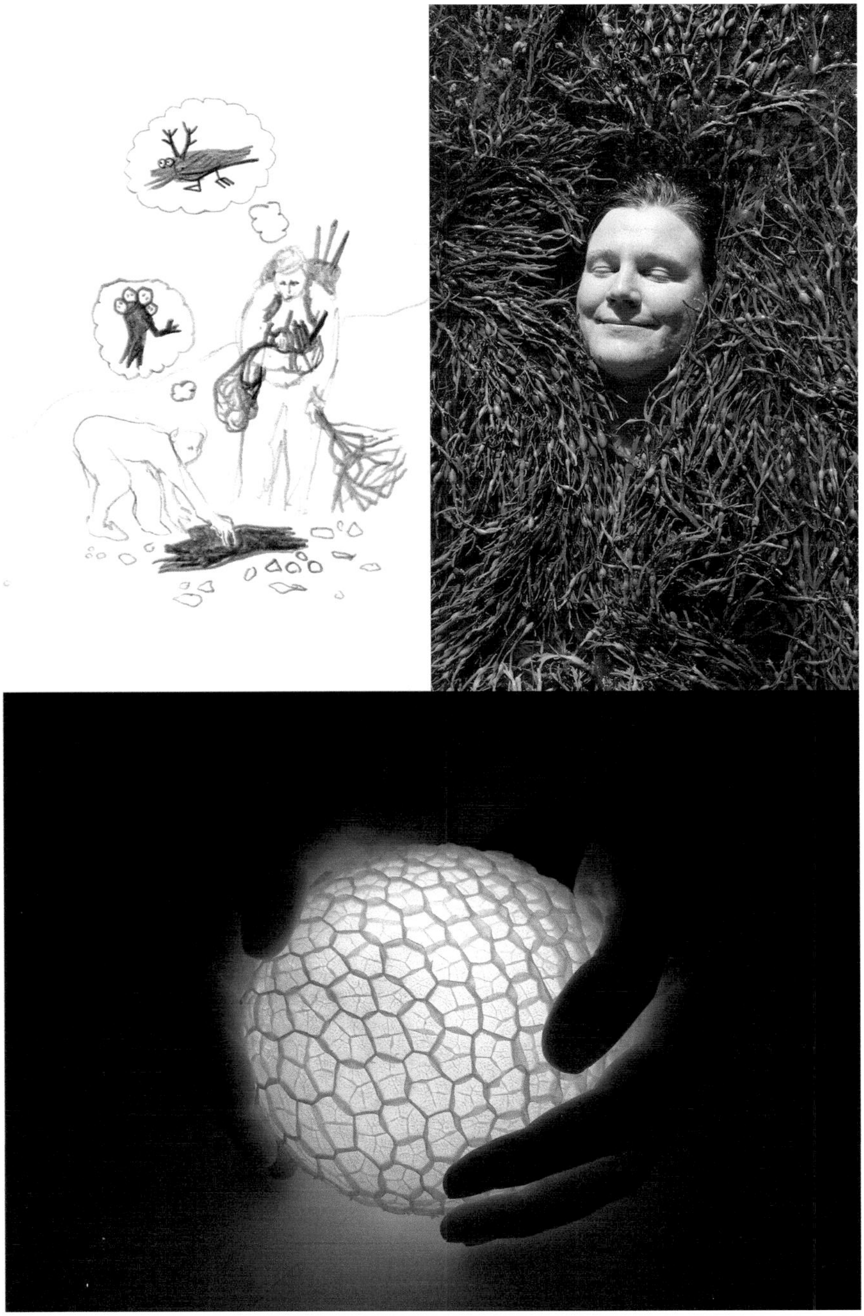

FIGURE 12.1 Top left: Childhood memories of gathering driftwood for flotsam sculptures. Illustration: Julia Lohmann. Top right: Immersion – Exploring the eco-systemic agency of seaweed in an Icelandic bladderwrack habitat. Photo: Anna van der Lei. Bottom: Probing ethics and value systems by re-contextualising animal materials – 'Ruminant Bloom' lamp made of a preserved sheep stomach. Photo: Julia Lohmann

- From convenience to care: Overcoming our single-species mindset, based on reflections on urbanisation;
- Seaweed: Material, muse and method, based on my community-of-practice-forming work with macroalgae as a material for making; and finally
- Distilling all of these aspects into the *Knowing, Caring and Acting* framework for regenerative multi-species design.

Early Encounters: Navigating the Natural and the Human-made World

Since childhood, I have been trying to understand how we connect with nature, feeling instinctively that we are just one of a myriad of species co-inhabiting the Earth. I felt at home playing in the forest, watching wildlife, befriending stray dogs on caravanning holidays, and helping my father collect flotsam for the sculptures he was making. I was baffled by narratives positioning humans as a species set apart and above nature. When I spent too long in environments that were built for a single species only, I felt displaced, restless, and wary and sought the non-human company of animals or trees to recalibrate my thoughts and emotions. I played that I myself was not a human.

I wanted to become a jungle explorer and later a veterinary surgeon until I was 12 and a school internship at a vet ultimately changed my direction: I loved interacting with the animals and was fascinated by observing operations, but the wide range of human-animal relations we encountered often jarred with my own perception. We treated lonely single rabbits and aged dogs that were beyond healing whose owners could not let go because they were their closest kin, whilst putting down healthy animals devoted to the humans who discarded them in favour of clean carpets.

Looking back, I see how these inter-species frictions and conflicts intensified my own reflections. At 19, I moved to the UK to study communication design where nature and animals again shaped my work: I explored multi-species themes through personal briefs, from probing perception and prejudice through macro-photography of mould to reflecting on growth and decay, as well as concepts of connectedness by creating artworks and performances in collaboration with maggots. As a postgraduate student I focused my practice-led research on re-linking biomaterials with their natural origins and probing the ethics and material value systems humans apply to nature.

Process-led Immersion: Proximity, Physical Interaction, and the Body as a Medium of Connection

If we want to re-imagine our place and role in the Earth's ecosystem, we need to conceive of ourselves as interconnected elements of a web of life, reliant on the ecosystems that nurture us. Then we can try to immerse ourselves in the otherness of the species we co-exist with, experiencing their lifecycles, appreciating and questioning our entangled webs of life, empathising beyond ourselves. We must question the fundamentals of what it means to be human in a complex ecosystem, instead of pursuing the 'paragon of animals' narrative at the cost of all other species. We need to reframe ourselves at eye level with the others. These considerations are illustrated by my experiences as an early career designer.

Iceland: An Immersive Inflection Point

After my bachelor's degree and six months working in an advertising agency in London, I needed to remove myself as far as possible from the city and recalibrate in proximity to the natural world. I spent three months working on a horse and sheep farm in Iceland. The sheep we bred roamed free for a summer in the highlands, before we rounded them up on horseback, driving them back down to the valley of their birth where half of them were slaughtered to provide the community with food and wool and the farmer with money to continue the cycle. The direct interconnectedness of human and animal lifecycles fascinated me, especially because ultimately it is the Icelandic climate that decides if, how and for how long this symbiosis continues. None of these aspects seemed to matter in London where I became acutely aware of people eating dinosaur-shaped bits of chicken, of whose origins or life they knew nothing nor seemed to care about, whilst treasuring and anthropomorphising their pets.

Therefore, probing the complexities of the ethics and value systems in human-animal relations, exemplified by those inherent in the production and use of animal-based materials, became the focus of my master's studies in Design Products at the Royal College of Art. What is the role and responsibility of a designer working with animal-derived materials? What might happen if I used materials made from animals that we commonly kill in an unexpected way? Would it make the known new to people, taking them – like Iceland had taken me – to a place beyond the veneer of acceptance that I felt coated our Western society, increasingly separating us from empathising with other species and the natural world?

In my encounters in Iceland, I had experienced the birth and lives of sheep but (deliberately) excluded anything related to their killing and slaughter. To speak about materials, I felt I needed to experience this part as well, to be with a dead animal, to feel empathy and reconnect animal materials with their origins with an immediacy that theoretical study cannot provide. I asked a taxidermist whether I could observe him and help with basic processing methods, and he agreed to let me join him in his studio.

Merleau-Ponty's frequently quoted notion that "the body is our general medium for having a world"[1] describes the importance of sensual perception in philosophy. Sensual perception and immersion are equally vital in creative processes. Immersive encounters connect us physically and emotionally with our topic, heightening our perception of nuances and complexities. They help us condense intellectual thoughts into a lived and deeply felt connection, described by feminist scholar Donna Haraway as "becoming with",[2] suggesting how we may attune ourselves to a non-human life. My aim was to extend 'becoming with' to a dead animal as it is processed into materials. It made me want to find ways to hold on to the connection, to link leather back to the cow, and value every part of the animals we kill to sustain ourselves.

Reframing: Developing Context Sensitivity

I took some tanning chemicals back to South London, wondering where to find skins to process. In the display of a local Halal butcher, I saw the most intricate, lace-like structures and learned they were sheep stomachs. A skin, I thought, is a permeable membrane between a body's inside and outside that absorbs some things and rejects others. By this definition of 'skinness' the lining of a stomach could be considered an interior skin. I bought some stomachs and experimented with preserving them. They possessed a beautiful parchment-like translucency that glowed warmly in the light. I turned the preserved stomachs into lights and light installations called 'Ruminant Blooms', distancing them from their original context and function. At a time when 3D printing technology was emerging, people thought they might have been manufactured digitally. In admiration of presumably human ingenuity, they asked: "Is this 3D-printed? Is it rapid-prototyped? What algorithm did you use to craft such an intricate surface?" As soon as they found out the lamps' ruminant origin, there was a moment of disgust, often followed by a pondering of the culturally ingrained clash of value systems, oscillating between fascination and revulsion. This is what I had hoped for. Not for shock value but to see if materials used outside of established material value systems could enable reflective experiences.

In the system 'sheep', every organ is of existential value. Sheep cannot live without their four stomachs. When we kill a sheep, we dissect this system, removing its elements and imbuing them with different levels of value, based on our needs and preferences. These reflect where we have come to position ourselves in relation to the natural world, viewing it from a distance that we justify in terms of objectivity, rationality and efficiency whilst omitting descriptors like ownership, extraction, exploitation and commercial gain. This is perhaps why, whilst we wear sheepskin on our skin and feed our flesh with that of the animal, we cringe when it comes to sheep stomachs. They seem to breach the societal constructs that foreground how far we have evolved beyond our hunter-gatherer origins, instead highlighting the intimate, life-sustaining processes in both human and ruminant bodies that mark us as fellow mammals, as dependent on what nature provides as ever. We can craft contexts where these connections reveal themselves, in which we care again. Encountering the 'Ruminant Blooms' dislocates perceived certainties and echoes my own process of discovery. I hope that triggering these moments of inter-species proximity through my work engenders in us a sense of responsibility and stewardship towards other lifeforms and the Earth's ecosystem as a whole.

Beyond Comfort: Reflecting on Transgression

When we want to design with a multi-species mindset, we need to perceive these cultural, societal and experiential complexities and perspectives, and enmesh our emotional and intellectual insights. But we can only do this if we actively gather them. If we watch maggots become flies, go to the abattoir, work with the taxidermist, if we invest ourselves into the entire process, not shying away from the moments that could taint our innocence.

Immersion, or 'staying with the trouble',[3] is essential, and every discovered answer will open new trajectories of inquiry. The lens through which I viewed my next research question were cattle, and the leather and leather products we make from them. I wanted to probe what happens to our emotions in the process of transformation from mammal to material. Where and when do we cease to *care*? Where do we engage in which way? I drew up three lists, firstly of animals we commonly kill, secondly of materials we make from these animals, and a third of objects we commonly make from these animal materials. I saw these three lists as distinct, conflicting emotional territories that interconnect through the design process. A designer is often responsible for the phase where a material becomes an object and considers the object's future user and needs to inform their decision-making process. In circular design we extend our scope of concern beyond this future. However, we can also extend the temporal scope back in time, into the prior lives our matter led, and their eco-systemic entanglements. Herein resides a deeper layer of the material's meaning that reaches beyond aesthetics and functionality to our understanding of who we are in the world and what impact our choices have.

Cowbenches: Multi-sensorial Access to Material Ethics

I wanted to see whether I could pull the sequence of conflicting emotional territories into a single object until, like in a temporal purse-seine net, they can all touch us at the same time.

I wanted to make a leather sofa that becomes a vessel for the memory of a living cow, her death, the cowhide, the leather sofa – and the user who is considering how to react when encountering the piece in the built environment, intellectually and body to body. These processes, qualities and responses should coexist within the piece, compressed yet layered and shimmering in a way that viewers might perceive the thing as a narrative of prior life and death – at times alive and dead at others – as cow, leather, sculpture or sofa. I hand-sculpt the torso of each *Cowbench* from furniture foam to look like an animal reclining in the field, however, without a head or lower limbs since the hide is stripped off after these the body parts have been removed. With paper-cord and wooden dowels I replicate bones visible under a cow's skin and those felt within the body when touching it. The *Cowbench* is then upholstered spine on spine with a single, high-quality tanned cow hide that shows the scars, wrinkles, and markings that the animal had gathered in her lifetime.

I had never built furniture before, so Ron Arad connected me to Alma Leather, a leather furniture company that allowed me to use their workshop and work alongside and guided by their Polish upholsterers. Krzysztof Siutkowski, in particular, embraced this challenge. He explained to me why the project excited him: he had previously worked on a cattle farm and loved the animals in his care. Siutkowski had immersed himself in his work. He knew their anatomy from calf to fully grown cows and could now transfer his lived experience gained in close proximity to the animals to upholstering the benches. He was able to craft with knowledge, muscle memory and a multi-species mindset. Together we developed a bespoke method to upholster the *Cowbenches,* each of which was distinctly different in shape and posture. We worked side by side, at eye-level, negotiating the placement of a minimum of seams on each bench to create a leather surface that was as continuous as possible. Each finished bench became memento mori of a single cow's lived experience instead of hiding the lives of six to eight cows by cutting away all the traces of life deemed to be 'imperfections'.

I had the opportunity to show the *Cowbenches* internationally and found that there were some responses that transcended cultures. Children would instinctively read the benches as animals despite their incomplete bodies, run up to them, cuddle and ride on them. Adults seemed more reserved in their approach, trying to establish what they were looking at. They would first circle the *Cowbenches*, gradually step closer and then pat them on the neck before sitting down, as though they were seeking permission to do so. A few people suggested that it would be more efficient to build a mould and cast the benches in self-skinning foam so that I could manufacture them in larger numbers. I explained to them that the cows the leather was made from had been numbers in an industrial farming process and that it was precisely this anonymity that I wanted to counteract, the reason why each bench was different and why I had given the benches names. To become what they are, the benches need my emotional and bodily immersion, my time and my empathic investment to connect with the life of one animal I never knew.

> Attachments to habitual ways of thinking are especially hard to break. More than this, that which is familiar and taken for granted fades into the background. We simply do not see, feel, or think about that which has become embedded in our mode of being.[4]

Tony Fry's words really resonate with me because this is what I am trying to do through my work. To make a carelessness that we enact in our daily life perceivable as such. If touching the scars in the leather of a *Cowbench* connects people with the animal that existed in this skin, they may also see the animals that became their shoes and belts, bags and sofas. Perceiving

FIGURE 12.2 Top left and right: Re-connecting leather with its animal origins – *Cowbenches* 'Luzie' and 'Antonia'. Photos: Julia Lohmann. Bottom left: Marking thresholds of empathy – Fleece, carcass and plaster cast of a sheep's internal body cavity, taken at a knacker's yard. Photos: Julia Lohmann. Bottom right: A volume all mammals share – The sculpture *The Lasting Void*, cast from a calf's internal body cavity. Photos: Fabrice Gousset/Galerie Kreo

them helps us consider our choices and, if we still decide to acquire them, build an empathic, longer-lasting human-material relationship with our animal-based objects.

The Lasting Void: Crossing Thresholds and Decoding Experience for Discourse

We know that something in our inter-species relationship ruptures with the death of an animal, and this point of departure tends to become the focus of our reflections. However, faced with a dead cow, we still empathise. We see it as another being, a fellow mammal, whose body reminds us of our own. How then does this trajectory of empathy lead us, if we follow it, from the dead cow to the steak on our plate? Which thresholds do we traverse? When do we leave our empathy behind and start seeing body parts as consumable *matter*? Of course, this process is highly subjective and dependent on our prior life-experience, the choices we make and the context in which we encounter the dead cow. I set out to discover when my own perception shifts occur.

The taxidermist introduced me to an acquaintance of his who ran a knacker's yard. When farm animals need to be 'put down' or die of natural or unknown causes, they are categorised as unfit for human consumption, and he collects them for incineration or to turn them into animal feed. I wanted to witness the processing of an animal into materials to identify and document this incremental loss of empathy and connection and the emergence of material utility value, a shift from body to carcass to cuts of meat and fleece to wear. I had to immerse myself again.

Encapsulating the Remains of Empathy

Two sheep and a calf had just been brought in and we watched the proprietor skin the first animal. We then processed the other sheep in the same manner by ourselves, engaging in a series of transgressions that our society condones and that remain unspoken of in everyday life. I reflected on my response to each processing stage, from skinning the sheep and removing her head and lower limbs to when we removed her organs and divided the cadaver into cuts of meat. It was when we started taking the animal apart that I recognised the change in perception that I had been looking for. In the bodies of all mammals, there is an internal space holding their organs. When an animal is alive, they fill this volume, and only in one moment of the slaughtering process it exists in an empty state – after the organs are removed and before the carcass is dismembered. I wanted to preserve this point of departure and loss of connection by making a cast of the dead calf's body cavity. Using an animal as a mould felt like another transgression, but it would also make tangible an unseen space that we share with many other species. I called the object *The Lasting Void.*

Theorising Practice: Converting Immersively-made Objects back into Words

The Italian designer, architect, and author Alessandro Mendini was horrified by *The Lasting Void*, and wrote a considered but stark open letter to my gallerist and myself, explaining his reasoning and response.

> I know full well that research in design is on the decline mow and I am myself part of that trend. I also know that design, in this fleeting era, is not motivated by ethics or by values. I know design now is not an idyll and I also know that one is forced to take extreme action to find innovations and new languages. However, Julia's creative energy, in the case of your stool, truly seems to be badly directed. I do not understand what so much unpleasantness is supposed to demonstrate. If the photograph of her stool enters into the history books of design, this will be one of the most bitter examples, an extremely sad moment in the history of objects.[5]

I was still a very young designer then and it shocked me that Mendini's interpretation of *The Lasting Void* could be so drastically different to what I had wanted to communicate through the object. At the gallery press view, people waved Mendini's letter at me, and a journalist of

Le Figaro urged me: "Please respond! Please respond!" I replied that my French wasn't good enough, that I would not respond without prior reflection on what Mendini had written. I had invested myself so profoundly into researching and making *The Lasting Void*, reflecting deeply and refining what I wanted to express, encapsulated in object form. Now Mendini's accusation highlighted the need to decode my insights back into language to position my work within this discursive space. Together with my husband and friends, we crafted an open letter response to Alessandro Mendini's arguments, about the ethical aspects of using animal materials and the need for designers to take responsibility for their actions and the effects that these may have on the natural world.

We posed the question: "Is an object that has the death of an animal as its starting point more ethical if it hides its origin as best as it can?"[6] and explained that in my view the reasoning underpinning the creation of objects needs to encompass concepts beyond ergonomics and decoration, efficiency of manufacture and commerce:

> Design has to be more than merely "pleasant". Our lives are increasingly mediated through objects and revolve around consumption. It is the responsibility of the designer to embed in objects an added emotional and ethical functionality. Design should stop us from becoming numb to the world and instead prompt us to rethink how we lead our lives.[7]

The exchange with Mendini was vital to the development of my practice. I realised that it is not enough to discover new knowledge individually, but that it needs to be shared and connected back to the bigger body of knowledge. Mendini's provocation gave me the impulse to learn how to turn a work into words, which set me on course towards the collaborative, practice-led PhD studentship I completed at the Royal College of Art and Victoria & Albert Museum in 2018.[8]

To enable discourse, we need to find a material, form and context that transports our insights to an audience. We must also develop the ability to translate and transpose our findings from one context into another, from one field of knowing to another one so that the idea becomes conductive.

Rethinking Ourselves as Multi-Species Ecosystems: From Human Beings to Creatures Collectively Being Human

The Wellcome Trust commissioned me to create a window installation about the co-existence of organisms within the human body, based on the finding that only forty per cent of the living cells in our bodies are human. The remaining sixty per cent are other species living in symbiosis with our body or inhabiting it – organisms without whom we could not exist. Thinking of our bodies as multi-species ecosystems is mindboggling and challenges our notions of identity, self-control and independence. *New Materialism* scholar Jane Bennett describes the shift in thinking that occurs:

> Each human is a heterogeneous compound of vibrant matter. If matter itself is lively, then not only is the difference between subjects and objects minimized, but the status of the shared materiality of all things is elevated.[9]

Accepting that we are not alone within ourselves can help us make sense of the physical, emotional and sensory complexities that make us human and may in turn help us relate better to complexity in other fields and our societies at large. To bridge scales and perspectives, I created

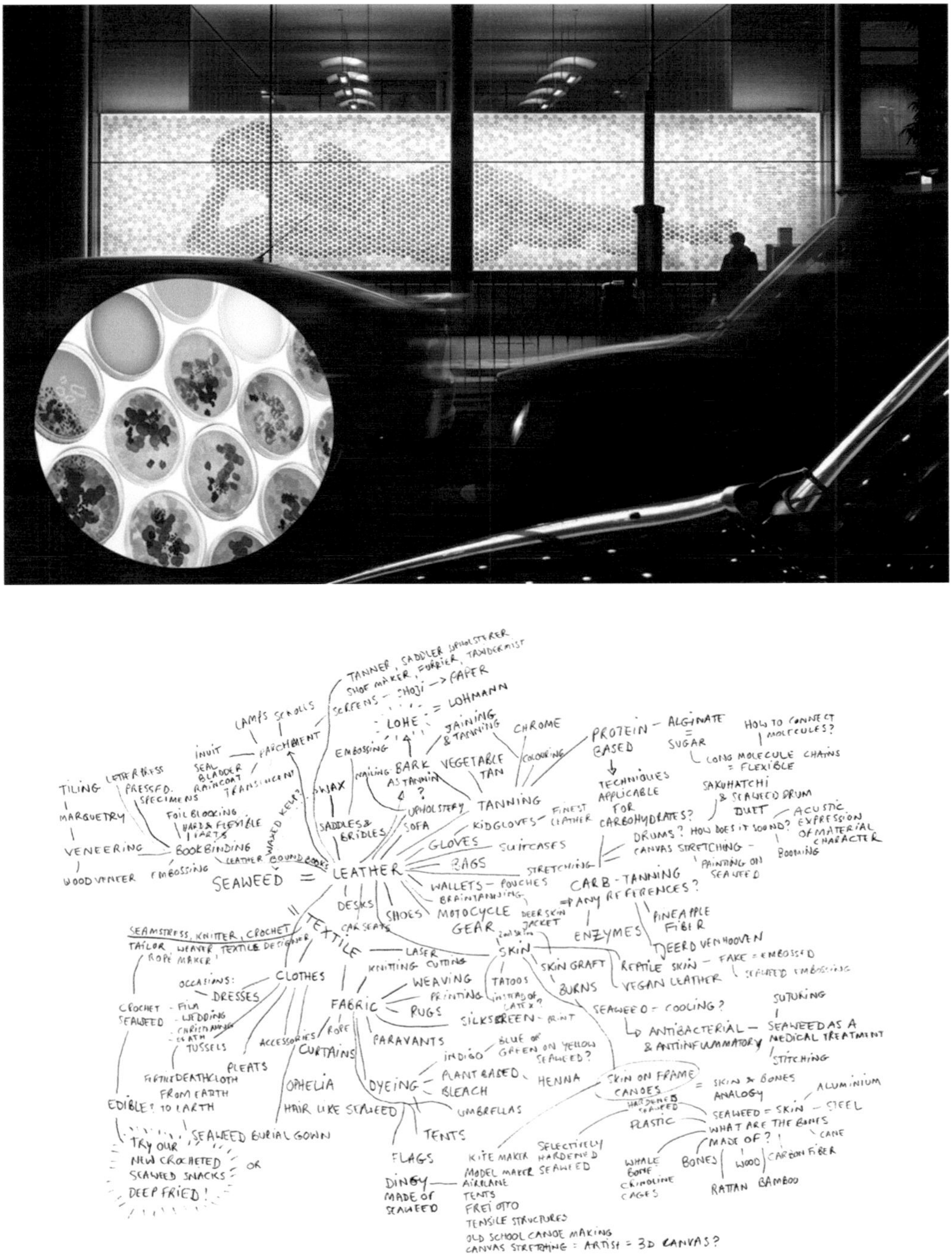

FIGURE 12.3 Top: The installation *Co-existence* on the co-existence of organisms in and on the human body, consisting of 9,000 Petri dishes mapping images of microorganisms according to their human habitats. Photo: Wellcome Trust. Bottom: Mindmap of material analogies between leather and seaweed, used to explore seaweed as a material for making. Illustration: Julia Lohmann

images of human bodies consisting of thousands of Petri dishes, acting both like pixels in a digital image and windows into our microcosms. Each Petri dish contained images of bacteria and other organisms that inhabit us, that I had mapped out with a microbiologist corresponding to the location of their habitats inside or on us. From afar, viewers would see depictions of reclining humans whilst close-up, they might see that we are not who we are after all.

Michael Levin and Daniel Dennett describe all intelligence as "collective intelligence",[10] as it is created through the interrelation of multiple, multi-species elements together forming a body. We are already multi-species actors on our insides, ecosystems nested within ecosystems.[11] Instead of human beings (a single species) we are multispecies collectives being a human. Can this shift in mindset help us engage less destructively in our environments?

From Convenience to Care: Overcoming our Single-species Mindset

The part of Helsinki where I live is dominated by construction work. The 400-year-old woodlands and rocks of Stansvik Forest, home to biodiverse webs of life, are felled and blasted to make space for housing blocks and shopping centres, to make room for humans. This isn't new. Already in 1965, the German social scientist Alexander Mitscherlich described the human-centeredness of our planning processes in his seminal book 'The Inhospitability of our Cities'[12] and the acknowledgement of unintended consequences of our actions led to the 'precautionary principle':

> When an activity raises threats of harm to human health or the environment, precautionary measures should be taken even if some cause and effect relationships are not fully established scientifically. In this context the proponent of an activity, rather than the public, should bear the burden of proof.[13]

Yet, the logging and pulverisation continues. Who gives us humans the right to own and shape land as we please, without considering the needs of our non-human co-inhabitants, whose ecological relationships we depend on? Sixty years on from Mitscherlich, our thinking remains human-centred, lacking in eco literacy and inclusion, and largely blind to humanmade *inhospitability* because the spaces we create seemingly serve us, humans, well. Next to single-species standards and interpretations of 'sustainability', conflicts of convenience and care are just as pervasive in the way we live. It is important to become conscious of convenience because it is a type of carefreeness that can easily lead to carelessness. Quite often, our design briefs and practices are geared towards convenience, not carefulness. I know that attempting to care in every situation would be overwhelming, however, we should speak out when something is designed for convenience to make us forget the design's potential negative implications and effects on others. After all, as stated by anthropologist, social scientist and cyberneticist Gregory Bateson, "The unit of survival is not the individual or the species, but the individual, the species and their environment"[14]

One of the exercises that I undertake with students is to think of a human societal concept, like democracy, and then attempt to consider it from different perspectives. It is wonderful what happens, because when you suddenly view democracy through the eyes of a chicken, scrutinise it as a chicken, you realise its limitations. Through the exercise, we reach a vantage point from where the limitations of our constructs become clearly visible, as well as paths towards reconfigurations that acknowledge non-humans. When we become aware of the challenging conflicting needs and wants, we have even more reason to base our actions on the precautionary principle and slow down when we are not sure whether we are heading in the right direction.

Thinking Further at the Fringes

We may just be one species among many, but we have evolved to become a lifeform that defines our entire planetary ecosystem including the relationships with and among the other species it contains. The question I am posing through my work is how we can become a more empathic, caring and regenerative species in the ecosystems we're creating or affecting. The biologist, innovation consultant and author Janine Benyus said: "Life creates conditions conducive to life".[15] When we design, we often aim to create the opposite: antibacterial fabrics, wipeable surfaces, designing for comfort and human control. In regenerative design[16] we try to shift from "recycle, reduce and reuse" to "restore, renew and replenish".[17] In short, we are asked to shift our agency towards becoming a nurturing force wherever we are present, from the deep sea to the Earth's atmosphere. Scientist and indigenous thinker Robin Wall Kimmerer writes:

> We need acts of restoration, not only for polluted waters and degraded lands, but also for our relationship to the world. We need to restore honour to the way we live, so that when we walk through the world we don't have to avert our eyes with shame, so that we can hold our heads up high and receive the respectful acknowledgement of the rest of the Earth's beings.[18]

So how can we redirect our impact? Who can help us become conducive to life, enabling others and ourselves to flourish? Can we team up with organisms that can act as our counteragents? Can we design for and with them? For me, algae have become teachers that help me reflect on these questions.

Seaweed: Material, Muse and Method

As artist in residence at the S-AIR centre in Sapporo, Japan, in 2007, I fell in love with algae, specifically brown kelp, because of its materiality and its ecosystem-building abilities. Algae are vital to oxygen production on Earth, absorb excess nutrients in the sea, act as carbon sinks and generate habitats and food webs for a wide range of marine species. Brown kelp have a positive presence in their eco-system that enables other life to flourish and that we can learn from, not just for our relationship with the ocean but nature in her entirety.

When you pull a blade of *Saccharina japonica* out of the water, it feels like a piece of leather, and since I had worked with leather before and I knew about the destruction and pollution inherent in processing leather, I wondered if seaweed could provide an alternative. Contrary to the intensive land, feed and chemical use in industrial livestock farming and tanning, seaweed farming using native species in their appropriate habitats can have a positive eco-systemic impact. It can be farmed on lines suspended over the seabed, creating new habitats. In many locations it does not need to be fertilised and grows by absorbing excess nutrients washed into the sea through our rivers and waterways. If we harvest this seaweed, we can ameliorate some of the imbalances we created whilst gaining nutrients and biomaterials to replace some of our most harmful material relations. This is how I think we should investigate its use, prioritising and expanding its regenerative impact and viewing seaweed materials and products as a positive side effect.

Expanding from similarities between seaweed and leather, I developed a system of material analogies to explore what else we might make from seaweed, for instance alternatives to paper or plastics. Drawing on existing cultural knowledge of the reference materials I mapped out similarities and differences, materialising my findings through processing tests and prototypes. Seaweed has very little documented design and craft history, apart from Aboriginal Australian uses of bull kelp to make water carriers. Speculative making and material analogies helped me establish a craft that should and might have existed but had been forgotten and not developed further.

The more explorative seaweed-related objects I made, the more I realised that its development as a material for making would take time and that it should be conducted within a community-led and community-building framework, rather than by a single author-designer. To give seaweed the time that it required, I enrolled in a collaborative practice-led PhD studentship at the Royal College of Art and Victoria & Albert Museum in London, which included a six-month residency at the Victoria & Albert Museum, where I founded the *Department of Seaweed*[19] to claim that seaweed was just as important as the other materials that have or have had a department there: silver, metalwork, textiles, ceramics and glass.

I set up the residency as a community of practice,[20] a group of people from different fields with an interest in seaweed who collaborate at eye level, exchanging and building knowledge and skills across disciplines. The nascent state of seaweed as a material for making was an ideal opportunity to flatten hierarchies between learners and experts that exist in relation to established materials. None of us were masters in macroalgae crafts which freed us to focus on shared, open-ended experimentation and discovery. Through the hypothetical space we created and enacted within the physical studio in the museum, we became a very real Department of Seaweed. Museum visitors would say: "Oh, I didn't know there was a Department of Seaweed in the V&A". and I would say: "Yes, there is. Welcome!" They visited an experiential, multi-sensory space suggesting what seaweed might become.

The Temporal Agency

Biomaterials enable time-based reflections, enabling us to think with them, both into their own eco-systemic past and into multiple possible ecological and material futures. Reflections along these trajectories of transformation connect us with our impacts on the ecosystems and lifecycles that we derive our materials from, as well as those in which we use them. On an artefact level we shift our gaze into the city with its infrastructure and populations both supported by the natural word and degrading it through waste streams and pollution, taking us right back to the ocean. This temporal agency of biomaterials appears heightened in seaweed because it is so highly reactive to its surroundings. Watching dried seaweed rehydrate in water shows how dramatically it changes from being rigid and brittle to being supple and animate, as though it came back to life. The Victorians used the atmospheric reactiveness of seaweed to forecast the weather. The physical interaction between one organism and another, touching and smelling the scent of the seaweed, provided immediate physical feedback that a thermometer or barometer cannot provide.

Bodily Engagement

We can identify with seaweed as a body, an organism, as a being. The minute there is a piece of seaweed on the table, the conversation changes. We become more aware of our being, gaining in physical presence and empathy. Seaweed is slimy, it smells, it provides sensual cues that make us connect with our inner self and childhood stories, with immersive moments of lived experience. In my practice, it has become the most effective way I know of bringing nature into human-made contexts. If we understand the transformation from organism to material artefact and the changing and evolving inter-species relations involved in design processes, we can begin to reflect on them at a systemic scale. I observed this with many other biomaterials as well, especially in scientists researching them, who have formed strong bonds with the subjects of their study and establish intricate protocols of care, and to whom biomaterials have become familiar spirits. In old manuscripts and folk tales there are many descriptions of such spirits, of cats or frogs helping witches and wizards work their magic. In the research consortia I contribute to, I have seen very similar relationships between scientists and their willow, liver cells, water fleas or zebrafish. I believe it is these interrelations between different species that make us gain new knowledge and a better understanding of the world. My own familiar, seaweed, is associatively linked to ocean health coastal communities and material streams. In designing and making with seaweed, we adopt a contextual approach, exploring it from diverse perspectives. Working with algae introduces us to ocean literacy[21] and systems thinking as described by Fritjof Capra: "Systems thinking is contextual, which is the opposite of analytical thinking. Analysis means taking something apart in order to understand it. Systems thinking means putting it into the context of a larger whole".[22]

Transdisciplinary Ruminations

To connect with matter, we can stretch our reflections into different systemic contexts. Our enquiry can become infused with transdisciplinary knowledges from diverse fields and levels of complexity from the micro- to macro-world. However, these can be challenging to navigate if we don't have translators in the form of natural scientists or ecologists, or indigenous cultures with their cosmologies. Realising the immense value of these translations and exchanges, I developed a method at the Department of Seaweed that I call Co-speculation. It meant that instead of individually crafting a speculation and developing outcomes from a single perspective, I was crafting a lens through which I could co-speculate together with others and co-imagine different futures for seaweed.

My lens was my residency department, with its studio space populated by lots of smelly materials, with its opportunities for sensual bodily engagement and numerous unfinished seaweed objects that had no clear functionality in a conventional sense, such as ergonomics by which you might judge a chair design. They were objects that were uncomfortable and that did not quite fit into our daily lives, however, they presented a wide range of material connotations that enabled people to grasp their seaweed-led materiality. The idea was that we use the undefined or unfinished quality of these things to initiate a conversation about what we might make from seaweed and what its future role might be. This is a process that can be transferred to other materials and their particular qualities, contexts and types of engagements, enabling open conversations about futures and aims, about what is desirable and what is undesirable, about potentials and dangers.

Portraying Material Agency

At the end of my Department of Seaweed residency, I constructed the seaweed and rattan sculpture *Oki Naganode*, as a portrait of the materiality of seaweed – seaweedness – and my dialogue with seaweed. The sculpture still changes every day, reacting to the humidity of its surroundings

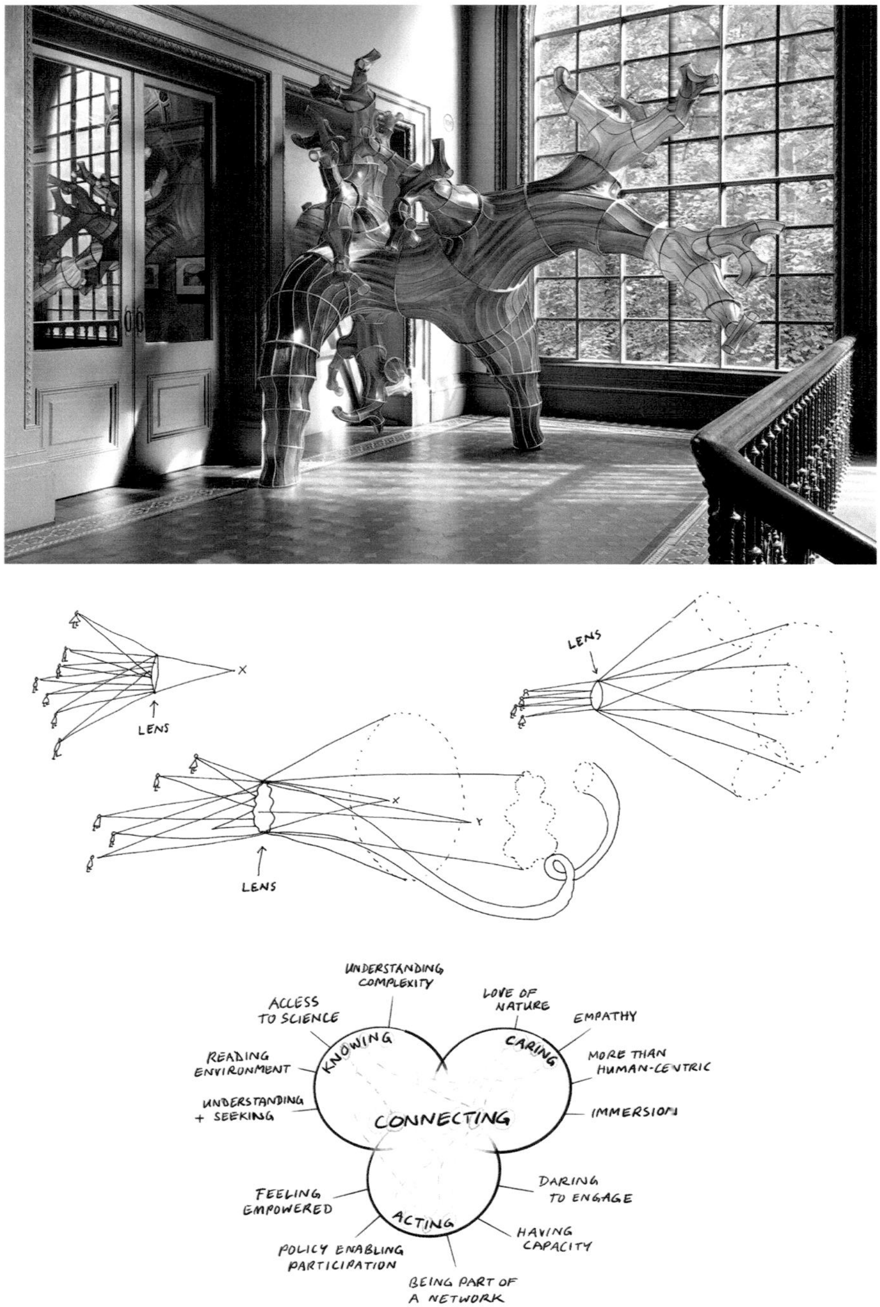

FIGURE 12.4 Top: Portraying the material qualities of seaweed – The 'Oki Naganode' seaweed and rattan sculpture at the Victoria & Albert Museum London. Photo: Petr Krejci. Centre: Different lenses used by communities of practice exploring a shared concern through Co-speculative Design. Illustration: Gero Grundmann. Bottom: Designing with a multi-species mindset by connecting knowing, acting and caring. Illustration: Julia Lohmann

and subtly changing its surface tension accordingly, as well as lightening in colour as the seaweed's chlorophyll fades away. I have made more seaweed sculptures since, using them as vehicles to talk about topics from biomaterials and ocean protection to reframing the expanding field of design with a multi-species mindset.

As creatives, we should be creative with our methods. This is something we should never give up and that we should fight for when we are entering the academic field. Because this is our *superpower*. That we can speculate and speak about topics before they become science, before they become political, projecting probable and possible scenarios and communicating them accessibly and engagingly to help people identify which of them are destructive and which are desirable – from both a human and eco-systemic perspective. Design is by its very nature future oriented and action oriented, as expressed in Herbert Simon's famous definition: "Everyone designs who devises courses of action aimed at changing existing situations into preferred ones".[23] When designing, knowledge about a situation is not the ultimate aim, but a means to find a pathway and understand the nature of the preferred situation we are aiming for. We are triangulating our future aim with all of the information present in the situation and seek further input when we do not yet understand enough. By designing we create a connective tissue between nodes of information and experience that are relevant to our action: we have methods that make us versatile, lateral and relational thinkers and doers, enabling us to embrace emergent discourses and complexity, connect across disciplines and articulate emotional and intellectual aspects. That is why for me it is so fruitful to collaborate with scientists from different fields because we can combine our ways of working to find new, more accessible and actionable expressions for new knowledge.

Biomaterials Are More than just Materials: Matter, Method and Muse

If we let biomaterials become our familiar spirits and embrace their influence on the creative process in its entirety, they become our material, our muse and our method. If we accept them as non-human actors or agencies we can converse with them, generating outcomes considered from a wider range of perspectives, contexts and temporal trajectories. The role of an organism in the eco-system guides how we should design alongside it. If an organism is a keystone species that struggles to survive because of shifting environmental factors (such as Bladderwrack in an over-fertilised sea), we need to treat it as a client: finding out how we can enable it to thrive again (for example by finding ways to remove nutrients). If the organism is superfluous or too dominant in a location (like *Cladophora glomerata*, that grows quickly and abundantly on top of the bladderwrack), we can devise methods to extract or collect and utilise it, devising a scenario which enables the removal process, for example by finding uses for the biomass or building alliances with others who will benefit from it. If the organism can potentially rebalance the ecosystem (like *Saccharina latissima* in the Atlantic coastal regions) or to create conditions for an abundance of life, we can employ it as an ally in the regeneration of a damaged environment. Here we can ask: Are there methods of scaling these benefits so that the outcome is desirable for both the ecosystem and the humans involved? If the organism is highlighting or reinforcing our human behaviours and imbalances it can become a symbol for a problematic relationship that helps communicate the plight of the seas and connect people across diverse groups in a common cause. Green-blue algae blooms are one example.

Knowing, Caring and Acting

As creative practitioners, we can connect knowing, acting and caring, much like in the 'head, heart and hand' approach found in transformative management. In any eco-social scenario we enter, the three elements knowing, acting and caring exist among the multiple actors present. We can map the knowing, caring and acting of the stakeholders we consider to be affected by the situation. We can identify who is present at the table, what we know, and project pathways to action. Then we can consider our blind spots and biases: What do we not know, whose voices should we also hear. Why are conflicts arising? We can then take stock of the situation and alleviate the weaknesses we have identified.

A situation dominated by care and agency will likely feel volatile and uncertain, and actions taken might not necessarily achieve desired outcomes because of a lack of the knowledge required to direct the energy. There is a risk of alienating potential allies. Realising this helps identify and invite collaborators from the fields of science we require to better understand the nature of what we are dealing with. Who might help us channel our energies into the right direction? How can we minimise our blind spots?

If care and knowledge are present, but we lack agency, our endeavours may stagnate and feel depressingly unsolvable. Eco-anxiety is a good example, with its contrast between actionable scientific knowledge on countering and adapting to the climate crisis and a lack of political will to take action that frustrates an increasing number of people. Here, we need to learn how to form and strengthen alliances, accessibly and actionably communicate our concern to a broad audience and constructively challenge decision makers. Locating and activating parties who did not previously consider themselves to be aligned with our purpose and uncovering hidden and unconventional pathways to a being heard and seen is important in this.

If agency and knowledge are strong but the concept of care is underdeveloped, we have a powerful scenario where the situation can quickly be changed, but it might not necessarily shift in the right direction. An example could be an influential company who acts solely for the benefits of their own shareholders. Here, we should aim to highlight the limitations of their care: inclusive design methods, long term thinking, considering future generations or giving a voice to the voiceless stakeholders can be methods for engagement in such situations. As Fritjof Capra describes:

> Care flows naturally if the "self" is widened and deepened so that protection of free Nature is felt and conceived as protection of ourselves.... Just as we need no morals to make us breathe ... [so] if your "self" in the wide sense embraces another being, you need no moral exhortation to show care.... You care for yourself without feeling any moral pressure to do it.[24]

We can only grow a holistic multi-species practice at the intersection of *knowing*, *caring* and *acting*.

I believe, as creatives, we are ideally placed to weave these together, to understand what is lacking, how to gather the complementary elements of knowledge and care, or the pathway towards effective agency.

- Knowing: Including scientific knowledge, unlocking science for interdisciplinary discourses;
- Acting: Building agency: methods of engagement, design activism and strengthen alliances;
- Caring: Extended empathy (Krznaric) and shift of mindset (F.Capra), long term thinking.

Conclusion

In this chapter I introduced the key insights I gained through practice and practice-led research over the past 20 years. They are: context sensitivity, process led immersion, reflection, enabling discourses, our bodies as eco-systems, thinking further at the fringes, materials enabling time-based reflections, bodily engagements, transdisciplinary ruminations, materials as muses and methods, and fostering knowing, caring and acting. Each of them, I hope, can nudge us to reflect more, connect further, instil care in ourselves and others and help us embrace a multispecies mindset and a widened understanding of our multispecies selves.

Notes

1 Merleau-Ponty, Maurice. *Phenomenology of Perception*. London: Routledge, 1962, 169.
2 Wirth, Jason, and Donna Haraway. *Environmental Philosophy* 5, no. 2 (2008), 165–170. http://www.jstor.org/stable/26167994.
3 Haraway, Donna J. *Staying with the Trouble: Making Kin in the Chthulucene*. Duke University Press, 2016. https://doi.org/10.2307/j.ctv11cw25q.
4 Fry, Tony. *Design as Politics*, London: Bloomsbury, 2011.
5 Lohmann, Julia, *Appendix B, The Department of Seaweed – Co-speculative Design in a Museum Residency*, PhD dissertation, Royal College of Art, 2018.
6 Lohmann, Julia, *Appendix B, The Department of Seaweed – Co-speculative Design in a Museum Residency*.
7 Lohmann, Julia, *Appendix B, The Department of Seaweed – Co-speculative Design in a Museum Residency*.
8 Lohmann, Julia, *The Department of Seaweed – Co-speculative Design in a Museum Residency*, PhD dissertation, Royal College of Art, 2018.
9 Bennett, Jane. The Force of Things: Steps Toward an Ecology of Matter. *Political Theory*, 32 (2004), 347–372. 10.1177/0090591703260853.
10 Cognition all the way down by Michael Levin and Daniel Dennett at aeon.co 13 October 2020.
11 Lohmann, Julia, *From Algae to Ocean Literacy: Knowing, Acting and Caring through Design*, keynote, Sustainability Science Days 2022, University of Helsinki, Finland, 19 May 2022.
12 Mitscherlich, Alexander. Die Unwirtlichkeit unserer Städte, Anstiftung zum Unfrieden. Frankfurt: Suhrkamp, 1976.
13 Staff, Science and Environmental Health Network. 26 January 1998. Wingspread Conference on the Precautionary Principle, archived at the Wayback Machine, 2 November 2005.
14 Bateson, Gregory. *Steps to an Ecology of Mind*. London: Paladin, Granada Publishing, 1973.
15 Benyus, Janine M. *Biomimicry*. HarperCollins, 2003.
16 Wahl, Daniel, *Designing Regenerative Cultures*. Triarchy Press, 2016.
17 Morton, Jennie. *4 principles of regenerative design*. www.buildings.com/resiliency-sustainability/article/10196146/4-principles-of-regenerative-design.
18 Kimmerer, Robin Wall, *Braiding Sweetgrass*. Minneapolis, MN: Milkweed Editions, 2013.
19 Lohmann, Julia. *The Department of Seaweed*, accessed 31 January 2024, www.departmentofseaweed.org/
20 Lave, Jean & Wenger, Etienne. *Communities of Practice: Learning, Meaning, and Identity*. Cambridge: Cambridge University Press, 1998. http://dx.doi.org/10.1017/CBO9780511803932.
21 *Ocean Literacy Portal*, UNESCO Intergovernmental Oceanographic Commission, accessed 31 January 2024, https://oceanliteracy.unesco.org/?post-types=all&sort=popular.
22 Capra, Fritjof & Luisi, Pier. The Systems View of Life: A Unifying Vision. In *The Systems View of Life*. Cambridge: Cambridge University Press, 2014, 1–498. doi:10.1017/CBO9780511895555.
23 Simon, Herbert. *The Sciences of the Artificial*, 3rd edition. The MIT Press, 1996, 111–167.
24 Capra, Fritjof & Luisi, Pier. The Systems View of Life: A Unifying Vision.

13

TEMPORALITIES OF CARE IN MORE-THAN-HUMAN DESIGN

Gizem Oktay, Minha Lee, Bahareh Barati and Ron Wakkary

Introduction

Traditional ways of viewing time as a linear trajectory toward infinite human progress and production cause overlooking how other living beings experience time, and how our relationship with them creates alternative understandings of time, beyond the coordination of clocks and calendars. Accessing plural understandings of time is crucial to tackling environmental injustice and climate crisis, as the dominant (e.g., capitalist) modes of production that are causing these issues are informed by conventional understandings of time (Birth, 2016). Time beyond the clock or a unit to be controlled and measured is an important avenue in design-oriented HCI research. Current examples include treating time as an experiential phenomenon that takes different shapes beyond clock time; as a contextual, everyday practice with socio-political underpinnings (Pschetz and Bastian, 2018), and as an element to be aligned across multiple beings of the Earth, who go through their own rhythms and subjective expressions of time (McKibben, 2003). More-than-human design (MtHD) as an emerging subdiscipline of design has taken an interest in this research avenue, through designers and practitioners exploring the multiplicity of time through the notion of MtH temporality. Noteworthy examples consist of design artefacts, objects (Ikeya, 2023), and fabulations (Søndergaard et al., 2023) that visualise/express more-than-human temporality through communicating how other beings (e.g., silkworms, microbes, dogs) might experience time.

Despite the growing body of work on MtH temporality, aspects of time from a care-centred viewpoint are understudied. Care, as a sensibility and practice, is crucial to decoding the relational aspects of time. Therefore, this chapter emphasises relationships formed between humans and nonhumans in studying time. From this perspective, MtH temporality not only means how other beings experience time (their temporality), but also points to time as a *cohabited* phenomenon that arises through relationships between different beings, whether human, living, or non-living. Temporalities of care are then proposed as a useful framework (Oktay et al., 2023), as it unravels the multiple agencies in constructing more-than-human care relations. Unravelling and acknowledging more-than-human agency in the construction of sustainable relations is crucial in design practice where the focus is on the well-being of not just humans, but the ecosystem as a whole.

DOI: 10.4324/9781003467731-16

Through a close look at the temporal dimensions of more-than-human care practices, designers can identify what kind of relations are the ones to design for/with, which is one of the main tenets of MtH design. As design with care in MtH design is becoming a fruitful research avenue, there is a need for enriched perspectives towards care that are informed by temporality. Understanding plural expressions of time and more-than-human temporality not only provides opportunities for multispecies flourishing via more-than-human design but also provides an alternative vocabulary to linear, quantitative forms of time that focus solely on ideas of infinite progress and accelerating futures, which result in catastrophic outcomes such as climate crises, environmental and social injustices.

This chapter responds to the need to provide generative definitions of time through the question: "How do care practices towards MtH beings bring forth a plural understanding of time?" and provides answers through an example of more-than-human care practices within human-dog cohabitation that focuses on care-time entanglements. The proposed contribution is a set of sensitising concepts (synchrony, sequence, repetition, asynchrony and latency) that reveal multiple agencies and dynamic relations at play. Bringing forth temporal concepts can shed light on care relationships that inform more-than-human care practices, with which we hope to restore, regenerate and repair within a cohabited world.

Time, Temporality and Care in Design Research

Care can be broadly defined as 'physical and emotional labour' on behalf of some other/s (Bowlby, 2012). Caring includes attending to need-based relationships and the interdependence between the carer and the cared-for, thus involving reciprocity (Noddings, 2010). Notions of interdependency and reciprocity are necessary thinking tools in MtH design, to help bring nonhuman entities to share the stage with humans (Wakkary, 2021). Care has been explored from numerous angles in design research that centres living beings, such as rewilding and composting activities for soil as an act of care for more-than-human bodies (Søndergaard & Woytuk, 2023), caring for bacteria living in the soil with the construction of microbial fuel cells (Ooms et al., 2022), exploring the essential elements of care such as affection, obligation, and maintenance through engaging with fermentation (Chen et al., 2021). In broader design research, tensions within care relationships (e.g., doing what is convenient instead of what is 'right') are taken up through autobiographical accounts of design (Helms & Fernaeus, 2021), as well as care as a design ethic for IoT at home (Key et al., 2021).

While previous definitions of temporality include the idea of a linear progression of past, present and future (Heidegger, 1927); recently, it has been referred to as the state of existing within time, or time as it is lived (Hoy, 2012). Research in HCI and design has dealt with multiple aspects of time and temporality. These vary from slow technology (Hallnäs and Redström, 2001) and slow design (Odom, 2012), which investigates longer-term human-technology relations (e.g., interaction beyond use), and designing for slowness, solitude and mental rest to other design-oriented approaches that centre temporality from an empirical perspective. Research in this realm includes designing with traces and patina (Rosner, 2013), as well as the afterlife, a 'non-use' state of things beyond immediate interaction. More recently, researchers have been dealing with underlying definitions of time. A well-known example is the conceptualization of 'temporal design' by Pschetz and Bastian (2018), where time is seen beyond dichotomies of speed (e.g., fast or slow) or duration (e.g., long or short) towards social, cultural and political dimensions of time. According to tenets of temporal design, time is multiple, and heterogeneous, and arises from the unequal entanglements between various social formations (2018, p.174).

The discourse on time, temporality and care are connected in the way that *we make time for each other through care practices*. These characteristics point to the construction of a time that arises through relations. In the following sections, through the example of an ongoing and intimate relationship between humans and dogs, we focus on the negotiation of time, push, and pull of agencies in care practices, as they reveal a diversity of temporalities that allow us to understand the plurality of time as a lived and situated experience.

Revisiting Care from a Temporal Standpoint

Care time arising through relationships between humans and more-than-humans opens new ways to see time as relational and contextual. This view of time entails situated materialities, practices and temporalities. Living with a pet, growing mushrooms, tending to a garden, and maintaining healthy soil all entail different practices, thus plural rhythms and engagements with time. These include the rhythms of motion and changes the cared-for goes through (e.g., metabolism, lifecycle, seasonality), as well as the relational temporalities that arise through the relationship between the caregiver and receiver. To illustrate, Puig de la Bellacasa states that "promoting soil care practices intensify the involvement of practitioners with soil's temporality" (2014, p. 1). Accordingly, ways of caring for soil follow the pace at which soil treats and renews itself. Engaging with soil care will mean involvement and attunement with soil's temporality, which will reveal interdependent temporalities, providing an alternative perception of time to predominant understanding based on progress and acceleration.

We unpack this invitation to engage with more-than-human temporality via care practices on multiple levels. First, the engagement with MtH temporality through caring for MtH entities allows an understanding of a time that is plural, consisting of many rhythms, paces and characteristics beyond clock/calendar time (e.g., within care, there will be different intensities, repetition, intervals). Second, when we engage in care through noticing, anticipating, and fulfilling more-than-human needs, we are not involved in the fast-paced and accelerated futures we are otherwise accustomed to. Instead, we enter an alternative temporality imbued with the temporality of the care receiver and one that stands outside the normative structures of time.

Through the engagement with more-than-human entities provided through care, we become aware of a temporality that is different from the generalised perceptions of nonhuman time, which is seen as cyclical and going at a slower pace than human time (McKibben, 2003), such as geological aeons that govern life through a scale of thousands of millions of years, or species with lifecycles much shorter and seemingly ephemeral and fleeting than our own (e.g., the example of the day-long life of a butterfly). Against this understanding which reduces nature to a timeless backdrop, geologist Marcia Bjornerud claimed instead that nature is time*ful*, not time*less*. That, at any given moment, there are plural temporalities governing life, inviting us to abandon any belief in the permanence of nature (Bjornerud, 2022). These temporalities are entangled, in various states that make them aligned, and in competition. Mismatches and conflicts ought to occur among these temporalities, which require negotiation, intervention and responsive attentiveness (Rose, 2013) given through care.

Within more-than-human care, plural temporalities are informed by the more-than-human moral relationship. Therefore, studying how care relations are formed, evolved, and changed through time, as informed by relations is important. Studying these relations reveals plural temporalities at play in constructing more-than-human care. This allows seeing temporalities in relation, rather than having a singular structure of time. In the following sections, we illustrate how we study these relations in the context of living with companion species through a diary study conducted with humans and dogs. The diary study focuses on the care practices enacted towards dogs as companion species and how these practices inform a plural understanding of time beyond clock and calendar time.

The Study: Human-dog Relations and Everyday Care Practices

To look at the plural temporalities of care amongst humans and dogs, we turn to Haraway in her Companion Species Manifesto (CSM). In CSM, Haraway states that we "inhabit companion species stories along with ours, cohabiting an active history where the smallest unit of analysis is a relationship" (Haraway, 2008, p. 20). Treating the relationship as our unit of analysis, we chose to conduct a diary study to capture the richness of care relations unfolding in everyday life. Czerwinski, Horvitz and Wilhite discuss the high ecological value of diary studies that are carried out in situ, in participants' environment as they unfold over time (Czerwinski et al., 2004). This aligns with Barbara Adam's view of everyday life as an 'intricate tapestry' (Adam, 1995) of 'intersecting and sometimes discordant times, rhythms and temporal regimes that are always more-than-human' (Power, 2009).

We recruited a total of twelve participants, comprising six humans and six dogs (see Table 13.1). Participants were given the choice of a digital document, a written notebook, or taking notes on their phone for the diary writing to make sure that writing was integral to their day-to-day lives. All participants chose a digital format (e.g., Google Docs or Microsoft Word), including pictures, narrative, duration and time-stamped information about activities such as walking, playing, feeding, and the time they spent together or separately. Additionally, we presented participants with 12 prompt cards, which they could pick one per day. These cards served two purposes: Stimulating participants to write about their day with specific attention given to time, and, to include more of their dog's perspective in the writing, such as through questions: "What's your dog's perspective of the day?" (see Figure 13.1).

The data gathered from the diaries were analysed through thematic (Braun & Clarke, 2006) and abductive analyses (Tavory & Timmermans, 2014), which revealed aspects of shared time (e.g., doing things together such as walking), competing agencies (e.g., waiting for each other), senses of responsibility, guilt (e.g., compensating for cut short walk time), and obligation experienced by humans through care. Our iterative coding process revealed how care practices were informed by both human and dog agencies (e.g., walk time and duration were defined by dogs as much as humans). As a result, we arrived at five sensitising concepts (synchrony, asynchrony, latency, sequence, repetition) that shed light on relational temporalities of more-than-human care that takes place in everyday life. Below we give examples of each concept from the study and provide a definition.

TABLE 13.1 Participant information

Participant code	*Age*	*Dog pseudonym*	*Dog's age*	*Breed*	*Time together*
P1	44	Button	5	Bichon Frisé-poodle mix	2.5 years
P2	34	Rickie	4	Spaniel-Chiuahua-Pitbull mix	3 years
P3	37	Sausage	10	German shepherd mix	8 years
P4	23	Biscuit	14	Terrier-poodle mix	Over 13 years
P5	29	Dundee	2	Maltese-Shih Tzu mix	Almost 2 years
P6	35	Brady	4	Pitbull	3 years

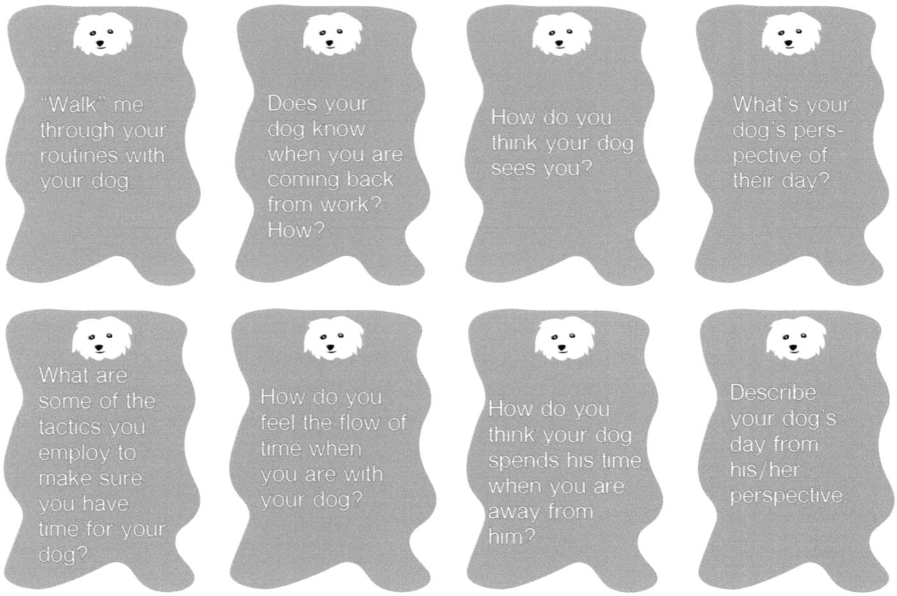

FIGURE 13.1 Prompt cards for participants.

Source: Image by the first author.

FIGURE 13.2 Sensitising concepts.

Source: Image by the first author.

Synchrony

Within care practices, synchrony is a state of temporal alignment brought about by enacting care practices. In our study, we observed synchrony through moments of humans and dogs doing things together, sharing the same space, and being close to each other. Most of our participants remarked on being with their dogs in their diaries, such as being with them 24/7 (P6), eating their food together (P4), working and studying from home to make sure they're almost always with their dog (P1), the dog being rarely home alone (P2), or to take specific actions to be together, such as working from a specific location to be closer to their dog (P5). In this case, we interpret synchrony as the ease and flow of being together, which the human and the dog achieve by being in each other's company. Our participants shared many moments of synchrony, reported either by their dogs' preference for being around them even when they are not doing anything together, such as P6's dog keeping them company and 'napping in the sun' while they work, or the dog sitting by their feet while they watch a movie (P2). We define this temporal alignment as synchrony as it does not require any time-related adjustment from either party. That, in moments of synchrony, the individual human and dog temporalities run parallel to each other. Synchrony is experienced by humans with a sense of ease and flow, such as in the words of P3 whenever walking with their dog in the vast forest: "Our time [together] is melting or like meshing".

Synchrony refers to instances where the movement or sensation of two beings overlap in time and form (Rennung & Göritz, 2016). Often referred to in human-to-human interpersonal contexts (e.g., synchronised sporting activities, military marches, collective rituals), the exploration of synchrony in more-than-human relations is understudied. Anthropologist Deborah Bird Rose brought up synchrony in her theorisation of an 'ethical time'. Within ecological relations, she positioned synchrony as a life-giving co-existence with overlapping temporalities. She gives the example of Australian flying foxes and the pollen and nectar they feed on: "*When* eucalypts start to flower, that's *when* flying foxes arrive" (2012, p. 133).

To sum up, synchrony is the matching of rhythms between different entities. Within humans and dogs, the matching of rhythms happens through the activities they engage in together, such as walking, playing, resting, sleeping and so on. An important aspect of synchrony to consider is how it manifests itself among entities that are not always in the company of each other: How would a synchronous human-fungi or human-soil relationship look? How could humans match their rhythms to that of a garden? These are important provocations that can inform the time-related aspects of more-than-human care practices.

Sequence

Within our study, we observed sequence as to how care practices inform each other. Taking place in the context of everyday life, care practices are informed by their precedent. Care practices may be altered in duration or intensity by the preceding (i.e., past) activity, such as how P5 does it: "I want to keep every day a little bit balanced. Because if we for example played with other dogs, then he's super tired, then next day we normally go for short walks instead of long walks because I know that he's tired from yesterday and that he's also less active". P6 and P2 paid close attention to this temporal element as well, in the ways that they adjusted their care activities based on what happened in the past: "The amount of walk we do depends on the last walk we had" (P6) or "No special food tonight because of what happened in the morning" (P2).

We interpret these accounts as human caretakers using sequence to achieve consistency and flow in the care practice. This consistency has a subjective understanding per participant that is the result of the ongoing, intimate relationship between the two. Through diary writing, participants became more aware of the sequential nature of care:

> Writing things down made me feel like: 'OK, this is how much I did yesterday, so today I should also do something similar. Then I will know that that much activity he did, will be enough for him and he will sleep a lot. And this diary made me realise that I know how much I should do every day'.

Sequence is broadly defined as an order of events preceding one another. This is what we're often used to hearing as the "cyclical" rhythm of nature. However, with sequence, there is a difference: things do not go back to their initial state such as in a cyclical state. They follow each other after another in a continuum. From this perspective, nature is not cyclical; seasons that follow each other are never the same way.

More importantly, this continuum implies that present care practices ought to be informed by their precedents, which mean adjustments, continuous decision-making and attuning to the results of care practices as they are enacted. Within more-than-human design practice, we invite designers to take advantage of sequence as a temporal element that speaks to the way care practices follow each other, and with which order they are practiced. An important question to ask is: "How can sequencing of care practices inform more-than-human care relations?"

Repetition

Within our study, we observed repetition in the construction of routines: sequential care practices repeated over some time. The perception of repetition flowed through our participants' narrative, through the extensive use of words such as *usual morning routine, walk routine, food routine, ritual, outdoors routine*, and so on. Routines are built alongside long-term cohabitation, reading each others' cues through nonverbal communication (e.g., P1 wrote that they correlate the dog's staring at them and the clock time to know its feeding time) and intimate relations. Participants relied on the existence of routines to ensure care was given and employed tactics to safeguard the continuity of care. P2 states that they have "such a pattern that the long walk happens at the same hour every day of the week", whereas P5 states that through routines, they can keep a consistent walk and feeding practice. When routines are not performed the way they take place, this is experienced by participants as an unusual situation and may result in inconsistency in the way care is given: 'I usually walk her as soon as I get up, but I didn't do it today like every day, I overslept and we went out late' (P6).

Repetition is a crucial temporal aspect of more-than-human care. In her plea to reconsider time and space in the way we care for 'country' (an Aboriginal term that holds all relations in a given place), Donna Haraway states: "there can be ongoing and effective care that stays alert to many sorts of history" (2011, p. 100). Similarly, Puig de la Bellacasa states that "expected repetition [...] is essential to ecological relations of care" (2008, p. 19). In design practice, what gets repeated and what is not, how many times, and over how long are important aspects to consider.

Latency

Within our study, we observed latency as the wait time that the cared-for endures, such as dogs waiting for their human companions to come back when they are left home alone. Several diaries included accounts of dogs waiting, resulting in latency, as the *delay in the interaction* between the human and the dog. One participant remarked on this, and a desire to compensate for it:

> I have early meetings on Thursdays, so Rickie got a very quick morning greeting. [...] Poor Rickie had to wait for [partner] to wake up to go for his first pee and poo. He went around 11:30 for about 20 minutes. Then we had playtime on the stairs with the fuzzy tug fox toy for about 20 minutes. I felt bad ending playtime so I gave him a Dentastix as a consolation prize.

Other examples include the dog having to go back to sleep if it is not time to go out yet (P5), or the dog howling endlessly (P1) when they are put in a crate for the humans to come back, waiting by the door while the human is out with friends (P4), or waiting by the window when the human is out shopping (P2).

Latency refers to the time that occurs between a cause and its effect (Floridi, 2021). In the digital technologies context in which it is often used, latency combines the concepts of communication and interaction. Delay, lag and wait time are all similar concepts of latency. Within more-than-human design, latency appears as an essential temporal element of care. Specifically, when care does not show its results immediately, requiring wait times to be endured, and requiring continuous attention to be given to the unfolding effects of care.

Asynchrony

Within our study, we observed asynchrony as the state of temporal misalignment. It is a time-related conflict between the human and the dog, in need of being adjusted. Our analysis showed that the reason for asynchrony was often a power differential in care practices, meaning the caregiver had more agency over the cared for and, therefore, made decisions on behalf of the cared for. Below we share an excerpt from the exit interview the first author had with P6, where the first author asked P6 about how their dog tried to get attention but was denied:

P6: Well, if I'm chilling and watching something, she would come up and try to engage and also try to be let on the sofa inside, but that's forbidden. So then, she usually puts her head on the sofa and looks at you. And this can last ohh, a long time. She can do about fifteen minutes of unbroken eye contact. Just making sad eyes.

Interviewer How long do you keep eye contact?

P6: Oh, I completely ignore her. I can give her a pet or something and tell her: 'Bruh, you're not coming up. *This is me time.* But yeah, it's usually ignored.

Another reason for asynchrony is anticipation: Both dogs and humans may anticipate events to unfold in a certain way, but then be wrong, such as when P5's dog is left alone:

> He stayed quite long home alone. We thought he would sleep. Instead of sleeping, we saw through the camera that he barked a lot. We usually take him out with us everywhere, so he begins to have a harder time watching us leave.

P6 also shared a humorous moment of how he anticipated their dog to excrete, but was wrong about it: "So I've been late to places because I thought that a 10-minute walk would be fine. And then she wouldn't cr*p". These instances show us that expecting the other to behave in a certain way (i.e., anticipation) can be a reason for temporal misalignment, which may or may not need adjustment.

Overall, asynchrony refers to mismatches, conflicts, and the push and pull of agencies. In the context of our study, it referred to moments of being wrong about predictions and anticipations, and conflicts between clock time and care time. In an ecological context, there are examples of asynchrony in the form of food-web (trophic) mismatches. The mismatch between when animals breed and when resources are available may result in the timing of breeding not quickly adapting enough to match the changing environment. This mismatch can have negative effects on the reproductive success of herbivores, resulting in trophic mismatches (Doiron et al., 2015).

To summarise the five sensitising concepts: *Synchrony* is the sense of temporal alignment, flow and ease that comes with the enactment of care practices performed towards MtH beings we are in a relationship with. *Sequence* speaks to care not being composed of a single and immediate act, but how they follow each other based on preceding acts of care. *Repetition* speaks to the way care is enacted within the everyday and the need for care to be ongoing and continuous. *Latency* is the delay of interaction between the caregiver and the receiver. *Asynchrony* is temporal misalignment, mismatches and conflicts between the beings that constitute the care relationship. These temporalities must be acknowledged, negotiated and intervened if needed through more-than-human design practice.

Discussion

Reflecting on the definitions of synthesising concepts and the study example, we discuss the implications of sensitising concepts for more-than-human care in design, as well as the fluidity of caregiving and receiving roles in care relations.

Attending to Plural Temporalities in Care Practices

The sensitising concepts we identified speak to the temporal nature of more-than-human relations, which are informative of care practices. While, within a human-dog care relationship, synchrony revealed itself as dogs and humans spending time in their vicinity with a sense of flow and ease, within a different human-nonhuman care relationship, synchrony (e.g., matching of rhythms, affect and behaviour) can mean something different. Our study revealed the concepts as they arose, as well as the situated and contextual care relations that are unique to each human-pet participant pair. We expect more implications and plural meanings of sensitising concepts to emerge through different relationships, e.g., relationships formed with service dogs, cats, plants, soil, fungi, water, bacteria, and so on. An important implication for more-than-human design is to unravel how temporal states may occur through care in more-than-human relationships.

More importantly, we urge designers to pay attention to not just moments of synchrony but also asynchrony and latency, as these point to how temporalities are negotiated and intervened in the care practice. Then, within care relations, synchrony is not the only outcome, and mismatches, conflicts and delays ought to occur. These conflicts are not to be dismissed, as they speak to the way plural temporalities are deliberately negotiated by the cared for and the carer, showing the push and pull of agencies. Examples from our study include how dogs go back to sleep if their human has not woken up yet (P5), or dogs not excreting during the allotted time to walk, causing conflicts in humans' expectations from care practices.

Synchronous and asynchronous relations take place in sequential and continuous ways, thus requiring ongoing (Haraway, 2011), repetitive (Puig de la Bellacasa, 2014) and responsive attentiveness (Rose, 2013) by the caregiver. Understanding how synchronous care relations occur in human-nonhuman relations requires developing an understanding of more-than-human temporality when we cannot rely on a long-term and communicative cohabitation such as the one between humans and their canine companions. Therefore, data-oriented approaches to noticing the environment (Rosén, 2022) and building kinship-based relations built over time with not-so-companion species such as fungi (Liu, 2018) are relevant. We invite designers to design more-than-human care practices with sensitising concepts as they inform the pace, rhythm, duration and timing of care practices.

To care is to make time for the other. This care time results in aligned and competing temporalities, within a state of pluralities. This means there will always be plural expressions of time, beyond the singular and abstract conventions of clocks, calendars, and schedules as coordination elements. The plurality of time within care practices urges us to pay attention to states of asynchrony, latency and repetition, sequentially and continuously. What is relevant for MtH design and designers is to pay attention to matches as well as mismatches between different temporalities. Acknowledging, highlighting, negotiating and intervening in plural temporalities will be of ongoing importance in more-than-human design as an essential component of more-than-human care relations. Through the sensitising concepts we identified (synchrony, sequence, repetition, latency and asynchrony) designers can better pay attention to more-than-human temporalities, which are essential to more-than-human care practices.

The Fluidity of Caregiving and Receiving

Through our study, we observed that while the conventional role of caregiving seemed limited to humans, through their diary reflections, humans acknowledged how their canine companions cared for them in turn. Participant narratives included accounts of dogs helping with the participant's seasonal affective disorder (P2), providing opportunities to socialise with others through organised walks (P1), accessing parts of the forest that they would otherwise not think of going to, and therefore discovering new areas (P3). These speak to the fluid nature of caregiving and receiving, where caregivers may become care receivers, and vice versa. The fluctuating nature of caregiving and care-receiving creates an imperative to closely look at power asymmetries and the notion of agency in the construction of care practices; care is not an exclusively human practice, but MtH entities also care for each other. How MtH beings care for each other is a relevant question that needs to be addressed. An inherent aspect of human-nonhuman relations is that they care for each other by giving each other time. Therefore, how MtH beings give and make each other time is a fruitful direction to take. Consequently, a strong implication for MtH design practice will be to decode who makes time for whom, when and how that time manifests itself.

Conclusion

In this chapter, we focused on temporalities of care as composed of five sensitising concepts, namely, synchrony, sequence, repetition, latency and asynchrony. These concepts speak to plural temporalities in more-than-human care practices, and the push and pull of agency in more-than-human care relations. Through the example of a human-dog diary study, we showed how sensitising concepts emerged through everyday enactment of care practices, in a way where caregiver and cared-for roles were subject to fluctuation. Going ahead, there is an implication that designers not reduce the myriad care relations to achieve synchrony, and that there will be asynchronies, conflicts and latencies to account for in more-than-human design. Accepting all aspects of plural temporalities will be an ongoing challenge for designers. However, this also poses an opportunity as sensitising concepts become helpful in decoding kinds of relations that are informative of care practices. We ask designers how they can construct synchronous relations between humans and other more-than-human entities, especially the ones that are not-so-companion, unlike the canine subjects of our study. We urge designers to design more-than-human care practices that open up plural expressions of time.

Acknowledgments

We are deeply appreciative of our editors Marie Louise Juul Søndergaard and Anton Poikolanen Rosen. We are immensely grateful for the participants, both human and canine. Lastly, the first author would like to dedicate this paper to their recently deceased canine companion of eight years. She is the main motivation for this research project and will be eternally missed.

References

Adam, B. (1995). *Timewatch: The Social Analysis of Time*. Polity Press; in association with Blackwell.

Birth, K. (2016). Calendar time, cultural sensibilities, and strategies of persuasion. In Hom, A., Mcintosh, C., Mackay, A., & Stockdale, L. (Eds), *Time, Temporality and Global Politics*. e-IR. www.e-ir.info/2016/07/15/edited-collection-time-temporality-and-global-politics/

Bjornerud, M. (2018). *Timefulness: How Thinking Like a Geologist Can Help Save the World*. Princeton, Princeton University Press.

Bowlby, S. (2012). Recognising the time: Space dimensions of care: Caringscapes and carescapes. *Environment and Planning A: Economy and Space*, 44(9), 2101–2118. https://doi.org/10.1068/a44492

Braun, V. & Clarke, V. (2006). Using thematic analysis in psychology. *Qualitative Research in Psychology*, 3(2), 77–101.

Chen, D., Seong, Y. ah, Ogura, H., Mitani, Y., Sekiya, N., & Moriya, K. (2021). Nukabot: Design of care for human-microbe relationships. *Extended Abstracts of the 2021 CHI Conference on Human Factors in Computing Systems*, 1–7. https://doi.org/10.1145/3411763.3451605

Czerwinski, M., Horvitz, E., & Wilhite, S. (2004). A diary study of task switching and interruptions. *Proceedings of the SIGCHI Conference on Human Factors in Computing Systems*, 175–182. https://doi.org/10.1145/985692.985715

Doiron, M., Gauthier, G., & Lévesque, E. (2015). Trophic mismatch and its effects on the growth of young in an Arctic herbivore. *Global Change Biology*, 21(12), 4364–4376. https://doi.org/10.1111/gcb.13057

Floridi, L. (2021). Digital time: Latency, real-time, and the onlife experience of everyday time. *Philosophy & Technology*, 34(3), 407–412. https://doi.org/10.1007/s13347-021-00472-5

Hallnäs, L., & Redström, J. (2001). Slow technology: Designing for reflection. *Personal and Ubiquitous Computing*, 5(3), 201–212. https://doi.org/10.1007/PL00000019

Haraway, D. J. (2007). *The Companion Species Manifesto: Dogs, People, and Significant Otherness*. Prickly Paradigm Press. https://books.google.nl/books?id=XzRTnQEACAAJ

Haraway, D. (2011). Speculative fabulations for technoculture's generations: Taking care of unexpected country. *Australian Humanities Review*, 50. https://doi.org/10.22459/AHR.50.2011.06

Heidegger, Martin (1976). *Basic Writings from Being and Time* (1927) *to The Task of Thinking* (1964). New York: Harper & Row

Helms, K., & Fernaeus, Y. (2021). Troubling care: Four orientations for wickedness in design. *Proceedings of the 2021 ACM Designing Interactive Systems Conference*, 789–801. https://doi.org/10.1145/3461778.3462025

Hoy, D. C. (2012). *The Time of Our Lives: A Critical History of Temporality*. MIT Press. https://books.google.nl/books?id=Q2bXQ36onvwC

Ikeya, Y., & Barati, B. (2023). Designing an algal relay computer: A critical orientation in exploring more-than-human temporality. *Extended Abstracts of the 2023 CHI Conference on Human Factors in Computing Systems*, 1–6. https://doi.org/10.1145/3544549.3585874

Key, C., Browne, F., Taylor, N., & Rogers, J. (2021). Proceed with care: Reimagining home IoT through a care perspective. *Proceedings of the 2021 CHI Conference on Human Factors in Computing Systems*, 1–15. https://doi.org/10.1145/3411764.3445602

Liu, J., Byrne, D., & Devendorf, L. (2018). Design for collaborative survival: An inquiry into human-fungi relationships. *Proceedings of the 2018 CHI Conference on Human Factors in Computing Systems*, 1–13. https://doi.org/10.1145/3173574.3173614

McKibben, Bill. (2003). “Worried? Us?” *Granta*, 83(Autumn), 7–12.

Noddings, N. (2010). Care ethics, caregiving, and global caring. In V. Moser & I. Pinhard (Eds), *Care – Wer sorgt für wen?* (1st ed., pp. 17–26). Verlag Barbara Budrich. https://doi.org/10.2307/j.ctvm2023j.4

Odom, W., Banks, R., Durrant, A., Kirk, D., & Pierce, J. (2012). Slow technology: Critical reflection and future directions. *Proceedings of the Designing Interactive Systems Conference*, 816–817. https://doi.org/10.1145/2317956.2318088

Odom, W., Wakkary, R., Hol, J., Naus, B., Verburg, P., Amram, T., & Chen, A. Y. S. (2019). Investigating slowness as a frame to design longer-term experiences with personal data: A field study of Olly. *Proceedings of the 2019 CHI Conference on Human Factors in Computing Systems*, 1–16. https://doi.org/10.1145/3290605.3300264

Oktay, G., Ikeya, Y., Lee, M., Barati, B., Lee, Y., Chen, Y., Pschetz, L., & Ramirez-Figueroa, C. (2023). Designing with the more-than-human: Temporalities of thinking with care. *Companion Publication of the 2023 ACM Designing Interactive Systems Conference*, 104–106. https://doi.org/10.1145/3563703.3591462

Ooms, D., Barati, B., Bruns, M., & van Dongen, T. (2022). From concern to care: A transformative reflection on designing-with the living. *Nordic Human-Computer Interaction Conference*, 1–15. https://doi.org/10.1145/3546155.3547285

Pauly, D. (1995) *Foundations of Socio-Environmental Research* (n.d.). Retrieved August 29, 2024, from www.cambridge.org/core/books/abs/foundations-of-socioenvironmental-research/daniel-pauly-1995/9501C9C8D9EA95F7E91362F46CD9B0CA

Power, E. R. (2009). Domestic temporalities: Nature times in the house-as-home. *Geoforum*, 40(6), 1024–1032. https://doi.org/10.1016/j.geoforum.2009.07.005

Pschetz, L., & Bastian, M. (2018). Temporal design: Rethinking time in design. *Design Studies*, 56, 169–184. https://doi.org/10.1016/j.destud.2017.10.007

Puig de la Bellacasa, M. (2015). Making time for soil: Technoscientific futurity and the pace of care. *Social Studies of Science*, 45(5), 691–716. https://doi.org/10.1177/0306312715599851

Rennung, M., & Göritz, A. S. (2016). Prosocial consequences of interpersonal synchrony. *Zeitschrift Für Psychologie*, 224(3), 168–189. https://doi.org/10.1027/2151-2604/a000252

Rose, D. B. (2012). Multispecies knots of ethical time. *Environmental Philosophy*, 9(1), 127–140.

Rose, D. B. (2013). Val Plumwood’s philosophical animism: Attentive interactions in the sentient world. *Environmental Humanities*, 3(1), 93–109. https://doi.org/10.1215/22011919-3611248

Rosén, A. P. (2022). Relating to soil: Chromatography as a tool for environmental engagement. *Proceedings of the 2022 ACM Designing Interactive Systems Conference*, 1640–1653. https://doi.org/10.1145/3532106.3533503

Rosner, D. K., Ikemiya, M., Kim, D., & Koch, K. (2013). Designing with traces. *Proceedings of the SIGCHI Conference on Human Factors in Computing Systems*, 1649–1658. https://doi.org/10.1145/2470654.2466218

Søndergaard, M. L. J., & Campo Woytuk, N. (2023). Feminist posthumanist design of menstrual care for more-than-human bodies. *Proceedings of the 2023 CHI Conference on Human Factors in Computing Systems*, 1–18. https://doi.org/10.1145/3544548.3581083

Søndergaard, M. L. J., Campo Woytuk, N., Howell, N., Tsaknaki, V., Helms, K., Jenkins, T., & Sanches, P. (2023). Fabulation as an approach for design futuring. *Proceedings of the 2023 ACM Designing Interactive Systems Conference*, 1693–1709. https://doi.org/10.1145/3563657.3596097

Tavory, I., & Timmermans, S. (2014). *Abductive Analysis: Theorizing Qualitative Research.* University of Chicago Press. https://books.google.nl/books?id=5OYvBAAAQBAJ

Wakkary, R. L. (2021). *Things We Could Design: For More Than Human-Centered Worlds.* (1st ed.) MIT Press.

14

ENVISIONING MULTISPECIES FUTURES IN MULTISPECIES ENVIRONMENTS

Methods, Outcomes and Learnings from a Future Scenario Workshop

Lotte Nystrup Lund and Shams Hazim

Introduction

This chapter delves into the design, outcome and learnings of a workshop located in a natural environment focused on envisioning future scenarios for multispecies cities. The workshop was conducted as part of an industrial PhD project titled *Biocities of the Future* (Nystrup Lund, 2024) and hosted by a Copenhagen urban development company. Before the workshop, most of Lunds Ph.D. research had been conducted in Copenhagen, where it had involved various urban planners and decision-makers. The workshop idea emerged as an opportunity to explore what would happen when changing research settings and participants by conducting a workshop in an experimental camp on the countryside with designers and researchers already familiar with more-than-human perspectives. A contrast to the everyday research involving urban actors in urban office spaces.

The workshop *Multispecies City of the Future* was integrated into a one-week research camp named *I.N.S.E.CT. Summer Camp 2022* which we, the authors of this chapter, joined (Keune et al., 2023). The camp was located in a scout camp in the Bidstrup forest in Hvalsø, Denmark, and based on co-creation; inviting all participants to organise an activity. The camp provided an interesting framework for a workshop on multispecies future scenario-making, as it was centered around collaborative more-than-human design and exploratory investigations into living-and-being-with more-than-human organisms (Keune et al., 2023). Our chapter captures and discusses learnings from the workshop, thereby contributing a concrete and detailed account of more-than-human design and futuring work in practice.

Background

The relevance of speculative practices targeting multispecies cities is manyfold. First, we urgently need to re-engage with our living environments to be able to re-understand and re-imagine our relations to radically changing surroundings (Haraway, 2016; Latour, 2018; Rosa, 2019). In this relation, imaginative practices focused on more-than-human needs in urban environments might inspire designers and others working with urban development to re-think what a city is

DOI: 10.4324/9781003467731-17

and could be. Second, most people in the world today live in urban areas, and as cities have a huge potential to better fulfill more-than-human needs, a focus on urban change potentials seems relevant across countries and species (United Nations, 2019; Zari, 2018). Third, research points to several benefits related to more-than-human urban design such as multispecies justice, habitat provision, air purification, mental/physical health and eco-literacy fostering care for the natural world (Beatley & Newman, 2013; Celermajer et al., 2021; Zari & Hecht, 2019) but radical imaginations of future cities seem scarce (Hopkins, 2021).

A hypothesis was that bringing imaginations of multispecies futures out-of-the-city, using other tools, and involving people from fields other than urban development might lead to more radical ideas compared to those developed in more conventional workshops with (most often) busy urban actors.

The workshop *Multispecies City of the Future* brought together participants from academic and artistic fields related to fifteen countries from four world regions. Participants had backgrounds in mostly creative and engineering disciplines like architecture, industrial design, electrical engineering and material sciences. Participants were predominantly affiliated with academic institutions as post doc's, lecturers, master and PhD students, in addition to some being professional artists and designers. People joined the workshop after an open call at the *I.N.S.E.C.T Summer Camp*.

In the following, we describe the context of the workshop, design and set-up, and reflect on learnings and challenges based on outcome, feedback, and insights from workshop preparing and facilitating.

Co-creative Workshops Based in Nature

Below, we introduce concepts of co-creative workshops in nature. We present three examples of workshops held in nature described in existing literature as well as the benefits and limitations of various workshop methods according to research.

We define a co-creative workshop as a collaborative and participatory event where a group of people come together to generate e.g., ideas, solutions, and strategies targeting a specific challenge or opportunity (Frow et al., 2015). Typical characteristics of co-creative workshops include participants working collectively to brainstorm, share insights, and develop prototypes through processes idealistically characterised by a sense of co-ownership, shared problem-solving and decision-making (Frow et al., 2015; Sanders & Stappers, 2008). Such workshops are often situated in lab-based environments, but in previous years there has been a tendency to move outside (Chamberlain et al., 2012). This shift into outside research/workshop methods is called in-the-wild research; a concept introduced by Altarriba Bertran et al. (Altarriba Bertran et al., 2022) and described as a method used for contextual research purposes, evaluation, and as a means for co-design and ideation practices (Altarriba Bertran et al., 2022). Workshops in nature aim at changing the experiences and perspectives of participants to gain different results and understandings in opposition to learnings gained in lab-based or urban environments (Altarriba Bertran et al., 2022; Haldrup et al., 2022).

Examples of in-the-wild future thinking scenario workshop methods are: 'Labs in the Wild', understood as prototyping, and employing the prototypes in the settings for which they are meant to be used (Wilde et al., 2014). In this methodology, social and physical aspects have a critical effect on the usage of prototypes. Another example is 'provotyping', an approach using artifacts, and experiences to stimulate reflection on the desirability of the future (Boer &

Donovan, 2012). Provotypes are employed as bridges between technology and human perceptions while exploring tensions between them. Finally, 'Speculative enactments' is a methodology where participants enact elements of possible futures through acting in a fictional narrative (Elsden et al., 2017). These enactments can be part of an event, service, or performance and can stimulate participants' actions and experiences of the workshop. Although such methods touch certain parts of future scenarios, they are not necessarily based in 'wilderness' environments, but rather in human-made ditto. Thus we saw an opportunity to apply aspects of the methodologies in our workshop to gain a greater understanding of the influence of natural environments on participant experiences and workshop outcomes.

In reaction to the described in-the-wild workshop methodology (Chamberlain et al., 2012), the work of Altarriba Bertran et al. aims to move the in-the-wild methods, to 'from-the-wild' i.e., more explicitly aiming to account for the perspective of other species. As mentioned, the limitations of in-the-wild methods often take place in human-made environments that offer qualities considerably different from nature's 'wilderness', and are thus not wild enough (Altarriba Bertran et al., 2022). Another limitation mentioned by Altarriba Bertran et al. is that through the constraints of design research, designers have a position of power and privilege which in turn compromises the connection with the context, skewing the aims and conclusions of the research.

Such limitations can be countered by placing design and research in nature as done by the Open Forest Collective. They collect "experiential forest data" by observing, smelling and touching surroundings, and turning them into stories (Botero et al. 2023). Similar research involves gathering sensorial impressions such as sounds, images and meanings to facilitate a relational inter-species interaction (Mitro, 2023). Relatedly, through imagining alternative futures in a Future Forest workshop of Collins et al. aimed to gain an understanding and seek an empathic interrelationship between humans and the forest (Collins et al., 2018). By narrating local stakeholder perspectives; a set of arts and humanities positions; and the ecological background of the forest, Collin et al. provided participants with information to contribute to a renewed cultural relationship with the woods. In this way, the immersion in the subject matter is greatly enhanced and offers new insights into experiencing the design challenges. By exploring ways of utilizing the from-the-wild perspective (Altarriba Bertran et al., 2022) we sought to gain knowledge on how a 'wilderness' workshop might enhance the ability for multispecies future scenario making.

Multispecies City of the Future Workshop Design and Set-up

The location of the workshop was defined by research camp possibilities, but we deliberately chose the farm area with a barn, surrounding meadows and forest due to the combination of a varied nature and a shelter for rain, sun and wind. By being in a semi-cultivated farm area marked by human activity (meadows were e.g., fenced due to grazing farm sheep) but with views to a more free-growing nature (the farm area was framed by forest on all sides), we sought to minimise the risk of disturbing fragile or threatened nature by our presence.

The workshop was planned using a design thinking approach recognised for stimulating creative and integrative solving of wicked problems (Dorst, 2015; Rind Christensen et al., 2014) and assessed as useful for imagining futures in nature (Altarriba Bertran et al., 2022; Collins et al., 2018). This approach guided us to design a co-creative workshop embracing prototyping, iterations, visualizations and reflections on the future.

Workshop Design

The three-hour workshop consisted of three main sessions: 1. An intro to prepare participants by setting a relaxed atmosphere, opening up their imagination, and connecting to surroundings. 2. A primarily silent session to activate and engage participants in exploring, co-creating and imagining a future where urban life is characterised by more-than-human needs. Leading to 3. A main session with team-based future scenario-making. The aim and activities of the three sessions are described below.

1. Intro – Preparing Participants by Using a Fictive Story Woven into a Meditation, and Underlined by Music to Generate a Relaxed Atmosphere and Opening up the Imagination

The intro aimed to prepare the participants for the workshop; embrace the theme of multispecies urban futures; create a common frame of reference stimulating radical scenario making (i.e., 'provotyping') and strengthen participants' engagement in fictional narration (i.e., the concept of 'speculative enactments'). In addition, the intro sought to stimulate senses and support participants in embodying a calm atmosphere. The overall aim was to stimulate a mental let-go of conventional urban design attitudes and signal design freedom in scenario making.

The intro consisted of four phases: First, the facilitator presented the aim and main task of the workshop: Respectively, to imagine radical scenarios for what urban life could be like, and create future scenarios in teams. Then, the facilitator invited the group to imagine that we all lived in 2024:

> I hope that we will transform ourselves into one big brain that will actually be living in the future city of 2040 in most of the workshops here. So, in a while, we will transform ourselves from the beings that we are now into the beings in another city that is very nice and beautiful and characterized by being good for everyone … by doing a lot of positive effects and synergy. And I rely on our big brain to create that city together and to sketch it and to write about our lives in that city.
>
> (Quote from kick-off talk by main facilitator)

Responding to this, a participant told an imaginative story describing an extreme event leading up to a radically different environment in 2040: A massive solar flare hit the planet in 2025 causing an extreme impact on Earth making people realise their deep dependence on biological systems. In consequence, radical reinventions of survival strategies were implemented including a total re-creation of urban life and design. Concluding the introduction, the facilitator told a story about a 2040 city where more-than-human species co-lived and thrived.

During storytelling a participant improvised guitar play with melodies of harmony and joy. Most participants laid down while listening to the story and guitar, with their eyes closed. We were surprised to experience how impactful the guitar play seemingly was. The combination of storytelling and music seemed to create a meditative (i.e., transformative) atmosphere which seemed to match the aim of opening up participants' imaginations. Though the meditative aspect was not planned, it felt appropriate and fruitful.

Participant contributions were agreed upon the day before as spontaneous ideas evolved during other camp activities. In this relation we find that openness towards co-creating a workshop design with participants and a 'go with the flow' attitude was useful for creating an engaging and creative atmosphere.

2. Silent Session – Activating and Engaging Participants by Inviting them to Explore Surroundings, Co-create and Imagine Future Living in a Diary Format

The aim of the second session was to help participants land in, and engage with the living environments; to stimulate new ideas and more-than-human perspectives stemming from interaction with a multispecies physical space (Chamberlain et al., 2012; Haldrup et al., 2022); and to activate participants as contributors. Thus, methods characterised by exploration, co-creation and individual imagination were chosen.

The second session consisted of three phases:

First, a three-step individual task conducted in silence:

1. Collect an item from the surroundings symbolizing life in a future multispecies city.
2. Place the item on a cardboard poster unrolled on the barn floor.
3. Anchor the item by writing or drawing.

When introducing the task, participants were invited to weave together personal ideas, and hopes with aspects from more-than-human agencies and perspectives stemming from outdoor observations, and a natural artifact that each participant selected, and reflected upon. An activity relating to theory highlighting that “direct contact with nature can promote ecological attitudes and behaviors” (Martin et al., 2020, p. 30). Figure 14.1 shows some of the contributions to the Collective Collage cocreated by participants at the workshop.

Second, participants were encouraged to make individual stories illustrating a personal envisioned future scenario (i.e., fictional narration) The task was to write and/or draw a diary describing a normal day living with more-than-human species in 2040. For the diary task, participants were again encouraged to work in silence on their own as we thought it might help keep focus on the more-than-human; trying to stretch the open-minded curious mindset from the outdoor exploration into the writing/drawing phase. The diary format was chosen to make an abstract concept of ‘multispecies cities’ tangible and relatable. Also, the diary method seemed useful for creating a space where everyone had a bit of silent time for concentrated work enabling mental/physical landing. Diaries were created in and outside the barn which had two large double doors wide open to the surrounding environment. Figure 14.2 shows photos from the diary writing activity and some of the diaries produced by workshop participants.

In the concluding step, participants shared a key insight from their diary. The sharing of diary snippets was done orally while participants sat in the barn in a circle. Hereafter, diaries were laid out on the barn floor as a visual linear gallery.

This last activity provided an informal opportunity to get to know each other a bit more; it created a ‘democratic space’ where everyone (among humans) had a voice; and invited all to get an impression of each other’s imaginations. Aspects we found useful for the following task; to self-organise in teams (explained below). In addition, by appreciating all sharings we could signal that all ideas were welcomed, no matter how different they were compared to today’s urban design practices, and urban living.

3. Group Session – Inviting Participants to Self-organise in Groups, and Providing a Storyboard and a Poster Listing Multispecies Inhabitants to Guide Co-creating of Species-inclusive Future Scenarios

This session aimed to allow participants to dive deeper into co-creating detailed scenarios and unfold various more-than-human perspectives on how species might live better together in

FIGURE 14.1 Contributions to a Collective Collage.

Source: Photos by Lotte Nystrup Lund

the future. Thus, methods supporting in-depth group work and a species-inclusive focus were chosen.

Participants were invited to self-organise in teams by identifying shared interests relating to multispecies futures. Self-organization took place outdoors with everybody standing on a grass field surrounded by meadows and a forest. A set-up that has the potential to stimulate flexible interaction. Participants wandered among each other, talked crosswise, and got together in groups.

FIGURE 14.2 Diary writing activity.

Source: Photos by Lotte Nystrup Lund

Thereafter, the four self-organised teams were invited to ideate ideas for a shared concept they could work on for one and a half hours; a time frame allowing for more in-depth scenario exploration compared to previous tasks. To steer scenario-making into a shareable semi-open format and guide co-creative imagination, each team was invited to a table with a large piece of cardboard with pre-made text and squares, and color markers. The card title said: *Multispecies City 2040,* a question asked "What is a day like living in a wonderful city of 2040?" and an empty 4-square storyboard was framed with the question "How are organisms thriving together?" The teams were invited to write and draw scenarios. Three teams worked inside the barn, one team moved their workstation outside in the farmers garden.

Before the workshop began, posters were installed at five work bases (one for each possible team). All posters had the title *Inhabitants of the Biocity*. Each poster highlighted a unique list of species covering various species-kingdoms (plants, mammals, insects, birds, humans, amphibians, microorganisms etc.). The posters were made to support creation of various future scenarios; and participants' focus on more-than-human inhabitants.

Wrapping-up of the workshop, the teams gathered outside the barn in the farmers garden to present scenarios for each other and us.This gave us an opportunity to videofilm scenario presentations hereby creating a data set for documentation and subsequently communication. Figure 14.3 shows a workshop team co-creating a future scenario on a grass area in front of the barn, and another team presenting a future scenario for urban co-living with more-than-human species, in the farmers garden.

Workshop Outcome: Scenarios Developed

Below, we describe the scenarios developed at the workshop. The presentation of scenarios was filmed by a camp member responsible for video documentation of the full camp. The videos and posters enabled us to summarise each scenario afterwards.

Future Scenario 1: Trees as Central for Living

In a post-apocalyptic future, architecture is grown rather than built. Genetically modified trees provide colossal collective homes and thriving ecosystems. Work life, energy and transport are redefined due to biotech innovations inspired by nature's intelligence enabling eco-friendly communities to live in harmony with nature and embrace a multispecies lifestyle. New challenges emerge, including ethical debates about the human use of trees.

Future Scenario 2: Human-insect Collective and Transformation

In a multispecies boot camp the human Stanley, the queen bee Avena, and a bacteria population co-live in a bee-hive-human-collective. Diverse nature/tech hybrid systems are imagined and include; a roof with bacteria-based bio-fill regulating indoor humidity for multispecies inhabitants; bees clustering and vibrating during winter hereby heating up the bed for Stanley and Avena, a spray can with pheromones attracting fireflies who lead you home when it's dark; and genetically modified trees growing lightening fruits hereby avoiding urban light pollution normally disturbing insects. The multispecies co-life includes learning processes that continuously transform humans, bees, ecosystems and technologies.

FIGURE 14.3 Co-creating and presenting future scenarios.

Source: Photos by Lotte Nystrup Lund

Future Scenario 3: Rhythms Embracing Ecosystems

In a multispecies future life is driven by natural cycles. Today's unwanted urban nature is welcomed and authentic rhythms of various species and identities steer life. Earth-based knowledge, permaculture and water flow play a crucial role. Water is a vital shared resource, with natural purification systems using e.g., urine and menstruation blood as fertiliser. Shared spaces, like a cocoon, cater to diverse species' needs and resources. Flexible nature-based residencies resonate with changing weather conditions. Humans focus on self-existence and rhythms, rather than monetary gain. The future ecosystem showcases harmony between species and eco-integrated humans.

Future Scenario 4: Rituals as Drivers for Change

A collective transition toward systemic change is ongoing and based on rituals allowing everyone to participate. The position of the sun and a hot climate define human activity throughout the day. Rituals help settle communities into new modes of living while companion species help humans engage in educative multispecies rituals. For example, humans are reflecting on the value and well-being of insects during quiet morning sessions. Human societies have divided into small groups gathering on the streets which now function as urban outdoor living rooms where cooking and decision making is done collectively.

Workshop Outcome: Participant Feedback

After the workshop, we conducted semi-structured interviews with five participants of which two also took part in workshop preparations. Our objective was to obtain learnings and identify challenges. A camp member of the documenting group conducted the interviews (one of the authors). The documenting group was responsible for recording, reporting and detailing experiences and results of camp events. In interviews, we asked questions about participants' personal and creative experiences during different parts of the workshop; about challenges encountered; and the usefulness of such a workshop in an environment with closer proximity to multiple species. The questions aimed at evaluating the workshop and discovering participant learnings and attitudes. Due to the open environment and character of the scout camp, three interviews were held hiking from the workshop's location to the main camp through a forest. During interviews, privacy was warranted by distancing interviewees from other participants hiking nearby. The remaining two interviews were carried out sitting outside the scout camp, to perceive opinions on the workshop in a more secluded and less hasty manner. Interviews ranged from six to 20 minutes. We live-recorded them, transcribed audiotapes and conducted a thematic analysis. Through the analysis, two topics arose; (i) envisioning multispecies futures, and (ii) belonging to a collective. In addition, we present feedback related to (iii) challenges and (iv) future thoughts.

Part 1: Exploring Nature and Envisioning the Multispecies Future

The first workshop session was centered around introducing the framework and atmosphere. While multiple participants mentioned the session as a starting point for initiating thoughts, and as a way of getting warmed up, the results were highlighted as varied. One participant mentioned that she "couldn't bring what she found" (Participant 5, 2022) into her final proposal. Another remarked that the initial exploration and collective collage creation made a helpful guideline for the later scenario-making: "I found some flowers and then I was putting them in the pine cone together and then I was like, oh, this looks kind of like architecture that is flourishing.... Then (the final scenario) became kind of a growing, flourishing city" (Participant 1, 2022). The participant mentioned that going out into nature and anchoring the natural elements with a sentence greatly inspired her to think about what kind of future she would want; and helped her create a basis for the following activities. Another participant mentioned that the natural elements in the environment inspired multiple ideas building on each other. "I grabbed a stock of grass with the little seeds tucked on the top and like the whole root system came out and that system is like seeing these three different parts, the roots and then the stock and then the seeds at the top. It was like this kind of three elements of connectivity" (Participant 3, 2022).

On the one hand this last quote illustrates how hands-on engagement with nature helped participants shift perspectives and consider phenomena novel to them. On the other hand the quote illustrates the challenge of balancing not entering more harmful or even extractivist relationships with the environment; as pulling out grass with roots kills the plant. From an eco-ethical perspective assessing all living organisms as equal and having the same rights to live and thrive, we could claim that removing plants and other natural elements from the living surroundings of the workshop was wrong, un-necessary and destructive. Furthermore, such activity seems to highlight (rather than challenge) the designers position of power and privilege. But, from a biological perspective such behavior might also 'just' reflect a more than million year-old ecosystem dynamic, as many (if not all) species use and modify other species for food, tools, habitat design etc. Birds eating insects and building nests of plants, beavers chopping down trees to construct dams etc. In this light, a designers' use of a plant to envision – and maybe later, work on realizing – environments where more-than-humans thrive, might be seen as okay, maybe even important.

To sum-up this dilemma, a workshop activity involving elements from nature can be assessed very differently depending on perspective, and risk to promote destructive behavior. In hindsight, we find it therefore appropriate to reflect on respectful ways of approaching the local multispecies context when locating a workshop in natural surroundings, e.g., by starting it with sharing perspectives on the issue, and agreeing on common guidelines.

Part 2 and 3: Belonging to the Collective

An interesting take on the workshop was the notion of belonging mentioned by a participant talking about the collective collage and diary creation: "It makes you feel like your ideas make sense. You are not like, alone... you know, and it's valued if you share it. So if you see another person is also thinking ... from out of nowhere the same way ... you feel like: I belong to this group" (Participant 4, 2022). The same participant reflected on her work, and how the workshop helped evaluate her research and thinking, which in turn made the feeling of belonging stronger: "It's working or reading all the time in your room, or thinking about (the topic of multispecies cities) sometimes. At least (during the workshop) I felt like: 'What am I doing, like, does it make sense or doesn't it?'" (Participant 4, 2022). The participant used the workshop activities as gauging points to aid in her thinking. Another participant said the social aspect of the workshop helped bring out new ideas and inspiration for future work. "The level of creativity that we were able to go to collectively; it seemed like it was very free. Just the whole thing felt very free" (Participant 3, 2022).

Challenges of the Workshop

Although the overall judgment of the workshop was described as "pleasant", "fun", and "well prepared", multiple participants stated that the workshop was too short to achieve a notable result. A participant said that the results were not realistic, having preferred the workshop to be over a longer period of time while acknowledging the constraints: "I would say it was a little hasty, but we do not have enough time to do [thorough research]" (Participant 2, 2022). Another participant mentioned that he conducted similar research but within a longer timeframe. Although the participant acknowledged that the limitation of time could also be a helpful factor for committing to ideas: "But I don't think that this kind of a workshop is actually useful in this kind of thinking. It is so useful in, like a longer period of time because it's so free form" (Participant 3, 2022). One participant described challenges related to the continuation of the workshop, expressing worries of "having to do something with it" (Participant 5, 2022) afterwards. A point shared by another participant who finds it difficult to integrate the generated ideas into her architectural work. "I always feel like I can't document them. Like I'm not doing anything with these ideas" (Participant 4, 2022).

Future Thoughts

An idea expressed for future workshops was to match activities to the people participating. A feedback highlighting that this particular workshop worked well due to the participant profiles. "So if it's like less designerly people, they need more structure, and they might need a bit more help with these steps. But because we are all creatives, we know how to deal with uncertainty. And that's why this worked really well as we knew how to deal with this process of design. So I guess for this audience, it was really good. But if you have an audience like decision makers. You would probably want to design the workshop a bit differently depending on the audience and their needs" (Participant 1, 2022).

Authors Reflections on Workshop Design and Outcome

Below, we would like to share the learnings obtained and challenges identified from conducting and evaluating the workshop. First, we acknowledge the profiles of participants (artists, designers, students, scientists) bringing unique viewpoints to the table, leading to a rich exchange of ideas that we believe had a strong impact on scenarios developed, and the ability to engage with more-than-human perspectives. Having stated this, we now present our learnings.

Three Learnings: How did the Workshop Help Envision Multispecies Futures?

1. *Challenging Human-centered Design*

From one point of view, a workshop based on classic design thinking seems creative and integrative. From another, it might stage participants as being 'in control' of ideas, materials, and scenarios thus representing human exceptionalism. After all, a designer's act reflects a strategy of mastering; a designer observes something, then tries to extract an essence and reproduce or reinvent such essence in a design, i.e., mastering (Negrotti, 2001). As our ambition was to envision multispecies futures in which humans did not put themselves on top of the species hierarchy (as design masters), we hoped that a workshop location in nature, and integration of nature-based inspiration, could stimulate a decentering of the human in a concrete and individually sensed way, compared to workshopping in urban/indoor environments. Not only is a natural environment recognised for enhancing creativity in design processes (Plambech & Konijnendijk Van Den Bosch, 2015). Also, the being-with a living environment might stimulate participants to be species-inclusive 'co-commoners' (Haldrup et al., 2022) rather than human-centered 'design masters' when envisioning multispecies futures. But, as elaborated on above, the potential of locating a workshop in nature also includes risks such as entering harmful relationships with the environment; as using natural elements might mean to destroy habitats kill more-than-human organisms.

Integration of living environments in the workshop did seem to stimulate first ideas (in collages and diaries) and then the creation of scenarios embracing more-than-human needs. In this relation, conducting an abstract task of 'imagining multispecies urban life in 2040' might have helped participants escape habitual ways of thinking, and opened minds to include needs and rights of more-than-human organisms. As all tools of the workshop aimed for storytelling rather than technical construction, participants might also have felt more free to invent radical scenarios that did not rely on current human-centred technical abilities (i.e., habits) of e.g., building, getting materials to work together etc. Also, participants were encouraged to base stories on inspiration from nature and each other, rather than on a specific urban environment, which might have supported an escape from everyday 'practical' thinking. In this relation, we find that a variety of open yet simple frameworks (such as a diary format, a collage, a storyboard) combined with rather basic tools (cardboard, pencils, natural artefacts) might support transgressive imaginations of future scenarios which are not 'stuck' in e.g., a computer's technical capabilities and related (human-centred) habits and perspectives. Also, simple frameworks might empower participants to create multispecies scenarios without getting overwhelmed by complexities of urban biodiversity and ecosystem thinking (Canepa et al., 2022; Zari, 2018).

We further find that the silent individual explorations of surrounding nature might have helped participants engage with the tasks and each other in a more sense-curious and less anthropocentric state of mind compared to a workshop kicked off in an indoor room where PowerPoint presentations and other similar activities are often expected. We experienced, that the initial reflection time where participants 'just' sensed and explored a living multispecies space which they were invited to be quiet in and relate to did set a special workshop mood, with tones of seriousness, reverence, mutual respect and curiosity - which we believe was fruitful for participants ability and curiosity for spotting and incorporating other species' needs and right to be. In hindsight, an explorative collection of inspiration from the surrounding nature by 'just' sensing (listening, viewing, touching, smelling) rather than extracting natural artifacts might have stimulated an even-more respectful and less anthropocentric state of mind.

2. *Flexible Perspectives*

Participants proved capable of envisioning alternative scenarios for co-living with more-than-human species in ways not staging humans at the top of the species hierarchy. This might be related to nature's impact on human creativity; such as stimulation of new ideas, playfulness, flexible shift of perspectives, and viewing a problem from different angles (Plambech & Konijnendijk Van Den Bosch, 2015). As the task was to imagine multispecies co-living, the importance of being flexible concerning the shift of perspectives (e.g., from a human to an insect) and seeing challenges from different angles (e.g., designing an urban habitat for several species) cannot be underestimated.

3. *Sharing Hope*

The workshop seemed to foster a sense of shared purpose and hope. We wonder, if the initial story about humans living in harmony with other species presented through spoken word and live music, in combination with surroundings characterised by the absence of urban disturbances (sensed through e.g., the sight of high-rise buildings, hearing traffic, smelling air pollution) might have helped to 'dare' keep up an atmosphere of optimism and create hopeful species-inclusive visions in a time otherwise often characterised by crisis and despair.

All in all, the workshop seemed to tap into a range of more-than-human perspectives, enabling the development of imaginative positive scenarios for future co-living with other organisms. As such, the workshop might be an exception proving the general picture of a (western) world that "simply has not developed a habit of imagining and sharing the long-range scenarios at issue in any concrete way" (Durfee & Zeiger, 2017, p. 45). This said, our learnings also point towards challenges, some being directly related to the location of a workshop in nature.

Three Challenges: What Challenged the Workshop?

1. *Nature location includes risks for more-than-human organisms, while an off-line atmosphere might block for fruitful digital activity, and scenario afterlife*

Planning a workshop located in nature, and with an ambition of extracting inspiration from the living surroundings into workshop activities, includes the risk of promoting a destructive human behavior. Depending on perspective, extracting elements from nature can be assessed as either harmful, and destructive, or as fruitful and okay. Thus, we find that a worksop located in nature should include reflections on ethical issues and ecological risks. Such reflections could be made in the workshop planning phase where guidelines for location and activities could be sketched, and in the initial phase of the workshop where more-than-human friendly guidelines could be co-created to provide a clear framework, and minimise more-than-human risks.

Another challenging aspect relates to the workshop promoting offline materials and use of various senses rather than the usual (artificial) tools of designers such as cell phones and computers. Setting such an overall offline atmosphere might have prevented participants from e.g., finding relevant knowledge online, or digitizing presentations on computers, which on the other hand could have qualified scenario content, and made it easier to take home learnings and ideas from the workshop. Also, cell phones could have been used to 'harvest' visual impressions from the natural surroundings through photographing and video filming, thereby minimizing risks of destructive behavior, and enabling participants to bring home essential inspiration from the workshop afterwards. Thus, working offline in nature seems to imply challenges concerning more-than-human risks, and re-communication afterwards with also the large size of cardboard posters being problematic to 'squeeze' into e.g., a PowerPoint or Word document. Perhaps, an offline atmosphere for a workshop prioritizing participants immersing in nature makes it challenging for participants to engage in continued multispecies envisioning afterwards. This being said, we find that the nature location and off-line atmosphere of the workshop did constitute a unique and fruitful framework for participants, which supported them in working visionarily on radical and more-than-human integrative futures – as elaborated on above.

2. No Preparation and Short Time Frame

Workshop tasks were not shared with participants beforehand, and no preparation was required. This meant that participants showed up unprepared. Engaging unprepared in tasks about imagining multispecies futures might be difficult compared to having had the chance to reflect on potentials and ideas beforehand. In addition, the workshop was done in just three hours which several participants assessed as being a short time frame, creating a hasty atmosphere resulting in scenarios that were not as detailed and fleshed out as participants wanted them to be.

3. Abstract Tasks

The workshop tasks could have been too abstract for some. Participants had no 'real city' to relate to, no specific challenges to overcome, and very few guidelines on what to develop. Wall posters suggested diverse species to integrate and a storyboard was intended to lead participants' imaginations into a presentable framework kicking off their scenario-making. But the workshop format depended heavily on participants' ability to transform abstract tasks, and impressions from nature, into shareable scenarios. A related limitation was to bridge ideas to the everyday working life of participants; with participants finding it challenging to use ideas afterwards.

Conclusion

We conclude that participants did engage in ideation, imaginative storytelling, and scenario making, with flora, fauna and fungi from the surroundings becoming sources of inspiration; and that these engagements led to (or, at least, did not block) envisionings of multispecies futures. We found that the scenarios developed represented radical and more-than-human centered urban environments compared to current city typologies as seen in e.g., Den Bosch and Copenhagen (where we, the authors, respectively live). In the co-created scenarios, humans were presented as more equal to other species compared to current urban biodiversity strategies in our cities. Furthermore, scenario presentations of 'a city' related more to natural elements such as trees, water flows and seasonal rhythms than to 'practical' construction techniques or hard urban infrastructure such as streets, squares and buildings.

All in all, we found that the workshop enabled participants to develop radical multispecies scenarios. In line with theory presented in this chapter, we suggest that a location in nature and integrations of natural elements might have positive effects on creativity, and integrations of more-than-human perspectives, though these are indicative suggestions. In addition, we suggest that workshop activities setting an atmosphere of hope and optimism (through stories and music), might help participants 'dare' create hopeful species-inclusive visions in times otherwise characterised by crisis and despair. Finally, we identified several challenges calling for improvements. Challenges related to the workshop location in nature potentially disturbing more-than-human organisms, and to the extraction of elements from nature, as such practices can be assessed as either fruitful or destructive, depending on perspective. A condition highlighting the importance of providing or co-creating guidelines for participant interaction with other living organisms, at a workshop located in a natural environment. Challenges also included lack of participant preparation and a short time frame, (too) abstract tasks and an off-line atmosphere blocking digital activity, and perhaps, scenario afterlife.

In conclusion, the workshop *Multispecies City of the Future* exemplifies one approach to explore the evolving nature of more-than-human design. Through engaging with designers and researchers already familiar with more-than-human perspectives various workshop methods were tested. On this basis, learnings and challenges point to development potentials. We suggest that further research and tests within this field could involve urban actors such as planners and decision-makers working with urban development but often lacking the time and opportunity to engage in radical scenario making.

References

Altarriba Bertran, F., Buruk, O. "Oz", & Hamari, J. (2022). From-the-wild: Towards co-designing for and from nature. *CHI Conference on Human Factors in Computing Systems Extended Abstracts*, 1–7. https://doi.org/10.1145/3491101.3519811

Beatley, T., & Newman, P. (2013). Biophilic cities are sustainable, resilient cities. *Sustainability*, *5*(8), 3328–3345. https://doi.org/10.3390/su5083328

Boer, L., & Donovan, J. (2012). Provotypes for participatory innovation. *Proceedings of the Designing Interactive Systems Conference*, 388–397. https://doi.org/10.1145/2317956.2318014

Botero, A., Dolejšová, M., Hee-jeong Choi, J., & Chewie. (2023). Open Forest. In *Creatures Co-Laboratory Catalogue* (pp. 71–84).

Canepa, M., Mosca, F., Barath, S., Changenet, A., & Hauck, T. E. (2022). Ecolopes, beyond greening. A multi-species approach for urban design. *Agathòn 11*, 238. https://doi.org/10.19229/2464-9309/11212022

Celermajer, D., Schlosberg, D., Rickards, L., Stewart-Harawira, M., Thaler, M., Tschakert, P., Verlie, B., & Winter, C. (2021). Multispecies justice: Theories, challenges, and a research agenda for environmental politics. *Environmental Politics*, *30*(1–2), 119–140. https://doi.org/10.1080/09644016.2020.1827608

Chamberlain, A., Crabtree, A., Rodden, T., Jones, M., & Rogers, Y. (2012). Research in the wild: Understanding "in the wild" approaches to design and development. *Proceedings of the Designing Interactive Systems Conference*, 795–796. https://doi.org/10.1145/2317956.2318078

Collins, T. M., Goto, R., & Edwards, D. (2018). A critical forest art practice: The Black Wood of Rannoch. *Landscape Research*, *43*(2), 199–210. https://doi.org/10.1080/01426397.2017.1318119

Dorst, K. (2015). Frame creation and design in the expanded field. *She Ji: The Journal of Design, Economics, and Innovation*, *1*(1), 22–33. https://doi.org/10.1016/j.sheji.2015.07.003

Durfee, T., & Zeiger, M. (Eds.). (2017). *Made up: Design's fictions*. Art Center Graduate Press.

Elsden, C., Chatting, D., Durrant, A. C., Garbett, A., Nissen, B., Vines, J., & Kirk, D. S. (2017). On Speculative Enactments. *Proceedings of the 2017 CHI Conference on Human Factors in Computing Systems*, 5386–5399. https://doi.org/10.1145/3025453.3025503

Frow, P., Nenonen, S., Payne, A., & Storbacka, K. (2015). Managing Co-creation Design: A Strategic Approach to Innovation. *British Journal of Management*, *26*(3), 463–483. https://doi.org/10.1111/1467-8551.12087

Hakio, K., Dolejšová, M., Mattelmäki, T., Choi, J. H., & Ampatzidou, C. (2022). Following seals and dogs: Experimenting with personal dimensions of transformative design. *Proceedings of the Participatory Design Conference 2022 – Volume 2*, 167–172. https://doi.org/10.1145/3537797.3537869 https://doi.org/10.1088/1755-1315/1122/1/012008

Haldrup, M., Samson, K., & Laurien, T. (2022). Designing for multispecies commons: Ecologies and collaborations in participatory design. *Proceedings of the Participatory Design Conference 2022 – Volume 2*, 14–19. https://doi.org/10.1145/3537797.3537801

Haraway, D. J. (2016). *Staying with the Trouble: Making Kin in the Chthulucene*. Duke University Press. https://doi.org/10.1215/9780822373780

Hopkins, R. (2021). *From What Is to What If*. Chelsea Green Publishing.

Keune, S., Ilgun, A., & Ludwig, C. (2023). I.N.S.E.C.T—Summercamp: Developing multispecies design perspectives, practices, and discourse through co-creating (in) community. In M. Mostafa, R. Baumeister, M. R. Thomsen, & M. Tamke (Eds), *Design for Inclusivity* (pp. 701–715). Springer International Publishing. https://doi.org/10.1007/978-3-031-36302-3_52

Latour, B. (2018). *Down to Earth: Politics in the new Climatic Regime* (English edition). Polity Press.

Martin, L., White, M. P., Hunt, A., Richardson, M., Pahl, S., & Burt, J. (2020). Nature contact, nature connectedness and associations with health, wellbeing and pro-environmental behaviours. *Journal of Environmental Psychology*, *68*, 101389. https://doi.org/10.1016/j.jenvp.2020.101389

Márquez Segura, E., Turmo Vidal, L., Rostami, A., & Waern, A. (2016). Embodied sketching. *Proceedings of the 2016 CHI Conference on Human Factors in Computing Systems*, 6014–6027. https://doi.org/10.1145/2858036.2858486 https://doi.org/10.1890/ES13-00075.1

Mitro, M. (2023). Open urban forest. In *Creatures Co-Laboratory Catalogue* (pp. 85–92). https://doi.org/10.1016/j.cities.2018.06.007

Negrotti, M. (2001). Designing the artificial: An interdisciplinary study. *Design Issues*, *17*(2), 4–16. https://doi.org/10.1162/07479360152383750

Nystrup Lund, L. (2024). *Biocities of The Future: A study of urban developers and protesting citizens unfolding ideas about biodiversity in the Copenhagen case Fælledby / Lærkesletten* [Phd thesis]. Royal Danish Academy – Architecture, Design, Conservation. https://doi.org/10.26687/archnet-ijar.v8i2.436

Participant 1. (2022, August). *Multispecies Future Scenario Workshop* [Personal communication].

Participant 2. (2022, August). *Multispecies Future Scenario Workshop* [Personal communication].

Participant 3. (2022, August). *Multispecies Future Scenario Workshop* [Personal communication].

Participant 4. (2022, August). *Multispecies Future Scenario Workshop* [Personal communication].

Participant 5. (2022, August). *Multispecies Future Scenario Workshop* [Personal communication].

Plambech, T., & Konijnendijk Van Den Bosch, C. C. (2015). The impact of nature on creativity – A study among Danish creative professionals. *Urban Forestry & Urban Greening*, *14*(2), 255–263. https://doi.org/10.1016/j.ufug.2015.02.006

Rind Christensen, P., Junginger, S., Design School Kolding, & Syddansk Universitet (2014). *The Highways and Byways to Radical Innovation: Design Perspectives*.

Rosa, H. (2019). *Resonance: A Sociology Of Our Relationship to the World* (J. C. Wagner, Trans.). Polity.

Sanders, E. B.-N., & Stappers, P. J. (2008). Co-creation and the new landscapes of design. *CoDesign*, *4*(1), 5–18. https://doi.org/10.1080/15710880701875068 www.unep.org/resources/making-peace-nature

United Nations, D. of E. and S. A. (2019). *World Urbanization Prospects: The 2018 Revision, Population Division: Vol. ST/ESA/SER.A/420*. United Nations.

Wilde, D., Underwood, J., & Pohlner, R. (2014). PKI: Crafting critical design. *Proceedings of the 2014 Conference on Designing Interactive Systems*, 365–374. https://doi.org/10.1145/2598510.2598603

Zari, M. P. (2018). *Regenerative Urban Design and Ecosystem Biomimicry* (1st ed.). Routledge. https://doi.org/10.4324/9781315114330

Zari, M. P., & Hecht, K. (2019). Biomimicry for regenerative built environments: Mapping design strategies for producing ecosystem services. *Proceedings of the TensiNet Symposium 2019*, 20.

15

PEERING THROUGH TIME

Harnessing Anticipation in More-than-Human Design

Camilo Sanchez, Anton Poikolainen Rosén, Antti Salovaara, Felix A. Epp and Tim Moesgen

The Innate Connection Between Futuring and More-Than-Human Design

The emerging field of more-than-human design has future-oriented underpinnings. It asks questions about the coexistence of humans and other species in possible futures and explores transformations that involve humans, technology, and a world of vibrant and lively matter. More-than-human design pushes designers to consider the entire globe as a stakeholder, emphasising feedback loops and interactions in large systems (Forlano, 2017). The future well-being of the planet depends not only on humankind's past and present actions, but also on dominant historical narratives and acknowledgement of neglected pasts, as well as on the inherent power of designerly intentions to evoke engaging and actionable future imaginaries.

Factors such as the ones listed above shape humans' relationships with the more-than-human world and its temporalities. They also connect more-than-human design with anticipation. To anticipate means to acquire knowledge from exploring the future and using this knowledge to act in the present (Poli, 2017). Therefore, anticipation is not about predicting more-than-human futures, but instead, it is about recognising how the assumptions about those futures are constructed. In this book chapter, we aim at identifying the missing procedures that would allow for a more thorough understanding of more-than-human perspectives and temporalities. Utilising this understanding to shape present actions makes anticipating an influential process (Miller, 2018b).

A substantial amount of work within more-than-human design has focused on speculative methods to generate scenarios that challenge anthropocentric worldviews and imagine radically different futures (Forlano, 2017; Tyszczuk, 2021). However, a challenge arises when speculation is done without being conscious of its embedded assumptions. Consequently, designers may unintentionally incorporate their worldviews and conceptions of desirable futures, without explicitly stating the process by which these speculative scenarios were formulated. This can be problematic because it may evoke misguided imaginaries and disregard essential viewpoints. Integrating an anticipatory approach in more-than-human design may help designers avoid such pitfalls and unpack the complex interactions of factors and motivations grounding future visions.

Inherently, engaging in more-than-human design endeavours can provide valuable learnings from tangible projects that tackle complexity and cultivate pluriversal perspectives. By

DOI: 10.4324/9781003467731-18

"pluriversal" (Escobar, 2018), we refer to committed attention to multiple worldviews in the anticipation processes. In contrast to pluralistic approaches, which tend to focus on human variance within the same worldview, pluriversal approaches forefront the co-existence of multiple ways of knowing and being in the world, also of other species. This requires acknowledging the limitations imposed by our human lenses. In this task, both anticipation and more-than-human design can collaborate to achieve a more comprehensive understanding of planetary futures.

In the following sections, we analyse how recent works in more-than-human design (Coulton & Lindley, 2023; Liu et al., 2018; Romani et al. 2022; Heitlinger et al., 2021) have adopted anticipatory approaches and addressed complexity, unveiled uncertainty and created new knowledge. We then outline the potential that Futures Studies methods offer for the anticipatory approach in more-than-human projects. We exemplify the adjustment of Futures Studies methods to more-than-human design by describing a workshop method that we have developed (Epp et al., 2022). Finally, we discuss other ways by which the anticipatory approach can be integrated into more-than-human design.

Anticipation in More-Than-Human Design

Given the complexity of the planetary ecosystem, the anticipation of possible futures and the design actions embedded in them benefit from thoughtful examination. A myriad of possible futures can be envisioned – utopian, dystopian, and all in between – all of which have merits to motivate their investigation. However, how should one determine which futures deserve closer attention? While a future's perceived likelihood or its desirability may be useful metrics for decision-making, anticipation's value resides in the insights gained from reflecting on the reasons why certain futures are contemplated (Chia, 2004; Miller, 2018b).

Among the different forms of futuring, more-than-human design's futures have especially reimagined humans' relationships with the environment. Reimagination is particularly evident in speculative methods that, for example, make use of prototyping to anticipate richer relationships between humans and fungi (Liu et al., 2018), understand the interrelations of IoT devices (Coulton & Lindey, 2023), participate in multispecies debates (Romani et al., 2022) and reconfigure food as more-than-human commons (Heitlinger et al., 2021). The anticipatory approach in these works not only illustrates the complexities of more-than-human futures but also highlights the intricacies – power dynamics, normative assumptions, attention to multi-species interests – that appear when humans seek to pursue their goals which however involves frictions (Pierce, 2021) with the environment.

The noteworthy form of anticipation in the studies above is their analysis of a possible future (i.e., an imaginary) in a way that has uncovered networks of interrelationships between humans and the environment. Through speculation, researchers have anticipated complex interdependencies between organic and digital environments, relations within such systems, and transformations that have implications from one system to others. This newly gained awareness has foregrounded aspects that are relevant to consider already in the present world. Subsequently, the papers written from these studies have described how phenomena in possible futures have implications for our actions in the present.

For example, Coulton and Lindley's (2023) study on networks of human and non-human actors in IoT systems has uncovered imagined futures' environmental, logistical and ethical implications that should be attended already in the present-day design considerations. Their speculative ontography has revealed hidden actors in IoT networks and questioned the notions of agency when this network is no longer controlled by humans. Heitlinger et al. (2021), in turn, used the blockchain to materialise and enact possible political, legal and social structures that are necessary for equitable interspecies food to be established. These examples used existing technologies to enact the complexities of more-than-human futures in the present. This method aligns with Guston's (2013) description of prototyping as an anticipatory practice whereby interdependencies and inherent complexities of a future can be materialised and therefore anticipated. These examples illustrate how more-than-human imaginaries can be studied with current means so that their scale and relationality can be appreciated, and actions can be enabled towards them.

When considering more-than-human life, we attend to lifespans and forms of survival different from ours (Liu et al. 2018). Engaging with more-than-human perspectives also reveals uncertainties that transcend diverse temporalities. Thus, exploring these uncertainties implies considering prolonged periods, such as twenty years or more into the future (Sardar & Sweeney, 2016), as well as reflecting on the continuity of historical contexts (Bendor et al., 2021), such as the "200-year present" extending from 100 years in the past to 100 years in the future (Boulding, 2017). Understanding what might have been and remains unknown is an anticipatory approach that pinpoints overlooked perspectives, mutable factors and direct action (Adam & Groves, 2011).

Speculation can be complemented with other foresight methods, which allows for broader exploration of timeframes that we are uncertain about, or which are challenging to relate to. Analysis of the evolution of trends, for example, can inform about the sequential unfolding of events (Romani et al., 2022) and facilitate the construction of holistic future archetypes such as growth, collapse, discipline and transformation (Dator, 2019). While different human cultures have different concepts of time – linear, circular, unidirectional and static (Bell, 2009) – we should be mindful that events also precede humans and will postdate us. This recognition enables us to design for temporalities that arise from environmental transformations deeply rooted in the interconnectedness of long-evolving systems.

Addressing multiple perspectives and temporalities requires abilities to contest implicit assumptions. The more-than-human design examples discussed here have been reflexive on how futures are constructed, and this has allowed them to consider synergies between pluriversal futures. For example, designers may build "provotypes" to provoke reflection on the implications of designing for multiple species (Romani et al., 2022); employ feminist and speculative ethics of care to examine how different interspecies relations dynamically shape conceptions about relevance and value (Heitlinger et al., 2021); and use new technologies to embed collective intraspecies knowledge in future-oriented reflection (Liu et al., 2018).

Above we outlined how elements of anticipation are already implicitly present in more-than-human futuring work. We now move to describe established Future Studies methods that can amplify the anticipatory capacities of more-than-human design. Many designers are familiar with scenario development (Schoemaker, 1995; Van der Heijden, 2005), but it is less frequently discussed how these scenarios can be combined with other methods to avoid falling prey to narrow-minded and short-sighted thinking. We have selected the following methods based on their alignment with the complexity, temporalities, collaboration, and decolonization aspects inherent in more-than-human design.

Menagerie of Postnormal Potentialities (Sardar and Sweeney, 2016) is an umbrella term for metaphors that can remind designers of the escalating pace of change and the possibilities of impactful events. "Black swans" are uncertain events of undetermined probability; "black elephants" are predictable but deliberately ignored events; "black jellyfish" are acknowledged critical events with outcomes of unforeseeable scale. These metaphors can help recognise human biases and blind spots when imagining more-than-human futures. For example, attending to "black swans" reminds us of the increasing evidence of intelligence and culture in a broad range of animal species (Whiten & Van Schaik, 2007) and helps us question the solipsistic beliefs about human uniqueness. The notion of "black jellyfishes" allows us to recognise how more-than-human agencies beyond our control shape the future, e.g., how the rising sea temperatures lead to the extinction of coral reefs, consequential loss of marine diversity, and the coral reefs' role as natural sea barriers (Kolbert, 2014). An awareness of "black elephants", finally, helps identify anthropocentric assumptions that need more attention and a broader scope: deforestation's effects, for example, are well-known and visible, yet it is permitted to continue at an alarming rate on a global scale. The awareness of postnormal potentialities such as these allows us to approach the complexities of the more-than-human world with humility, adaptability and a broader scope of consideration.

Backcasting (Robinson, 1982) is a method where participants adopt a desirable future as their starting point and reason what it would take to plausibly arrive at that future, working backwards to the present. Backcasting can identify future challenges and implications of co-existing more-than-human networks and has therefore been popular in sustainability research in which analysis of desirable sustainable futures requires careful reflections (Bibri, 2018). Backcasting can help us realise that desirable futures are not universal, and how therefore historic insensitivities and injustices – including more-than-human ones – have led the planet to the unsuitable circumstances of today. Vice versa, backcasting is also amenable to inclusive analyses that involve non-experts as well as non-Western and marginalised communities in the construction and evaluation of more-than-human futures. This can be done by orchestrating processes that gather multiple perspectives and counterfactual narratives (Lewis, 1973) whereby under-considered cultural narratives and values can be integrated into the co-creation of a collectively desired future. Backcasting is cognitively demanding, but this can be alleviated by making use of artefacts and narratives that render the complexities of plausible causal influence chains more comprehensible (Wilde et al., 2021). This helps make the process more inclusive and engaging for participants.

The Three Tomorrows (Sardar and Sweeney, 2016) looks at complexities and uncertainties from the perspective of degree, space and temporality. These perspectives appear on a different scale on the Three Tomorrows – extended present, familiar futures and unthought futures. As an example, more-than-human futuring on Artificial Intelligence's consequences can situate its thinking in the "extended present" to analyse current socioenvironmental concerns such as data bias and AI energy consumption. Here, we know the socio-environmental impact of AI to a certain degree; the issues of data bias and energy consumption operate on a space of collective acknowledgement; and the worry for these concerns extends from the near-past to the near future. "Familiar futures" outlines the development of current concerns and reflects on the socio-political expectations of AI. This second tomorrow explores data bias and AI's environmental footprint consequences separating the plausible ones from unlikely ones on a timeframe we can predict or are familiar with due to our collective future imaginaries. Lastly, "unthought futures" consider yet unknown AI socio-environmental implications, including changes in values, threats, environmental catastrophes and socioeconomic instability. In the third tomorrow, we do not know what the implications are nor the impact level of known events. It represents a space where we question our assumptions about AI and consider critical uncertainties; for this tomorrow, temporality is extended to 20 years in the future and beyond. By projecting a more-than-human endeavour across the three tomorrows, designers engage in an insightful exploration of complexity and uncertainty, their degrees of intensity, the knowledge space that projects them and the different temporalities of the ever-emerging more-than-human world.

Causal Layered Analysis (Inayatullah, 1998) was created to enable a critical reflection on the discourses, power structures and history that generate normative futures. The objective is to contest dominant assumptions and incorporate other ways of knowing, thereby decolonialising future narratives. This goal is addressed by critically pondering trends, socio-political issues, worldviews and myths. For example, Causal Layered Analysis can be used to deconstruct the metaphor of a "network", which often refers to complex interconnected systems – both natural and artificial. However, an analysis of the "network" metaphor on a "litany" level – statistics, trends and media – may reveal that complexity may be represented in an inherently technocentric way (August, 2022). Looking at the metaphor at the second, sociopolitical and structural layer, reveals how networks are often described by agencies that wield power, legitimacy and authority, thereby carrying implicit political ambitions, even if the result comes close to obscurity. At the third, worldview level, we may unravel underlying assumptions about efficiency and infinite growth. Finally, moving on to the fourth level consisting of myths, we might recognise that ecosystem networks have emotional underpinnings. We may notice a dichotomy between violence and harmony or detect interactional values such as balance and respect in symbiotic structures. By inviting designers to exercise a critical approach and reflection using several layers, causal layered analysis can be useful in promoting non-Western perspectives, which is a central concern in more-than-human design.

Futures Workshops (Jungk and Müllert, 1987) are originally an action research method. It aims to democratise futuring through collective, democratic and inclusive stakeholder engagement. To lower the participation threshold, these methods employ multimodal techniques to encourage active engagement and participation: sticky notes, performances and visual representations (Alminde & Warming, 2020). A characteristic feature of the methods that build on future workshops is the use of three phases (Glenn, 2009; Candy & Dunagan, 2017; Epp et al., 2022): 1) critique or current issues, 2) fantasy or futuring, and 3) implementation or practical steps. Futures workshops may seek pluriversality by giving "voice" and agency also to non-human organisms and systems. This can be done by electing representatives for non-human actors (Romani et al., 2022) or by including live-action roleplaying activities that enact futures of more-than-human scenarios (Heitlinger et al., 2021). In this chapter's later sections, we will present how one futures workshops method – Future Ripples (Epp et al., 2022) – can be adapted to more-than-human design projects.

Lastly, Futures Studies often make use of a heuristic list of important future-contingent factors. This list is called STEEPLE[1] which is an acronym of its contents – social, technological, environmental, economic, political, legal and ethical. Bearing these factors in mind can protect futuring from solely techno-centric, market-driven, Western-dominated, or other biased considerations. STEEPLE can be aligned with the posthuman perspective to overcome artificial divides and tries to forefront connections and complexity (see Ulmer, 2017, and the presentation of the Future Ripples method below).

We introduced the above-presented methods because of their compatibility with a more-than-human design's multifaceted approach. However, not all Futures Studies methods attend to more-than-human design endeavours. The futures cone (Taylor, 1993), for example, depicted in Figure 15.1, has been an influential representation of how futures can be thought about (Gall et al., 2022), but has been criticised for several shortcomings (Howell et al. 2021):

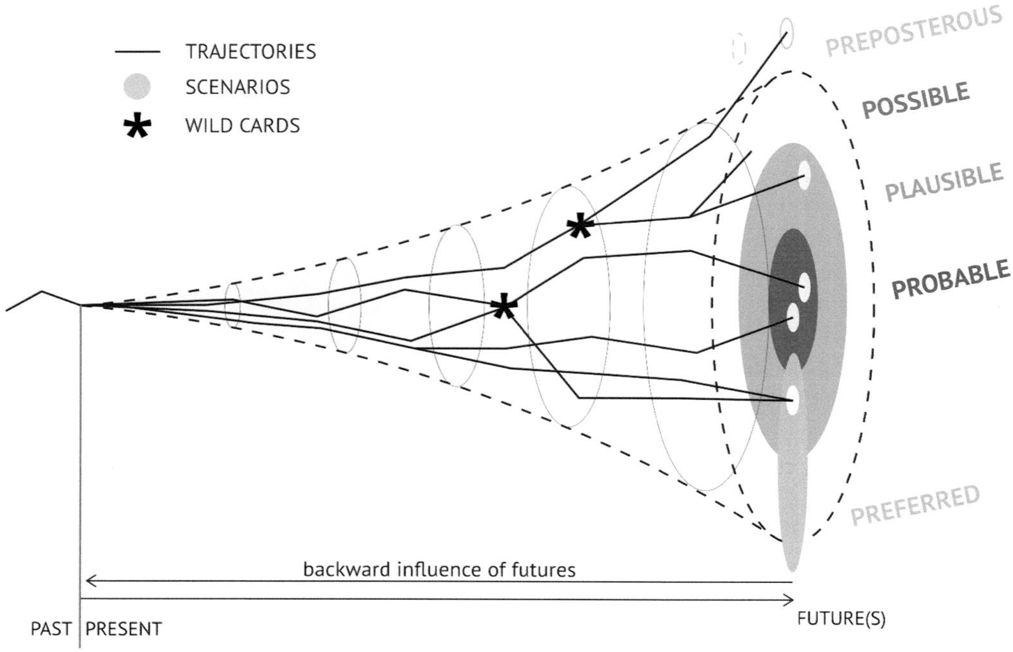

FIGURE 15.1 A simplified visualization of the futures cone, adapted from Gall et al. (2022). While the futures cone effectively portrays concepts such as intertwining trajectories, high-impact wild cards, exponential growing uncertainty, and a porous boundary of possibilities, it does not fully capture the richness of diverse perspectives, overlooked histories, and non-linear conceptions of time (Howell et al., 2021).

the traditional futures cone presentation of possible futures has the present world as its origin and neglects the past, enforces that a linear representation of time is adopted, and categorises futures technocratically, based on their plausibility.

Adopting an anticipatory approach in more-than-human design aids in acknowledging that there is no definitive "right" method for understanding the future. Anticipation is an intricate task with no guarantees as the future can only be known once it unfolds (Poli, 2017). However, the process of anticipation – characterised by introspection on individual assumptions and biases as well as extrospection on the broader social, political, cultural and environmental contexts (Miller, 2018a) – amplifies our awareness of the inherent dynamics and power relations shaping the discourse around future scenarios. This is particularly crucial in more-than-human design where understanding and negotiating such dynamics holds significant implications for the different temporalities that are shaped with the more-than-human world.

Future Ripples Method: A Tool for Bringing Multiple Voices to the Futuring Process

To extend more-than-human designers' opportunities in embracing an anticipatory approach, we present a practical case of adapting the Futures Ripples workshop method (Epp et al., 2022) to a more-than-human design case. The rationale for using this method is its explicit aim of collecting multiple perspectives, tracing asynchronous future trajectories, and enhancing the anticipatory capacity of its participants.

We developed the Future Ripples as an agile futuring method for designers, aiming to support reflection on possible futures within their practice. The method builds on a well-known Futures Studies method, the Futures Wheel (Glenn, 2009). The participants start by identifying weak signals[2], trends, or other future-oriented phenomena, and continue with a discussion on their future relevance from personal or professional standpoints. Through this collaborative exercise, the participants select a starting point whose first-order, second-order and third-order consequences will be then envisioned collaboratively and in parallel. The creation of each ripple consists of silent brainstorming where each participant considers the contents of the preceding layer and generates ideas about immediate consequences from some of them. These ideas are then presented to others, possibly with a requirement of selecting only those that seem intuitively most important. The expanding, concentric ripples that result from this process let participants lay out possible future trajectories (Figure 15.2). To ensure higher quality in this anticipatory process, the participants use STEEPLE as the set of future-contingent factors that must be attended to. To avoid focusing solely on a linear sequence of consequences, it is crucial to consider a variety of alternative paths. This can be achieved by prioritizing the completion of one full ripple of envisioning at a time. Finally, the fact that participants need to justify and discuss why they map a particular event to a ripple provides a way of eliciting and contrasting assumptions about the future. While originally aimed at designers, the simplicity of the method facilitates a collaborative and systemic examination of future trajectories that can encompass a wide range of perspectives.

Adapting the Future Ripples Method for More-Than-Human Design

To illustrate how anticipation might be integrated into more-than-human design, we present how we have used the Future Ripples method for more-than-human concerns. So far, we have used the method for futuring climate wearables, multi-sensory communication, novel smart materials (Epp et al., 2022) and design for the Baltic Sea. Based on these experiences, we discuss two ways the method can be used for – and adapted to – the relational and systemic interactions intrinsic to more-than-human design. Notably, this approach can be beneficial both in projects directly focusing on designing for and with other organisms, and in more conventional projects where it may be important to identify unintended harmful consequences of human intervention in the more-than-human world.

The first possibility for adapting the Future Ripples method for more-than-human design is to use the perspective of a more-than-human entity as the initial prompt. As illustrated in Table 15.1, different types of more-than-human prompts may lead to thinking about potential interventions and their consequences from different perspectives in the anticipation process. These examples highlight the possible diversity of more-than-human prompts, not an all-encompassing typology.

When we used a more-than-human entity's perspective as the initial prompt, we found that it is essential to remember how the scanning process, i.e., the analysis of trends and weak signals, tends to be anthropocentric due to the human-centred nature of information outlets. This requires futuring participants to place extra effort into finding information that covers more-than-human perspectives. However, representing pluriversal ecosystems can be challenging due to limited knowledge of diverse species. This challenge should be seen as an

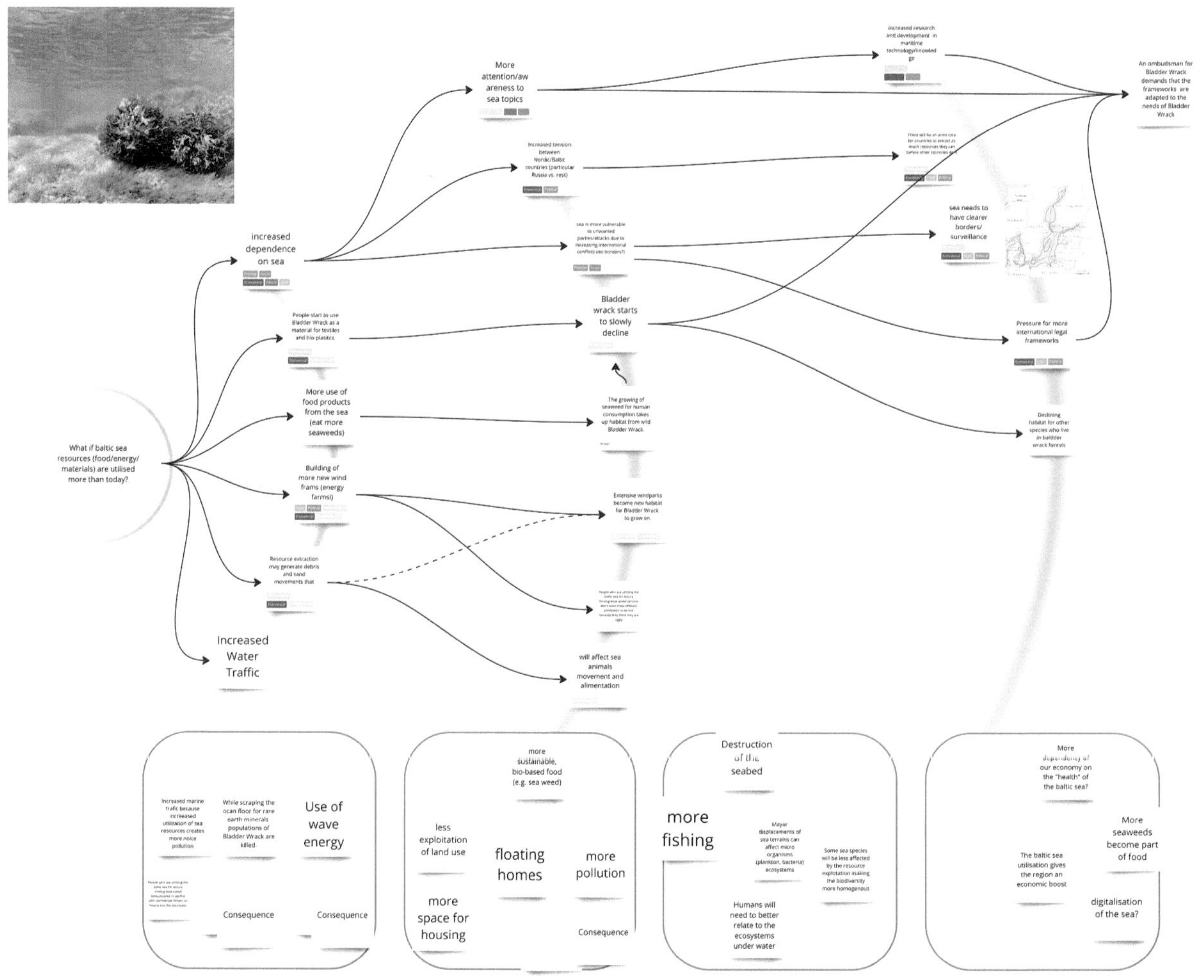

FIGURE 15.2 The outcome from a future ripples session that mapped the consequences of the Baltic Sea's increased resource utilization. In this image, consequences are laid out according to their causal order.

TABLE 15.1 Examples of more-than-human prompts that can be used in anticipatory processes

Type	*Example prompt*
The internal experience of an animal	"You are a coastal porpoise whale that faces increasing navigation problems and has stress symptoms due to noise pollution from ships in the heavily trafficked Baltic Sea Archipelagos"
Technological agency	"A new electrical boat motor is 12dB quieter than an oil-based motor providing the same horsepower"
Meeting the observed need of an animal	"Humans need to decrease noise pollution from traffic in order to not disturb porpoise whales"
Observed challenge for a whole ecology	"Increased water temperatures due to global warming affect the living conditions of all species in the Baltic Sea"
An environmental disruption	"A Norwegian oil tanker runs aground near the Curonian Lagoon biosphere polygon, Lithuania"
A new environmental protection regulation	"Motors that are louder than 90dB are forbidden in the Baltic Sea"

opportunity for reflection, knowledge gathering and uncertainty recognition. Lastly, when addressing more-than-human futures, it is already explicit that we are touching on environmental and ethical issues. As other works have demonstrated, STEEPLE categories can be reconfigured to expand their scope (Dumit, 2014). For more-than-human design, this requires rethinking how the STEEPLE categories could account for more precise and relevant more-than-human factors, such as Cultural + Floral + Faunal + Cybernetic, to give an example.

The second possibility for adapting Future Ripples to more-than-human concerns is assigning a representative to advocate for the interest of a particular species throughout the process. The idea of humans representing the interests of a specific more-than-human entity is established in more-than-human design research (Wakkary, 2021; Light, 2022; Clarke et al., 2019; Romani et al., 2022). Some call this representative a 'spokesperson' to highlight active advocacy of the species. Others call this 'ombudsmen' to highlight that human language – and perception in general – is not necessarily the most efficient way to know another species. Yet others assign 'species diplomats' to shift perspectives to the human effect on the species, thus asking how humans should behave to leave the best impression on the respective species. While being mindful of the limits of assigning an interspecies representative, such representation can help identify indirect consequences of design interventions on a particular species or whole environment.

In our Futures Ripples on the Baltic Sea, we assigned a spokesperson for the bladderwrack seaweed (Figure 15.2). We chose this species since it is a cornerstone species, creating living conditions for a broad range of other species in the ecology. We found that the processes benefited from making the represented species more present through images or other tangible material acting as a constant reminder to consider their perspectives. Additionally, we found that an introductory round where everyone defined their perspective improved this process. It clarified both the position of species represented, as well as the various human viewpoints: subjective (based on personal feelings or opinions); objectivistic (based on facts and evidence); and detached (seeking to maintain a sense of impartiality and neutrality).

When attending to more-than-human perspectives in the futures ripples, the consideration for other species enhanced creativity by requiring us to think from a novel perspective. However, we also believe that our limited knowledge about a given species poses challenges for further ideation. Representing more-than-human perspectives requires comprehensive background research to prevent participants' perspectives from being based on their own experiences and assumptions only.

Anticipation in More-Than-Human Design: Nurturing Relationality, Embracing Ways of Being and Knowing, and Fostering Transparency

In this chapter, we have discussed how more-than-human design might benefit from approaches of anticipation – and vice versa – as the more-than-human perspective intrinsically explores diversity, complexity and uncertainty. Merging more-than-human design with anticipation may broaden our understanding of futures by allowing us to reflect on the meaning of desirable futures and evaluate the alternatives that emerge from futuring processes. To crystallise the crucial aspects of this process, we provide the following considerations for further integration of anticipation within more-than-human design.

Relationality, Cause and Consequence

A central aspect of more-than-human design is the quest to understand how relationality affects emerging phenomena in design. In this context, Future Ripples – and other methods of anticipation – can be used to think of cause and consequence in systems identified in more-than-human design processes. Here, anticipation supports non-linear consideration of causalities, thereby supporting pluriversal possibilities and addressing diverse temporalities, both of which are essential perspectives when managing complex interspecies networks that transcend human lifespans (Adam & Groves, 2011).

Emphasising the Transformative and Experiential

The experience of engaging in an anticipatory process is transformative because it suggests that the present time is malleable and how we consider futures represents how we are in the present (Poli, 2017; Epp et al., 2022). Transformative experiences enable us to develop a deeper sense of empathy towards the more-than-human world, rooted in current realities and enacted through thorough contemplation. This approach aligns with more-than-human design's emphasis on situated knowledge (Haraway, 1988), where the tenet is that knowledge is embedded in and thus affected by the concrete historical, cultural, linguistic and value context of the knowing person. This implies that assumptions about the future emerge from this baggage and anticipation happens from a particular situated perspective. The transformative quality emerges through sharing exchange and discussion of subjective knowledge, facilitating a "cross-pollination" of understanding among workshop participants (Halskov & Dalsgård, 2006).

Externalising Thoughts Facilitates Collaboration

More-than-human design is highly collaborative and co-creative as it emerges through the co-constructive capacities of people, technologies and other species (Wakkary, 2021). This process requires the externalisation of interconnected chains of arguments, as well as explicit consideration and integration of multiple and potentially conflicting perspectives. For example, since all participants of Future Ripples workshops were tasked to research a specific topic and formulate insights based on it, a diversity of perspectives was already introduced at the very onset of the method's execution. This diversity then branched further out as the process continued (see Figure 15.2). Overall, the maintenance of multiple perspectives is more successful when their expression, exploration and documentation are at the heart of the process.

Transparency, Trust and Flux

Few things are more important than the collective imagination of radically better futures, yet such efforts are often dismissed as wild speculation, idealism, or unrealistic utopia. Considering this, the Futures Studies methods introduced in this chapter have the potential to make more-than-human designs and future visions more persuasive. This is because a more formal and accountable process for breaking down arguments and chains of thoughts facilitates comprehension of events' causality and environmental transformations. However, such activity demands being transparent on how futures are constructed and from what situated knowledge they are extrapolated. The methods proposed in this chapter aim to acknowledge own assumptions, consider various perspectives, and ascertain the extent to which unknown elements have been accounted for. Thus, futures ought to be revisited and reassessed consistently and iteratively.

Conclusion

Anticipation offers an approach for more-than-human designers to peer through time and enriches the understanding of designing for pluriversal futures. As Manzini and Tassinari (2023) highlight, this process enables a paradigm shift towards new modes of existence and action, thus generating new modes of design. These new modes of design involve attending to the different temporalties experienced by human and more-than-human organisms and systems. To attend to those temporalities, we need to be aware of our role in the process, the values we embed and the factors we ignore. By placing this emphasis on the exploration of futures, anticipation aims at highlighting our situatedness in the process of attending more-than-human endeavours.

Futures exploration is not only forward-driven, but it also involves questioning the past and overall, our position in the present. Moreover, outlining expectations, beliefs, methods, and lines of action used and identified throughout the anticipatory process helps external stakeholders better understand the relevance of more-than-human imaginaries. These are important points for the nascent field of more-than-human design, which is still searching for its place in the academic/design community and seeking wider acknowledgement.

However, much work remains to be done in the conscious integration of anticipation within more-than-human design. We have presented some Futures Studies methods that could support this integration. Yet, representing more-than-human perspectives in futuring activities is still a matter of research. To encourage further work, we suggest considering the following questions:

- What methodological approaches could support a more nuanced representation of more-than-human beings' perspectives in anticipatory processes sensitive to the complexities of what is WEIRD (Western, Educated, Industrialised, Rich, and Democratic)?
- How do we shape more-than-human future imaginaries by including specific more-than-human perspectives in anticipatory practices, given that resource and capability constraints may prevent other perspectives' consideration?

Acknowledgements

Although we the authors represent different nationalities, we work in the same Finnish university. We acknowledge that our shared academic context has the potential to both enhance and limit our viewpoints, as well as to influence the narrative in this chapter. Therefore, with this chapter, we call for further introspection of our position as designers and/or researchers in more-than-human endeavours. This work has been supported by the Research Council of Finland grant 330124.

Notes

1 An extension of the ETPS framework by Aguilar (1967).
2 Opposite to trends, weak signals are seemingly insignificant events that hold the potential to shape the future (Kamppinen et al., 2002, as translated in Holopainen & Toivonen, 2012).

References

Adam, B., & Groves, C. (2011). Futures tended: Care and future-oriented responsibility. *Bulletin of Science, Technology & Society*, 31(1), 17–27. doi:10.1177/0270467610391237

Aguilar, F. J. (1967). *Scanning the Business Environment*, 1st ed. New York: Macmillan.

Alminde, S., & Warming, H. (2020). Future workshops as a means to democratic, inclusive and empowering research with children, young people and others. *Qualitative Research*, 20(4), 432–448. doi:10.1177/1468794119863165

August, V. (2022). Network concepts in social theory: Foucault and cybernetics. *European Journal of Social Theory*, 25(2), 271–291. doi:10.1177/1368431021991046

Bell, W. (2009). Assumptions of futures studies. In *Foundations of Futures Studies: Human Science for a New Era. Volume 1, History, Purposes, and Knowledge* (pp. 115–164). New Brunswick: Transaction Publishers.

Bendor, R., Eriksson, E., & Pargman, D. (2021). Looking backward to the future: On past-facing approaches to futuring. *Futures*, 125, 102666. doi:10.1016/j.futures.2020.102666

Bibri, S. E. (2018). Backcasting in futures studies: A synthesized scholarly and planning approach to strategic smart sustainable city development. *European Journal of Futures Research*, 6(1). doi:10.1186/s40309-018-0142-z

Boulding, J. R. (2017). Expanding our sense of time and history: The 200-year present, from building a global civic culture (1988). In J. R. Boulding (Ed.), *Elise Boulding: A Pioneer in Peace Research, Peacemaking, Feminism, Future Studies and the Family* (Vol. 6, pp. 155–158). doi:10.1007/978-3-319-31364-1_13

Candy, S., & Dunagan, J. (2017). designing an experiential scenario: The people who vanished. *Futures*, 86, 136–153. doi:10.1016/j.futures.2016.05.006

Chia, R. (2004). Re-educating attention: What is foresight and how is it cultivated. Managing the Future: Foresight in the Knowledge Economy, 21–37.

Clarke, R., Heitlinger, S., Light, A., Forlano, L., Foth, M., & DiSalvo, C. (2019). More-than-human Participation: Design for Sustainable Smart City Futures. Interactions, 26(3), 60–63. https://doi.org/10.1145/3319075

Coulton, P., & Lindley, J. (2023). Designing technology for more-than-human futures. In P. A. Rodgers & J. Yee (Eds), *The Routledge Companion to Design Research* (2nd ed., pp. 112–125). doi:10.4324/9781003182443

Dator, J. (2019). Alternative futures at the Manoa School. In *A Noticer in Time* (Vol. 5, pp. 37–54). doi:10.1007/978-3-030-17387-6_5

Dumit, J. (2014). Writing the implosion: Teaching the world one thing at a time. *Cultural Anthropology*, 29(2), 344–362. doi:10.14506/ca29.2.09

Epp, F. A., Moesgen, T., Salovaara, A., Pouta, E., & Gaziulusoy, İ. (2022, June). Reinventing the wheel: The future ripples method for activating anticipatory capacities in innovation teams. *Designing Interactive Systems Conference*, 1, 387–399. doi:10.1145/3532106.3534570

Escobar, A. (2018). *Designs for the Pluriverse: Radical Interdependence, Autonomy, and the Making of Worlds*. Duke University Press.

Forlano, L. (2017). Posthumanism and design. *She Ji: The Journal of Design, Economics, and Innovation*, 3(1), 16–29. doi:10.1016/j.sheji.2017.08.001

Gall, T., Vallet, F., & Yannou, B. (2022). How to visualise futures studies concepts: Revision of the futures cone. Futures, 143, 103024. https://doi.org/10.1016/j.futures.2022.103024

Glenn, J. C. (2009). The futures wheel. *Futures Research Methodology: Version, 3*, 19.

Guston, D. H. (2013). "Daddy, can I have a puddle gator?": Creativity, anticipation, and responsible innovation. In R. Owen, J. Bessant, & M. Heintz (Eds), *Responsible Innovation* (pp. 109–118). doi:10.1002/9781118551424.ch6

Halskov, K., & Dalsgård, P. (2006). Inspiration card workshops. *Proceedings of the 6th ACM Conference on Designing Interactive Systems* – DIS '06, 2–11. doi:10.1145/1142405.1142409

Haraway, D. (1988). Situated knowledges: The science question in feminism and the privilege of partial perspective. *Feminist Studies*, 14(3), 575–599. doi:10.2307/3178066

Heitlinger, S., Houston, L., Taylor, A., & Catlow, R. (2021). Algorithmic food justice: Co-designing more-than-human blockchain futures for the food commons. *Proceedings of the 2021 CHI Conference on Human Factors in Computing Systems*, 1–17. doi:10.1145/3411764.3445655

Holopainen, M., & Toivonen, M. (2012). Weak signals: Ansoff today. *Futures*, 44(3), 198–205. doi:10.1016/j.futures.2011.10.002

Howell, N., F. Schulte, B., Twigger Holroyd, A., Fatás Arana, R., Sharma, S., & Eden, G. (2021, May). Calling for a plurality of perspectives on design futuring: An un-manifesto. *Extended Abstracts of the 2021 CHI Conference on Human Factors in Computing Systems*, 1–10. doi:10.1145/3411763.3450364

Inayatullah, S. (1998). Causal layered analysis. *Futures*, 30(8), 815–829. doi:10.1016/S0016–3287(98)00086-X

Kamppinen, M., Kuusi, O., & Söderlund, S. (2002). Tulevaisuudentutkimus: Perusteet Ja Sovelluksia (Vol. 896). Suomalaisen kirjallisuuden seura.

Kolbert, E. (2014). *The Sixth Extinction: An Unnatural History*, 1st ed. New York: Henry Holt and Company.

Lewis, D. K. (1973). *Counterfactuals*. Cambridge: Harvard University Press.

Light, A. (2022). Ecologies of subversion: Troubling interaction design for climate care. *Interactions*, 29(1), 34–38. doi:10.1145/3501301

Liu, J., Byrne, D., & Devendorf, L. (2018). Design for collaborative survival: An inquiry into human-fungi relationships. *Proceedings of the 2018 CHI Conference on Human Factors in Computing Systems*, 1–13. doi:10.1145/3173574.3173614

Manzini, E., & Tassinari, V. (2023). Anticipations of more-than-human futures. In *Design for More-Than-Human Futures* (1st ed., pp. 51–57). doi:10.4324/9781003319689-3

Miller, R. (2018a). Futures literacy: Transforming the future. In R. Miller, *Transforming the Future* (1st ed., pp. 1–12). Routledge. https://doi.org/10.4324/9781351048002

Miller, R. (2018b). Sensing and making-sense of futures literacy: Towards a futures literacy framework (FLF). In *Transforming the Future* (1st ed., pp. 15–50). doi:10.4324/9781351048002

Pierce, J. (2021). In tension with progression: Grasping the frictional tendencies of speculative, critical, and other alternative designs. *Proceedings of the 2021 CHI Conference on Human Factors in Computing Systems*, 1–19. doi:10.1145/3411764.3445406

Poli, R. (2017). *Introduction to Anticipation Studies*. doi:10.1007/978-3-319-63023-6

Robinson, J. (2003). Future subjunctive: Backcasting as social learning. *Futures*, 35(8), 839–856. doi:10.1016/S0016–3287(03)00039-9

Romani, A., Casnati, F., & Ianniello, A. (2022). Codesign with more-than-humans: Toward a meta codesign tool for human-non-human collaborations. *European Journal of Futures Research*, 10(1), 17. doi:10.1186/s40309-022-00205-7

Sardar, Z., & Sweeney, J. A. (2016). The three tomorrows of postnormal times. *Futures*, 75, 1–13. doi:10.1016/j.futures.2015.10.004

Schoemaker, P. J. H. (1995). Scenario planning: A tool for strategic thinking. *Sloan Management Review*, 36(2), 25–50.

Taylor, C. W. (1993). *Alternative World Scenarios for a New Order of Nations*. US Army War College Press.

Tyszczuk, R. (2021). Collective scenarios: Speculative improvisations for the anthropocene. *Futures*, 134, 102854. doi:10.1016/j.futures.2021.102854

Ulmer, J. B. (2017). Posthumanism as research methodology: Inquiry in the anthropocene. *International Journal of Qualitative Studies in Education*, 30(9), 832–848. doi:10.1080/09518398.2017.1336806

Van der Heijden, Kees. (2005). *Scenarios: The Art of Strategic Conversation* (2nd ed.). Chichester: John Wiley & Sons, Ltd.

Wakkary, R. (2021). *Things We Could Design: For More Than Human-Centered Worlds* (K. Friedman & E. Stolterman, Eds). Cambridge, MA: MIT Press.

Whiten, A., & Van Schaik, C. P. (2007). The evolution of animal `cultures' and social intelligence. *Philosophical Transactions of the Royal Society B: Biological Sciences*, 362(1480), 603–620. doi:10.1098/rstb.2006.1998

Wilde, D., van Gaalen, S., Dolejšová, M., Ravan, P. G., Trahan, S., & Karyda, M. (2021). Backcasting [better] futures. *Proceedings of the Nordes 2021 Nordic Design Conference*.

16

TEACHING FOR MORE-THAN-HUMAN VALUES AND PERSPECTIVES IN TECHNOLOGY DESIGN

Eva Eriksson, Elisabet M. Nilsson, Anne-Marie Hansen, Daisy Yoo and Tilde Bekker

Introduction

In this chapter, we challenge the dominating paradigm of teaching technology design, which often primarily focuses on human values (Friedman & Hendry, 2019; Nørgård et al., 2022). We do so by foregrounding systemic perspectives and planetary values. As we have grown to understand our entanglement with complex social-ecological-technological systems, we are becoming increasingly aware of the shortcomings of human-centrism prevalent in current technology design research and practice. We argue that *things* are more than a mute, passive backdrop to human activities, ready to be extracted and *used* as resources. We need to shift towards embracing other-than-human beings, entities, materials and technologies as actors or companions with which we are intimately entangled and share the world (Haraway, 2016; Poikolainen Rosén et al., 2022; Frauenberger, 2019; Wakkary, 2021). We see that integrating more-than-human values and perspectives in the design of technologies requires a continuous process of increasing awareness also in education (Nijs et al., 2020; Hansen et al., 2022). When examining how to integrate such perspectives in design curricula, it is relevant to explore how this links to elements of signature pedagogies of design, such as the design process, and design critique within a design project (McLain, 2021).

In answering the question "What topics and skills should be taught in design curricula?", Brosens et al. (2022) concluded that students should learn human-centred design and sustainability. In this chapter, we extend this view by arguing for integrating more-than-human approaches in design. How can we educate responsible designers in addressing more-than-human values and perspectives? What has to be unlearned to change our perspectives and attitudes? Providing a complete curriculum on more-than-human values and perspectives in design would require an extensive exploration of 1) what *concepts, theories*, and 2) *methods* should be taught, and 3) how these can be embedded in various design *practices*. In this chapter, we zoom in on describing a collection of teaching activities primarily concentrating on *methods*.

Three concrete activities for teaching more-than-human values and perspectives in design are presented: *Council of All Beings in design* inspired by Macy (2023), *More-than-human*

DOI: 10.4324/9781003467731-19

Ethnography inspired by Bennett (2010), and *More-than-human pattern thinking with AI* building on Selvitelle and Raskin (2023) and Sevaldson (2022).

These teaching activities may serve as inspiration for educators who are about to embark on this journey of teaching for more-than-human values and perspectives in technology design. The examples also serve as illustrations to some of the discussions we raise in this chapter. The main takeaway from this chapter is two-fold. Firstly, considerations in teaching for more-than-human values and perspectives in technology design, and secondly, concrete teaching activities to get started.

Background

As seen in earlier chapters of this book, research offers many different approaches to a more-than-human perspective and points out the need to think differently about design and technology development. However, research and pedagogy do not always move at the same speed; many questions about *how* to teach more-than-human values and perspectives in technology design education are yet to be addressed (Beach & Fox, 2021). The conceptual complexity that arises when thinking in assemblages of humans and non-humans might be challenging for students who are used to focusing on simplistic user-technology relationships (Bekker et al., 2023). Since we live in a time of environmental exigency, there are needs and opportunities for both education and educational research to respond in democratic ways and with a more-than-human perspective (Ulmer, 2017).

The explorations of more-than-human perspectives in technology design span a wide spectrum, which Yoo et al. (2023) summarise in four categories: 1) more-than-human *species* and living forms such as, animals, plants, fungi, algae, eco-systems, 2) more-than-human *things* beyond living species such as robots, artificial intelligence (AI), smart products, digital platforms and applications, 3) more-than-human *designers* where new alignments move past the blind spots of human-centred design and address the expanding universe of algorithms, forms of intelligence, and forms of life that are entering design practice, casting them as partners in more-than-human design practice, and finally 4) more-than-human *design education* which calls for an ontological shift from human-centrism to embracing our deep interdependencies and entanglements with the world (Frauenberger, 2019), in turn, leading to questions on how this might change the design curriculum and what to teach the designers of tomorrow (Yoo et al., 2023). This also leads to a shift from understanding designer responsibilities as something that can be ascribed or attached to some individual or collective entity, toward "doing responsibilities", that is, a framing of being responsible as a shared, hybrid activity rather than an attribute (Fuchsberger & Frauenberger, 2023).

Teaching Activities for More-Than-Human Values and Perspectives in Technology Design

The teaching activities presented in this chapter have been captured and turned into shareable and adaptable formats using a modification of a pedagogical design pattern method (Laurillard, 2012). The teaching activities are designed for appropriation (Dix, 2007) and allow educators to tailor the exercises to fit different courses in various educational settings, across different levels and disciplines. They will be part of a forthcoming collection of teaching activities published as an online open educational resource targeting educators in technology design programmes in higher education (MOVA, 2023). The goal of the teaching activities is to create conditions for students to develop the knowledge and skills required to consider the broader context and implications of design and design processes. This is achieved by going beyond human-centred perspectives, which currently are the prevalent perspectives in technology design research and practice.

The teaching activities presented represent three different takes on how more-than-human values and perspectives can be addressed in design work. The first teaching activity takes inspiration from indigenous spiritual practices, more specifically from Native American cultures (Macy, 2023). The second teaching activity is inspired by Bennett's (2010) work on vital materialism and draws from a set of new genres of ethnography inspired by posthumanist theories (Kirksey & Helmreich, 2010; Murray-Rust et al., 2019; Giaccardi et al., 2016). The third teaching activity builds on systems oriented design (Sevaldson, 2023) and the work by the Earth Species Project (Selvitelle & Raskin, 2023).

We will first introduce the core elements of the teaching activities and the rationale behind them. We will then discuss the teaching activities based on the results from piloting and reviewing them. The insights gained will serve as input to the final discussion on how to educate responsible designers of tomorrow.

Teaching Activity: Council of All Beings in Design

This teaching activity builds on the work of Saxgren (2023) and Macy and Seed (2023) who developed the Council of All Beings as a ceremony inspired by Native American cultures. The purpose of the original Council of All Beings is to identify with and speak on behalf of all beings and entities (direct or indirect stakeholders) that are affected by design decisions made to change the environment that everyone is part of. For example, if a river is affected, a person would speak on behalf of this river, becoming familiar with the river as a matter of concern. The Council of All Beings in Design teaching activity is thus an empathizing exercise where students get a chance to visit nature and attune with a more-than-human being or entity to emphasise with and understand (as far as possible) the perspective of this more-than-human.

Beyond empathizing with more-than-human beings or entities in the environments, the teaching activity explores speaking on behalf of – through "attuning with" and eco-phenomenology (Riley & White, 2019). The idea of "attuning with" in the context of this exercise is to role-play how it feels to be in the position of another being in the natural environment. It encourages students to reflect critically on the Rights of Nature and what it means to "speak on behalf of". Recommended reading to prepare for the activity: Fitz-Henry (2021), Garver (2020), Gilbert et al. (2023), Riley and White (2019), and Rzedzian (2023).

Why this Teaching Activity?

When working with more-than-human values and perspectives in technology design, it may be difficult to speak on behalf of more-than-human stakeholders that affect or are affected by the design. In some cases, this can be done by carefully studying the scientific knowledge and data associated with other species, mountains, rivers, ecosystems etc. This offers a rational approach to understanding what is at stake when decisions are made about how we might live with more-than-human beings and entities. However, we need other ways of speaking for those which are not rational (Riley & White, 2019).

Organising a Council of All Beings is a way to bring up the perspectives of nature and the environment when we make design decisions that concern the environment. It is also a way to bring in new perspectives in the ongoing negotiation about how our human (urban) environments should be integrated with the biosphere. The teaching activity is based on the view that humans should not act by themselves since they are part of an assemblage consisting of different entities; humans, nonhumans, things, and matters that are linked together to form a whole (Latour, 2005).

A Council of All Beings is an empathizing practice and method inspired by Macy's (2023) communal ritual with the same name. According to Macy,

> [t]he Council of All Beings is a communal ritual in which participants step aside from their human identity and speak on behalf of another life-form. A simple structure for spontaneous expression, it aims to heighten awareness of our interdependence in the living body of Earth and to strengthen our commitment to defend it. The ritual serves to help us acknowledge and give voice to the suffering of our world.
>
> (Macy, 2023)

The aim of the teaching activity, is to create conditions for students to explore and test how they can pay attention to and open up communication with more-than-human beings or entities that allow us to sense and understand more-than-humans. The teaching activity should be staged outside in a natural environment to create a deeper connection with nature, where students can express various perspectives and concerns about specific elements of the environment and explore their ecological awareness with each other. To be able to engage in such an activity, students might need to prepare themselves by studying specific more-than-humans, for example, rivers, plants, bacteria, to build an understanding of their situations.

Teacher Instructions

The teaching consists of three activities taking place both in the classroom and outdoors, individually and in groups.

Activity 1: Preparation (2h)

Each student picks a more-than-human being or entity that might be affected by the design to empathise with (it can be a design developed by the students, or an existing design). They are asked to reflect and contemplate on the (living) conditions of the living being or entity. They do this by first doing some desktop research on the living being or entity to find background knowledge. Then they are instructed to contemplate on what they learned by going for a walk in nature, thinking about this being or entity. Lastly, they pick an object in nature that can represent this more-than-human. This could for example be a leaf from a plant, a stone, a feather.

Activity 2: The Council of All Beings (2h)

The students and the teacher meet outside in nature, preferably around an object as the centre of attention (such as a bonfire) in the environment under exploration. Depending upon the size of the class, the students might be split up into smaller groups to ensure that everyone is allocated enough time to speak. Each student brings their own found objects and when they enter the circle around the bonfire (or what they chose to gather around), they announce to the teacher who or what they represent. The teacher acts as a human being, who learns from what the other beings or entities have to say. The teacher first asks how they are doing, and if they have a message that they would like to deliver. Based on the knowledge students have about the more-than-human entity that the students speak on behalf of, there might be something that they would like to say to humanity about their living conditions. A talking stick is then handed over to other students representing another more-than-human being or entity, who choose to speak. When all speeches are listened to, the teacher representing the human asks the students representing various beings and entities of the more-than-human world what they dream of, and what the world should be like for them to thrive. The council is concluded when all voices have been heard.

Activity 3: Reading and Reflection Seminar (2h)

The seminar should be arranged shortly after the Council of All Beings. In the seminar, students critically reflect on the council as a way of empathizing with more-than-human beings and entities and giving voice to direct or indirect stakeholders that affect or are affected by the design. The students should reflect upon how taking the Rights of Nature into account in a design process might influence a design. Potentially, students could also reflect upon who – if not themselves and the scientific knowledge – could speak on behalf of more-than-humans? For example, potentially people that already live in close connection with specific more-than-humans. These people might be considered important stakeholders.

Learning Outcomes

The intended learning outcomes of the Council of All Beings in design teaching activity are:

- *Describe* an individual and collective experience related to the Council of All Beings.
- *Express* their feelings and emotions related to empathizing and "speaking on behalf of".
- *Relate* personal experiences to "attuning with" and eco-phenomenology as well as the Rights of Nature.
- *Reflect* on what it means to speak on behalf of the more-than-human beings or entities in ecological environments as a method for stakeholder research for design.
- *Discuss* how designers can pay attention to the matters of concern that involve more-than-human beings and entities.

Piloting the Council of All Beings in Design Teaching Activity

The teaching activity was piloted in a class with four second-year master students in interaction design. The results from the piloting demonstrate the challenges of addressing more-than-human values and perspectives in a teaching situation. In this version of the activity, the students were given twenty minutes for the preparation part, which consisted of reflecting on which more-than-human being or entity they would like to associate themselves with. That is, they did not have time to conduct this in a natural environment beforehand. From the discussions, it became evident that this approach did not immerse the students enough to encourage them to enact the more-than-humans to a level where they could engage. Students stated that scientific knowledge combined with local knowledge about more-than-humans is necessary to establish an emotional connection. Students needed something to be at stake to emotionally engage in a way where they felt that they could give voice to something.

An important observation is that enough time and solitary space must be dedicated to the first part of the activity, that is, the walk in nature and thinking about a more-than-human being or entity. As Riley and White (2019) point out, this kind of teaching activity requires what they call "slow pedagogy" and "wild pedagogy" where there is "affective entering of bodies" (Riley & White, 2019, p. 264). To run this teaching activity successfully, the teacher should preferably arrange an excursion to a place with enough natural surroundings to create the right context. It is highly recommended to allow enough time and space for the activity to unfold in natural environments where students can attune to nature. This goes both for the preparation stage of this activity and the actual Council of All Beings. A variation of this activity could be to attune with living beings and entities in a local environment.

The feedback given by the students participating in the Council of All Beings showed that they struggled with empathizing because they lacked knowledge about the specific more-than-human that they chose to represent. They thus called for more information and more time to study the being or entity. The students also stated that there is a danger that individual utterances might suffer from cliches that are caused by a lack of knowledge about the more-than-humans they speak on behalf of. Furthermore, students mentioned that there can be a tendency to speciesism, where species that are very popular in the news are considered, while others "less interesting" are overlooked.

Teaching Activity: More-Than-Human Ethnography

This teaching activity is inspired by the "nonhuman turn" (Grusin, 2015) in ethnography. Traditionally, ethnography is rooted in colonial anthropology, and thus primarily focused on understanding human cultures (Daynes & Williams, 2018). More recently, new genres of ethnography have been introduced that are grounded in posthumanist theories such as multispecies ethnography (Kirksey & Helmreich, 2010) (see also chapter 9, "Multispecies Ethnography in Design Research and Practice" in this book), thing ethnography (Giaccardi et al., 2016), and entangled ethnography (Murray-Rust et al., 2019). Such more-than-human ethnographies draw attention to the creatures and things that previously appeared only on the margins of anthropological research – small organisms such as insects, fungi and lichens, and mundane objects such as a mug or a piece of garments – and bring them to the centre stage of conversations.

Beyond considering social justice for marginalised communities, this teaching activity encourages students to critically reflect on the meaning(s) of marginalisation and socio-environmental justice by examining the *things* that are neglected and pushed to the margins of society. Drawing from Wakkary (2021), we refer to things as "nonhumans made by both humans and nonhumans" (p. 10). Thus, marginalised things can be both artefacts (e.g., public toilets, trash bins, sewage) and things in nature (e.g., feral pigeons, stinky ditch, piles of dirt). Inspired by Bennett's (2010) work, students will learn to recognise marginalised things as animate, vital materials that deserve to be cared for. Recommended reading to prepare for the activity: Bennett (2010, Wakkary (2021), and Tsing et al. (2017).

Why this Teaching Activity?

Ethnography is a fruitful way of exploring situated and embodied knowledge. One of the defining features of ethnography is the "field" in which the researchers immerse themselves into. Ethnography offers methods and techniques for how to situate research in a "field site" in which researchers (students) use their own embodied experience to investigate the complexity and messiness of the life-worlds they are studying. In particular, ethnography is effective in uncovering unarticulated or latent needs that more-than-humans cannot express explicitly through careful observation, paying attention to minute details of the atmosphere, behaviours, and interactions in the field.

By drawing attention to the marginalised things in public spaces, this teaching activity aims to address more-than-human topics beyond engaging with nature, reflecting on issues of marginalization and care (or lack thereof) around technological artifacts and infrastructures that we bring into the world as designers. The activity provokes students to critically examine their everyday surroundings and to start questioning what this thing is, why is it here, and how it became systematically damaged and neglected by society. In doing so, students are led to reflect on what marginalisation is, how it takes form, and the role design can play in either reinforcing or mitigating marginalisation.

Teacher Instructions

The teaching consists of three activities taking place both in the classroom and outdoors, individually and in groups.

Activity 1: Mini-Ethnography on Marginalised Things (1h Lecture + 1 week Self-study)

The teacher introduces ethnography as a research methodology, its roots in anthropology and colonialism, and the emergence of more-than-human perspectives in the field of ethnography such as multispecies ethnography, thing ethnography, and entangled ethnography. Then each student is asked to identify and analyse a local socio-ecological challenge. Specifically, each student is asked to find a marginalised *thing* (as opposed to a marginalised community) within their local neighbourhood as their focus and carry out a mini-ethnographic fieldwork around it for a week (if not longer). Students are instructed to keep a meticulous log of observation through handwritten notes.

Activity 2: Secondary Research (1 week Self-study)

In addition to mini-ethnographic fieldwork, each student is asked to collect secondary data by reviewing related literature, reports, and news to become knowledgeable about

a) the history and genealogy of the *thing* at stake – Where does it come from? How did it start? How long has it been there?
b) the (material) life cycle – What is it made of? When will it become obsolete, and where will it go? and
c) the underlying policy and infrastructure – What are the regulations and tacit rules? What technological and financial infrastructures are in place for maintenance?

Activity 3: Joint Reflection and Mapping Workshop (2h)

The result of the individual mini-ethnography and secondary research assignments is synthesised and visualised as a map. Each student brings the more-than-human ethnography map (an interpretation) as well as their field notes (raw data) to class. In the joint reflection session, the teacher encourages students to reflect on their experience of being in the field doing more-than-human ethnography. Students do so by exchanging and reading each other's field notes. Depending upon the size of the class, the students might be split up into smaller groups to ensure that everyone is allocated enough time to discuss their case.

Thereupon, a collective entangled ethnography map of all marginalised things is developed together. By engaging in a joint mapping activity combining the individual mini-ethnography case studies, students reflect on how different things, practices, actors and actions hang together in entanglement. They critically reflect on the complex processes of marginalisation from historical, material, infrastructural, policy perspectives and more.

Learning Outcomes

The intended learning outcomes of the more-than-human ethnography teaching activity are:

- *Recognise* and *apply* more-than-human ethnographic research methods.
- *Describe* and *illustrate* complex entanglements of humans and more-than-humans in shaping marginalised things and unsustainable practices.
- *Reflect* and *discuss* the mechanisms of marginalisation and the roles design and designers play in this process.

Piloting the More-Than-Human Ethnography Teaching Activity

The More-than-human ethnography teaching activity was piloted in a class with 41 masters students in industrial design, with most (if not all) students being trained in human-centred design. An important insight gained from this piloting activity was the challenges of "unlearning" the human-centred perspective that is so deeply ingrained in students' research skills and attitudes. The students, who are used to observing and analysing human behaviours in their user studies, found it difficult to centre their observations around marginalised things because they were simply "not interesting" – these things don't talk, they don't move, and they are not even particularly beautiful to watch. For instance, a student, who chose to investigate fake trees often found in shopping centre corridors, reported that "taking field notes about a tree is rather unspectacular" since nothing ever happens. Similarly, many students shared their experiences of feeling lost during their field observation, not knowing what to pay attention to when humans are out of sight. As such, this pilot revealed a blind spot in how we teach design research, which primarily comes down to "user" studies. With human-centred design education, students learn various methods to gain a better understanding of their users' behaviours, needs and desires. At the same time, anything other than their target users becomes rather irrelevant and uninteresting for the students. They are trained to auto-focus on user behaviours and to prioritise user needs while filtering out the rest.

The students in this pilot had to unlearn their way of doing design ethnography as a form of user study to learn more-than-human ethnography. It calls for fostering a new ability to sense the livelihoods of the nonhuman, which Anna Tsing refers to as the "arts of noticing" (Tsing et al., 2017). Accordingly, the teacher instructed the students to re-do their mini-ethnography exercise guided by the following questions: What does the thing look like? What is it made of? (e.g., materiality and embodiment)? What do you hear (e.g., sound, noise)? How does it feel (e.g., touch, smell, taste)? How does it move (e.g., in subtle scales and delayed temporality)? Where is it located? How does it fit into the surrounding space/other things (e.g., situatedness, infrastructure)? Last but not least, to question, what is missing, what is not there?

FIGURE 16.1 Examples of mini-ethnography on marginalised things including (left to right) vandalised electricity boxes, lamp posts covered with stickers, and an unknown object on the campus lawn, which later turned out to be a geothermal storage box.

Teaching Activity: More-Than-Human Pattern Thinking with AI

This teaching activity builds on the work by Earth Species Project (Selvitelle & Raskin, 2023) and offers students the possibility to conceptualise and even experiment with artificial intelligence (AI) in relation to human and more-than-human entanglements. Students will get an introduction to systems oriented design methods, for example, giga mapping used for investigating relationships between categories of beings and entities from a large to small scale (microscope, telescope, wide-angle views) and ZIP (Zoom, Innovation, Potential) analysis for finding potential areas for interventions, innovations and ideas (Sevaldson, 2023, 2022). The students are also introduced to different AI concepts related to data analysis, such as pattern recognition algorithms used for finding hidden information in large data sets by recognizing certain patterns from those sets. Data sets could for example consist of data from Movebank – an online free database of animal tracking data hosted by the Max Planck Institute. The students are asked to identify and select specific entanglements of humans and more-than-human beings and entities of concern (e.g., an urban ecosystem, a food production system, a global broadcasting network) that they thoroughly investigate by using the system-oriented design methods introduced. Based on the outcome of their analyses, the students sketch scenarios and concepts where AI might be implemented and intervene in these entanglements – possibly to establish regenerative relationships between humans and more-than-human beings and entities. That is, concepts in which humans are actively revitalising the health of, for example, their local ecosystem and of themselves, as they incorporate both human needs and more-than-human needs. Finally, students will get to implement and explore a few simple AI pattern recognition algorithms concerning live data generated by more-than-humans (e.g., animal vocalizations, movements, temperature shifts) to discover patterns in data that could be of relevance. Besides being a tool for exploring how humans and more-than-human beings and entities are interconnected, AI is here intimately entangled with human thinking, and scaffold thinking which would not have been made possible without the AI. Recommended reading to prepare for the activity: Selvitelle and Raskin (2023), and Sevaldson (2023; 2022).

Why this Teaching Activity?

Students might not see the potential of AI in terms of how AI can be used to analyse and describe how humans and more-than-human beings and entities are entangled, and how humans might establish ways of life that are regenerative. It can be very complex to design for more-than-human beings and entities because designers need to take many perspectives and related parameters into consideration when they design. Systems oriented design methods offer ways in which designers can comprehend complexities in entangled worlds (Sevaldson, 2022).

Teacher Instructions

The teaching consists of three activities taking place in a classroom setting.

Activity 1: Workshop I (3h)

The teacher picks a few systems oriented design methods from for instance the chapter in this book *Systems Thinking in More-Than-Human Design* or from Sevaldsson (2022) and the accompanying systems oriented design website (Sevaldson, 2023). In the workshop, the students sensitise themselves to the complexities of the entanglements of humans and more-than-humans by using of one or several of the methods (e.g., giga mapping, ZIP analysis). Research from the fields of environmental sciences, biology, botany, zoology, and oceanography might also be relevant to include.

Activity 2: Lecture (2h)

Students are introduced to different concepts of AI related to data analysis, such as various pattern recognition algorithms. This should happen at a level where design students can understand the overall concepts without needing an advanced level of math. The teacher may, for example, ask the following question: What kinds of AI could aid humans in understanding entanglements and complex eco-systems?

Activity 3: Workshop II (3h)

Students identify specific entanglements of humans and more-than-human beings or entities that they investigated through the methods introduced. Then they sketch scenarios of how specific AI's might be implemented and intervene in these entanglements. This could be to establish regenerative relationships between humans and more-than-humans. Realistic, utopia and dystopia scenarios can be explored, and students create small experiments where they visualise and analyse live data that is generated from more-than-humans. The generative questions for the workshop could be: What aspects of our ecological, biological and technological environment could AI be concerned with? How might we utilise pattern thinking and recognition to support regenerative ways of life?

Learning Outcomes

The intended learning outcomes of the More-than-human pattern thinking with AI teaching activity are:

- *Recognise* systems-oriented design methods.
- *Apply* a selection of systems-oriented design methods.
- *Describe* different entanglements of humans and more-than-humans.
- *Illustrate* how a selection of AI concepts might be used in understanding human, more-than-human entanglements.
- *Create* scenarios where different kinds of AI play a role in human, more-than-human entanglements.
- *Reflect* and *discuss* design scenarios where AI plays a role in human, more-than-human entanglements.

Reviewing the More-Than-Human Pattern Thinking with AI Teaching Activity

The teaching activity was peer-reviewed by teachers in technology design. As highlighted by the reviewers, for the students to benefit from this activity they must possess a certain level of understanding of AI concepts. This might not be the case of design students, which makes the preparatory phase where the students are introduced to systems-oriented design methods and AI a crucial part. Also, for teachers in design who have a light technical background, it can be a challenge to introduce theories related to pattern recognition algorithms etc. A suggestion brought up was to invite teachers in, for examples, computer science or engineering studies as co-teachers.

As also pinpointed by the reviewers, using AI as a tool for developing an understanding of human and more-than-human entanglements and complex systems holds great potential as a pedagogical approach since it could challenge us to think beyond human-centred perspective. Most students, as well as teachers have been taught to experience the world from anthropocentric, and user-centred perspectives. Bringing in AI as a "co-researcher" in a design process may support students in "unlearning" and finding new ways of looking upon and sensing the world.

Another potential of this teaching activity is the opportunity offered to students to learn to collaborate with AI, as a more-than-human *thing* and in this case also a more-than-human *designer*. The students co-design with the AI by engaging in a creative and dynamic process driven by both parts. However, for humans to benefit from such a process also demands a critical awareness of potential bias and the negative consequences of inviting AI. Both teachers and students must learn to take a critical standpoint towards the use of AI to counteract unwanted effects and consequences of involving AI, which may become a profound risk to society and humanity (Future of Life Institute, 2023).

Educating Responsible Designers of Tomorrow

As evident, there are a myriad of aspects of more-than-human perspectives that need to be considered when developing teaching activities targeting design students. In the examples of teaching activities presented in this chapter, we focus on one of the questions raised by Hansen et al. (2022), namely, "How might we listen to the more-than-human and define what more-than-human values are?" The Council of All Beings in design addresses this question by empathising with more-than-human beings and entities that might affect or be affected by a design, while the More-than-human ethnography is more of a hands-on systematic way of identifying various entities. The More-than-human pattern thinking with AI teaching activity is about how designers can sensitise themselves to different entanglements by engaging with large data sets and AI methods. The activity also offers an opportunity to develop a critical awareness of biases and unwanted consequences of involving and collaborating with more-than-human things and designers, which is claimed to be of utter importance to the future of humanity (Future of Life Institute, 2023). These three examples of teaching activities represent different ways of both listening to and defining what more-than-human values and perspectives are.

Another question to consider in the design of teaching activities that go beyond the more traditional human-centred design is what we, in this case, design students, ought to "unlearn" to change our perspectives and attitudes toward more-than-human worlds. Although this is an important question, we believe that we instead need to extend our repertoire of means and methods in design rather than unlearn. While some of these methods are variations of traditional ethnographic methods, such as More-than-human ethnography, others are more distant from our traditional way of approaching design, such as the Council of All Beings in design. We firmly believe that a mix of both is needed.

Making an ontological shift from human to more-than-human centeredness is not only important for future research paths, but also for rethinking what we teach the designers of tomorrow in the design curriculum (Hansen et al., 2022). How can we educate responsible designers with regard to more-than-human values and perspectives in technology design (Yoo et al., 2023)? How can we develop teaching so that designer responsibilities in terms of more-than-human perspectives shift from something that can be had or attached to some individual or collective entity, toward "doing responsibilities", that is, a framing of being responsible as a shared, hybrid activity rather than an attribute (Fuchsberger & Frauenberger, 2023)?

We experience a gap between the socio-environmental problems that need to be solved and the human-centred design methods that are professionally practised and taught in technology design education with a particular focus on users (e.g., user experience design, user research). The examples of teaching activities presented in this chapter do not bridge this gap but are a starting point for further elaboration, and hopefully, many more teaching activities will be developed and shared by the design community. We also see a need to further examine how incorporating a more-than-human approach in design will influence signature pedagogies for technology design (as described in McLain, 2021), to detect the characteristic forms of teaching and learning, and as they define what counts as knowledge in the field and how things became known (Shulman, 2005).

References

Beach, M. W., & Fox, T. (2021). Value sensitive speculative design: Exploring more-than-human relations in the age of climate catastrophe. *Interaction Design and Architecture(s)*(51), 111–131. doi:10.55612/s-5002–051–005

Bekker, T., Eriksson, E., Fougt, S. S., Hansen, A.-M., Nilsson, E. M., & Yoo, D. (2023). Challenges in teaching more-than-human perspectives in human-computer interaction education. In *Proceedings of the 5th annual symposium on HCI education* (p. 55–58). New York, NY: ACM. doi:10.1145/3587399.3587406

Bennett, J. (2010). *Vibrant Matter*. Durham, NC: Duke University Press.

Brosens, L., Raes, A., Octavia, J. R., & Emmanouil, M. (2022). How future proof is design education? A systematic review. *International Journal of Technology and Design Education*, *33*(2), 663–683. doi:10.1007/s10798-022-09743-4

Daynes, S., & Williams, T. (2018). *On Ethnography*. Oxford: Polity Press.

Dix, A. (2007). Designing for appropriation. In *Proceedings of the 21st British HCI Group Annual Conference on People and Computers: HCI... But Not As We Know It – Volume 2* (p. 27–30). Swindon, GBR: BCS Learning & Development Ltd.

Fitz-Henry, E. (2021). Multi-species justice: A view from the rights of nature movement. *Environmental Politics*, *31*(2), 338–359. doi:10.1080/09644016.2021.1957615

Frauenberger, C. (2019). Entanglement HCI the next wave? *ACM Transactions on Computer-Human Interaction (TOCHI)*, *27*(1), 1–27. doi:10.1145/3364998

Friedman, B., & Hendry, D. G. (2019). *Value Sensitive Design*. London: MIT Press.

Fuchsberger, V., & Frauenberger, C. (2023). Doing responsibilities in entangled worlds. *Human-Computer Interaction*, 1–24. doi:10.1080/07370024.2023.2269934

Garver, G. (2020). Are rights of nature radical enough for ecological law? *From Environmental to Ecological Law*, 90–103. doi:10.4324/9781003001256-10

Giaccardi, E., Cila, N., Speed, C., & Caldwell, M. (2016). Thing ethnography: Doing design research with non-humans. In *Proceedings of the 2016 ACM conference on designing interactive systems*. ACM. doi:10.1145/2901790.2901905

Gilbert, J., Soliev, I., Robertson, A., Vermeylen, S., Williams, N. W., & Grabowski, R. C. (2023). Understanding the rights of nature: Working together across and beyond disciplines. *Human Ecology*, *51*(3), 363–377. doi:10.1007/s10745-023-00420-1

Grusin, R. (Ed.). (2015). *The Nonhuman Turn*. Minneapolis, MN: University of Minnesota Press.

Hansen, A.-M., Nilsson, E. M., Eriksson, E., Yoo, D., & Norgaard, R. T. (2022). Teaching for more-than-human perspectives in technology design – towards a pedagogical framework. Retrieved from https://cumulusdetroit2022.org/

Haraway, D. J. (2016). *Staying with the Trouble*. Durham, NC: Duke University Press.

Kirksey, S. E., & Helmreich, S. (2010). The emergence of multispecies ethnography. *Cultural Anthropology*, *25*(4), 545–576. doi:10.1111/j.1548-1360.2010.01069.x

Latour, B. (2005). *Reassembling the Social: An Introduction to Actor-Network-Theory*. Oxford University Press.

Laurillard, D. (2012). *Teaching as a Design Science: Building Pedagogical Patterns for Learning and Technology*. Florence, KY: Routledge, Taylor Francis Group. Retrieved from www.routledge.com/books/details/9780415803878/++/

Macy, J. (2023). *The Council of All Beings*. rainforestinfo.org.au. Retrieved from www.rainforestinfo.org.au/deep-eco/Joanna%20Macy.htm

McLain, M. (2021). Towards a signature pedagogy for design and technology education: a literature review. *International Journal of Technology and Design Education*, *32*(3), 1629–1648. doi:10.1007/s10798-021-09667-5

MOVA (2023). *The MOVA Project*. Malmo University. Retrieved from https://mova.uni.mau.se/

Murray-Rust, D., Gorkovenko, K., Burnett, D., & Richards, D. (2019). Entangled ethnography: Towards a collective future understanding. In *Proceedings of the Halfway to the Future Symposium 2019*. ACM. doi:10.1145/3363384.3363405

Nijs, G., Laki, G., Houlstan, R., Slizewicz, G., & Laureyssens, T. (2020). Fostering more-than-human imaginaries: Introducing DIY speculative fabulation in civic HCI. In *Proceedings of the 11th Nordic Conference on Human-Computer Interaction: Shaping Experiences, Shaping Society*. New York, NY: ACM. doi:10.1145/3419249.3420147

Nørgård, R., Eriksson, E., Nilsson, E., & Yoo, D. (2022). The vase pedagogical framework: Teaching for values in design in higher education. In R. Rocco, A. Thomas, & M. Novas-Ferradás (Eds), *Teaching Design for Values* (pp. 194–217). Netherlands: TU Delft Open. doi:https://doi.org/10.34641/mg.54

Future of Life Institute (2023). *Future of Life Institute*. Retrieved from https://futureoflife.org/

Poikolainen Rosén, A., Normark, M., & Wiberg, M. (2022). Towards more-than-human-centred design: Learning from gardening. *International Journal of Design*, *16*(3), 21–36. https://doi.org/10.57698/v16i3.02

Riley, K., & White, P. (2019). "Attuning-with", affect, and assemblages of relations in a transdisciplinary environmental education. *Australian Journal of Environmental Education*, *35*(3), 262–272. doi:10.1017/aee.2019.30

Rzedzian, S. (2023). Divergent environmentalisms, conflicting counter-hegemonies: Lessons from the rights of nature movement. *Environment and Planning E: Nature and Space*, doi:10.1177/25148486221148646

Saxgren, J. (2023). *Earthways.* Himmelbjerggaarden. Retrieved from https://earthways.dk

Selvitelle, B., & Raskin, A. (2023). *Earth Species Project.* Earth Species. Retrieved from www.earthspecies.org

Sevaldson, B. (2022). *Designing Complexity: The Methodology and Practice of Systems Oriented Design.* Common Ground Research Networks. doi:10.18848/978-1-86335-262-8/cgp

Sevaldson, B. (2023). *System Oriented Design.* Earth Species. Retrieved from https://systemsoriented-design.net

Shulman, L. S. (2005). Signature pedagogies in the professions. *Daedalus*, *134*(3), 52–59. Retrieved 2023-11-15, from www.jstor.org/stable/20027998

Tsing, A. L., Swanson, H., Gan, E., & Bubandt, N. (2017). *Arts of Living on a Damaged Planet: Ghosts of the Anthropocene: Monsters of the Anthropocene*. University of Minnesota Press.

Ulmer, J. B. (2017). Posthumanism as research methodology: Inquiry in the anthropocene. *International Journal of Qualitative Studies in Education*, *30*(9), 832–848. doi:10.1080/09518398.2017.1336806

Wakkary, R. (2021). *Things We Could design: For More Than Human-Centered Worlds*. Cambridge, MA: MIT Press.

Yoo, D., Bekker, T., Dalsgaard, P., Eriksson, E., Fougt, S. S., Frauenberger, C., Friedman, B., Giaccardi, E., Hansen, AM., Light, A, Nilsson, E., Wakkary, R., Wiberg, M. (2023). More-than-human perspectives and values in human-computer interaction. In *Extended Abstracts of the 2023 CHI Conference on Human Factors in Computing Systems.* New York, NY: ACM. doi:10.1145/3544549.3583174

17

STEPPING OUT OF THE CLASSROOM AND INTO THE WORLDS OF OTHER SPECIES

Daniel Metcalfe

Introduction

The widespread recognition of the devastating effects of unsustainable design on the natural world and other species we share this planet with has led to attempts at developing more sustainable design approaches. Efforts to reduce the negative impact of design on the environment typically prioritize reducing, reusing, and recycling material and energy. However, in recent years there has been a growing interest also in a more active role for design in addressing biodiversity loss and our growing alienation from nature – leading to the development of more-than-human design approaches (see for example Mancini et al., 2023; Metcalfe, 2015; Root-Bernstein & Ladle, 2010 and previous chapters in this book). To facilitate the development of this emergent design discipline and help disseminate it, there is a need for design education programs that introduce students to the prospects of designing with and for more-than-human worlds. Programs that nurture a view of design that is not separated from the natural world but seeks integration and connection, along with methods for studying and relating to nonhuman entities in the context of design. In this chapter, I reflect on my experience in teaching Multispecies Design to outline some possible characteristics of such a design pedagogy. While more-than-human design approaches tend to be broader in their inclusion of more-than-human entities, Multispecies Design, as described here focuses more strictly on other species and predominantly on wild animals.

Teaching More-Than-Human Design

In 2014 I taught a series of four workshops on Multispecies Design for the first time to Arts and Design students. The workshops were a way to evaluate and further develop a set of Multispecies Design Tools I was putting together at the time (Metcalfe, 2015) and reflect on design processes that address the needs of wild animal species interacting with anthropogenic systems. Following a short introduction to emergent conservation strategies that view human-dominated habitats as potential sites for biodiversity conservation (Marris, 2011; Rosenzweig, 2003), and the subsequent need for design strategies capable of understanding and responding to the needs of nonhuman species – the students were presented with some initial tools and asked to develop a concept of a project for a selected wild animal "client".

DOI: 10.4324/9781003467731-20

One of the things that struck me, as the workshops unfolded, was that this new approach to design seemed to be as much about unlearning old *ways of doing* as it was about learning new ones. The simple logic of *Identifying and responding to the client's needs* revealed itself as insufficient and at times problematic as the groups were brainstorming for solutions to the needs they hypothesized. For example, in many of the initial ideas the groups generated, *Design* revealed itself as a taming and domesticating force, posing, a risk of creating dependency of the animal on the designed artefact. The good news was that, even in the short period of the workshop, many participants were able to identify the shortcomings of conventional design thinking in dealing with nonhuman species and gradually shift their thinking as the workshop evolved.

It became clear that more-than-human design requires a shift in perspective that goes beyond just considering nonhuman species as potential stakeholders in the processes of design. A new literacy in the field needs to develop – drawing less on our personal experience as humans in the world and more on alternative ways of imagining and experiencing the world from multiple points of view. A couple of years later, as I was developing a course on Multispecies Design I decided to focus more on the shifts in thinking that enable us to design for and with nonhuman species, than on the actual outcomes of the design. The result was the course Human-Animal-Machine, which has been running since 2018 at the Industrial Design Department at the Technion-Israel Institute of Technology. The course focuses on the meeting points between humans, animals and technology and each year is dedicated to a changing theme. For example: rehabilitation after a fire, urban wildlife, the ecological role of dead trees or animal-technology interactions in the sky. We start with some theoretical background from diverse fields regarding human-animal relations, nature conservation, animal-technology interaction and design theory. In parallel, we go through a series of exercises and explorations designed to bring forth new ways of seeing and interacting with the world, which feed into the development of a design response relating to the topic of the course.

Here, I draw on my experience from developing and teaching the course to reflect on various ideas and methods that have helped my students and me acknowledge and address the human-centredness of our practice and imagine new roles for design and technology in the intersections of humans and nonhuman species. Specifically, five gradual shifts are highlighted, starting from 1) increasing our interaction with nature, 2) looking at natural phenomena through varying modes of perception, 3) experiencing the world from the points of view of other species, 4) regarding design as an evolving, open-ended process, and finally, 5) looking back at design for humans with new insights.

1. Increasing Our Interaction with Nature

> We have lost a lot of nature in the past three hundred years – in both senses of the word lost. We have lost nature in the sense that much of nature has been destroyed... But we have also lost nature in another sense. We have misplaced it. We have hidden nature from ourselves.
>
> Emma Marris, 2011, *Rambunctious Garden – Saving Nature in a Post-wild World*

In 1978 Robert Pyle introduced the term the Extinction of Experience to talk about the growing alienation of people from nature, arguing that our detachment inevitably leads to disregard and indifference towards other species (Pyle, 2002 p. 312). Later studies have shown a strong correlation between interaction with nature, the attachment people feel towards nature and their motivation to preserve and protect it (see, for example, Schuttler et al., 2018; Soga & Gaston, 2016).

The disregard of nonhuman species within the context of design is one manifestation of the collective indifference Pyle talks about. It is a result, in most cases, not of ill intentions but of inherent blind spots in our profession. *We have hidden nature from ourselves*. Interactions with nature are on a decline globally (Soga & Gaston, 2016), and as a result, we are less aware and mindful of more-than-human entities also within the practice of design. To start addressing this, we need to find ways to increase their presence in our lives. According to Soga & Gaston (2016), the leading causes of the Extinction of Experience are the loss of opportunities to interact with nature and the loss of orientation and affinity to nature. In the course, we try to address both these causes by creating opportunities for ourselves to experience nature and engaging in activities that help us relate to and appreciate the presence of nature around us.

We start by stepping out of the classroom and immersing ourselves in environments richer in life. Here we seek to create opportunities for ourselves to interact with other species. To observe, listen, touch, smell, and experience nature first-hand and in a holistic way. Throughout the course, we attend the University's ecological gardens on various occasions, to enrich our thinking and conversations, explore nature, develop our sensory attentiveness, and practice alternative ways of knowing and experiencing the world. To help focus our attention, we often start class with a sitting or walking exercise that guides us to shift our attention between different sights and sounds in our environment. We turn our attention first inwards to our breath and bodily sensations and then outwards to the sounds of the garden. We scan the environment with our eyes and ears, looking to notice what draws our attention and captures our curiosity. In doing this we are reminded that nature is not just pristine wilderness somewhere "out there", it can be found all around us if we learn to notice and pay attention to our surroundings. I am often amazed at how sitting quietly and listening, even in a large group, opens a space for other species to get closer. One such memorable moment came on one of the occasions when we sat quietly in a group of around thirty people with our eyes closed and listened. After about a minute of silence, we started hearing our surroundings come to life: the chattering of warblers in the bushes and tree above us, the alarms of blackbirds in the distance and the beautiful call of a song thrush that seemed to get closer every time it stopped and started again. We opened our eyes to see the bird a couple of meters away from us looking at us curiously. As a teacher, I feel that these moments of cross-species connection are often more significant than anything I could talk about or show in the classroom. They form the emotional bond that is the basis for the motivation to engage with nonhuman species in the context of design.

Theories of embodied cognition teach us that our thoughts and perceptions are shaped by bodily experiences and interactions with the environment (Glenberg, 2015). The environment in which learning takes place influences the ways we perceive, understand, interact and converse (Schilhab, 2021). The simple act of stepping out of the classroom whenever we can (in many cases it truly is simple), puts us in direct contact with nature and other species and promotes situated and embodied models of learning. Engaging in activities of noticing and reflection is a way to increase our attentiveness and awareness of the myriad manifestations of life surrounding us so that we can start learning and responding to them in the next steps.

2. *Looking at Nature through Varying Modes of Perception*

> Oft the beholder marvels at the wealth
> Of shape and structure shown in succulent surface –
> The infinite freedom of the growing leaf.
>
> Johann Wolfgang von Goethe, 1790, *The Metamorphosis of Plants*

On one of our visits to the ecological gardens, we engage in an exercise designed to open new ways of perceiving nature and natural phenomena based on Johann Wolfgang von Goethe's *delicate empiricism* (Wahl, 2005). Goethe's approach to studying nature developed as both an alternative and complementary epistemology to the dominant scientific practice of his times based on rationalism, a dualistic separation of the observer from the observed and favouring representation over direct contact with the world (Roy, 2018). It emphasizes direct engagement with the studied phenomenon, the use of imagination and reflection together with sense-based perception, and encourages an interconnected and dynamic understanding of phenomena, acknowledging the interplay of various elements within a system.

The exercise follows the four steps of Goethe's science process as developed by the School of Life Science (The Life Science Centre, n.d.) and inspired by Isis Brook's writing on practical applications of Goethean Science (Brook, 1998). It involves observing, experiencing and documenting natural phenomena holistically. Paying attention to different manifestations of the phenomenon as they appear, both to our senses and to the *eye of the mind* (Goethe, 2009), through our imagination. Brook describes four stages of observation, following an initial preparatory stage, that together represent Goethe's approach to perceiving phenomena as a whole. Each stage represents a separate mode of perception, and illuminates the phenomenon in a different light, revealing its various manifestations.

In the preparation stage, we wander about with open senses, paying particular attention to what we hear, see and smell. We search for a phenomenon that captures our attention. This can be, for example, a group of birds in a tree, an interesting plant, ripples in the fishpond, footprints in the mud, or other signs of activity around us. As we do this, we acknowledge and make room for how we normally are, our personal history, our likes and dislikes and our sensations and feelings at present. Once we choose a phenomenon to study, we note our first impressions of it, together with everything we already assume to know about it. We acknowledge and note these in order to put them aside in the next stages.

Following the preparatory stage, we aim to put aside all prior knowledge and try to look at the phenomenon as if for the first time. We use our senses to experience it without interpretation. We pay careful attention to its size, borders, shapes, colours, textures and proportions. We capture these observations in a drawing that seeks to give an exact representation of the phenomenon (see Figure 17.1).

Next, we look back at the phenomenon, considering it now as a phenomenon in time. We are no longer observing a static object but a living dynamic system. We try to imagine what we cannot see – What are the driving forces behind the phenomenon? What relationships does it have with other phenomena? How does it change over time? We go back to our drawing to add these new layers of information.

After actively engaging with the phenomena in the first two steps – first by seeing and documenting its outer characteristics, and then by experiencing its dynamic processes – in the third stage we attempt to suspend active perception and allow the phenomenon to express itself to us to reveal something of its natural essence. As we do this, we observe what thoughts and feelings the phenomenon arises in us, what it tells us about its whole nature, about its unique place in the world. We add these layers of insight to our drawings and give a name to the phenomenon – as if it were just discovered for the first time.

In the final stage of the exercise, we look beyond the specific context in which we have experienced the phenomenon. We ask ourselves, as designers, what can we learn from it?

About materiality, relations, rhythm, function and shape? About other ways of being in the world? What about it is unique, or in contrast, universal? This last stage differs from the School of Life Science's final stage which focuses on *becoming one with the phenomenon.* I adopted this variation partly because I believe there are specific insights to gain as designers from Goethean Observation, and wanted to draw attention to these in a designated stage. The second reason for this variation is that later in the course we have an exercise dedicated to experiencing the world from the points of view of other species, which brings us back to the notion of *becoming one with* and is built into a more appropriate time frame for truly experiencing this sensation.

As various design scholars have emphasised (see, for example, Irwin, n.d.; Korenblat, 2021), training ourselves in the different modes of perception of Goethean Science and learning to apply them within the context of design, holds promising potential in understanding and relating more holistically to the world around us. Irwin (n.d.), for example, points out how the process can help us focus on dynamic and temporal relationships instead of isolated objects and look for unity among seemingly unrelated phenomena. When thought of in the context of more-than-human design, the process of Goethean Science offers us ways to expand and enrich our understanding of the entities we would like to design with and for. In addition, it may help foster a deeper understanding and appreciation for the interconnectedness of all living things by shifting our attention to the fluid and continuous nature of natural phenomena.

The exercise, together with the introduction at the beginning, takes about two hours. At the end of it, we gather back together to share our work and experience. Students often share that the exercise had a calming effect on them and helped them appreciate the complexity of other life forms and the ingenuity of nature. I notice a tendency to gravitate towards phenomena the students have prior knowledge of and some difficulty putting this knowledge aside in the first stages of the exercise. One way to overcome this is to instruct students to choose a phenomenon they have little or no prior knowledge of.

FIGURE 17.1 Drawings created during different stages of the Goethean Observation exercise in the course illustrate how various aspects of the plant are highlighted via distinct modes of perception in each stage. Work by Noga Mazover.

3. Experiencing the World from the Points of View of Other Species

> To do so, we must first blow, in fancy, a soap bubble around each creature to represent its own world, filled with the perceptions which it alone knows. When we ourselves then step into one of these bubbles, the familiar meadow is transformed. Many of its colorful features disappear, others no longer belong together but appear in new relationships. A new world comes into being.
>
> Jakob Von Uexküll, 1934, *A Stroll Through the Worlds of Animals and Men: A Picture Book of Invisible Worlds*

After turning our attention towards more-than-human worlds and learning to observe natural phenomena through varying modes of perception, our next shift in perspective aims to take us a step closer to seeing and experiencing the world from the points of view of other species. This stage follows a long tradition of empathic design, where designers use various techniques to get closer to the lives and experiences of existing or future users and gain a deep understanding of their experience in the world (Kouprie & Visser, 2009). To this end, we engage in an exercise called the *Umwelt* Apparatus, in which, students attempt to design a simulation device or protocol that would offer them a glimpse into the world through the unique perceptive abilities of a selected animal species. The name and inspiration for the exercise come from von Uexküll's *A Stroll through the Worlds of Animals and Men* where he invites us to take part in a thought experiment of experiencing the world through the sensory organs of other species, referring to this self-world of the animal as its *umwelt* (von Uexküll, 1992). In the exercise, we devise physical aids and movement protocols that would help us experience aspects of the animal's *umwelt* not only in our imagination but also in an embodied way (see Figure 17.2 for examples).

The process of designing such an apparatus pushes us to learn about the animal through a specific filter, focusing on the animal's experience of the world and how it is mediated through its unique perceptive abilities. When we do this, something changes also in the way we view the animal and our relationship with it. In the second stage of the exercise, we put our devices to use. Each apparatus exposes us to a new perspective and opens new possibilities of being in the world. Some change our physical point of view, taking us, for example, high up into a tree or into a maze of chemical traces left on the ground by ants. Some change our depth or field of view, or the colour spectrum of our vision. Some block out some senses, leaving room for others to take a leading role – inviting us, for example, to navigate through a dark space just by the sense of humidity on our skin. The collective work of the group highlights the multitude of ways of experiencing the world, reminding us that our human way is but one of many different forms of existence. We make sure to use the devices in both, natural and human-dominated areas. Using them in the latter is a good reminder of how these are designed predominantly for humans and largely ignore their influence on other species.

Any attempt to experience the world from the point of view of another creature, human as well as nonhuman, should be done with humility and care. Simulation devices tend to be over-simplistic, focusing on one or two aspects of the experience. They may give a false sense of confidence in our understanding of the experience we are studying. It is important to properly frame the exercise and to adjust our expectations from it. The point of the exercise is not to create an accurate representation of the animal's self-world – as this would undeniably be impossible. The point is to expand our own perception, to introduce ourselves to other possible points of view of the world and to observe what changes in us when we *try* to experience the world from different perspectives. We remind ourselves that the experience we get from using the apparatus is not that of the animal, it is our own experience using a device inspired by the *umwelt* of the animal. In 2022 I conducted a series of interviews with students following the *Umwelt* Apparatus exercise, in an attempt to describe the influence of the exercise on their approach and relation to the animal they studied. Rather than creating a precise sense of knowing the animal, the exercise seemed to generate a more holistic view of the animal, and an appreciation of all the things we don't know, and perhaps can't know about the animal. In addition, the exercise led, in many cases, to changes in sentiment and real-world interaction with the animal, together with increased curiosity and motivation for designing for the animal (Metcalfe, 2023).

FIGURE 17.2 Examples of *umwelt* apparatuses developed in the course, representing different approaches to the exercise. Clockwise from upper left: 1. Apparatus for experiencing the world from the point of view of the honeybee involving a visual filter and an ultraviolet light, developed by Ofek Raz and Dana Taub. 2. Apparatus for coding the common fruit fly's "olfactory language", developed by Erez Ezra in the D.Dlab research facility. 3. Simulation using role-play of typical scenes from the life of great tits developed by Orni Bloch and Yael Salem (video screenshot). 4. Simulation of different visual capabilities of different fish using digital filters, developed by Shelly Boneh.

4. *Regarding Design as an Evolving, Open-ended Process*

Next, we gather what we have learned in the previous stages to start thinking in terms of possible interventions. We observe the world not just to experience and document it but also to ask how it might be different. If we have identified, for example, places where human activity undermines the basic biological needs of certain groups of nonhuman organisms or places where it poses specific risks to them, we might choose to try and intervene. We embody our roles as designers while relying on the shifts in perspective from the previous stages to think with and through multiple points of view. Here we might begin to notice that a shift in our relationship with our own designs is also immanent.

The need to consider wild animals as potential ‘users’ of designed artefacts does not come from a deficiency in nature’s provision but from a systematic removal of nature from areas humans have claimed for themselves. If we want to reverse this and reintroduce nature back into our lives, then there is a lot we can learn from the way other species interact with their environments. Consider an Oak tree for example. As it grows, matures, dies and decomposes, it provides different functions for different species in different stages of its life cycle. For many birds, mammals, reptiles and insects it will only become a viable habitat once the processes of decomposition start breaking it from within, creating diverse niches, crevasses and material flows (Bunnell et al., 2002). The view of natural phenomena as dynamic processes in constant change, awoken by the Goethean observation exercise, may begin to also influence the way we view our design endeavours. When designing for more-than-human worlds we try to align our designs with such natural processes, to view them as becomings, rather than finite objects (see Figure 17.3 for an example). We accept that we do not have full control over them. We expect them to change over time as other species and natural processes interact with them. This blurs the lines not only between designer and client but also between process and outcome (a separation that does not exist in nature). Everything in nature is continually evolving, decomposing and recomposing cyclically, there is no waste or end of life, just a continuous flow of material and energy. We ask ourselves, is it possible as designers to cultivate this quality of constant change in our designs? To let go of the wish to freeze the artifice in a static stage. To design for processes rather than products.

When we imagine design as a fluid process, shaped by multiple perspectives and the agency of other species alongside our own intentions, we begin to accept that we will not have full control over its outcomes. Moreover, we do not wish to have full control. We wish for them to be modified and shaped by other species using them, we wish for them to follow natural processes of decomposition and succession. These entanglements are often also reflected in the material and visual aspects of the designs. More-than-human perspectives feeding into the process result in textures, colours and formations not always recognizable to our human eye. Looking at our work, we sometimes get the sense that *someone* or *something* else was involved in the process. Or that there is something not entirely comprehendible about their aesthetics, as these are intended to convey affordances to other species, based on their specific perceptive abilities. Surface textures on a seawall tile, for example, might include groves, steps, crevices and holes all designed to accommodate the biological needs of different marine creatures – the function of which might not be recognizable to people without specific explanation. Reflecting on this is an opportunity to talk about more-than-human aesthetics (French et al., 2020) and how they come into existence when we integrate diverse, more-than-human points of view and address the needs of different species in the design.

FIGURE 17.3 Photos of the process (four left) and outcome (right) of a group project developed in the course. The goal was to replicate the ecological significance of dead trees within the local ecosystem using advanced 3D printing technology of natural and locally sourced materials which would gradually decompose back into the earth. The printed element aims to emulate the complex geometrical patterns found in dead trees and provide comparable environmental benefits to those offered by dead trees, changing over time through the processes of ecological succession and catering to a diverse range of organisms including insects, fungi, vegetation and other forms of wildlife. Group members: Yoav Dabas, Alon Nisan, Amit Aidlin, Surayyn Uthaya Selvan, Avraham Cohen and Dafna Galeen.

Photos by the group.

5. Looking Back at Design for Humans with New Insights

> A great deal is at stake in such meetings, and outcomes are not guaranteed. There is no teleological warrant here, no assured happy or unhappy ending, socially, ecologically, or scientifically. There is only the chance for getting on together with some grace. The great Divides of animal/human, nature/culture, organic/technical, and wild/domestic flatten into mundane differences – the kinds that have consequences and demand respect and response – rather than rising to sublime and final ends.[1]
>
> Donna Haraway, 2007, *When Species Meet*

In the last part of the course, we look back at our habitual ways of thinking and doing design to see if something in our perspective here might also change. We ask ourselves, what have we learned from engaging in more-than-human design that we might take back to our everyday production of design? We often notice that we don't necessarily need to create hard divisions between human and more-than-human design. In every project we are engaged in as designers it is possible to integrate multiple points of view, in varying degrees, depending on the scope and frame of the project. Most acts of design have implications that extend beyond our species at some point in their life cycle: from procuring material resources, through manufacturing processes, maintenance and end-of-life. Students often share that new ways of thinking and increased sensitivity towards other species have found their way into their work in other courses or projects they are currently involved in.

The blurring of the hard divisions between humans and animals, natural and artificial, wild and domestic, raises questions also regarding the ways we view people as users of design. For example, around the theme of *dependency* on a designed artefact or system. When designing for wild animals, we often take extra care not to create dependency of the animal on the design, in order to safeguard its autonomy and wild nature. Do we do the same when designing for people? The topic often comes up in the course concerning the feeding of wild animals. On the one hand, it is understandable to want to respond to the basic biological needs of animals by providing food for them, on the other, anthropogenic feeding systems may create dependency of the animal on humans, exposing them to further risks. When the theme comes up, we discuss possible ways to address it and use it to reflect also on our relationship, as humans, with design. What would human technology look like if it were designed not to create dependency? What form would a GPS navigating system take, for example, if it was designed to enhance our navigating abilities and not navigate for us?

The last part of the course is also the part where tough questions regarding the real-world application of Multispecies Design are often raised: How to identify all the different multispecies stakeholders influenced by a design project? Which ones to consider, and to what extent? What to do when a design decision would clearly benefit some species over others? Many of these questions remain unanswered or only partly answered in the course. They are the questions I struggle with in my own practice, and, more often than not, do not have definitive answers.

Concluding Thoughts

My goal with the course is to help students develop a sensitivity to more-than-human perspectives within the context of design, hoping this will find its way into their future work. My approach to doing this is through experiential exercises, exploration and first-hand interactions designed to raise questions and highlight new possibilities of being in the world, rather than giving firm guidelines or answers. Assessing to what extent this sensitivity has been developed is not always straightforward. It is often reflected in the way students talk about the animals they work with, the considerations they raise when developing their projects and their ability to address complexity and multiple points of view within their work. Multispecies design projects often start from a naive and simplistic view of animals. I look to see to what extent students manage to develop their approach as they progress, to acknowledge the risks and unknowns in their projects and to define humble goals for it.

I assume this experiential and open quality to the course partly because I believe this is the best way to learn something new, and because, in most cases, definitive answers do not exist. As we expand the scope of design to include more-than-human perspectives, we will undoubtedly also make mistakes. We will misunderstand and misinterpret the needs of other species, we will favour some species while disadvantaging others, and we will create new dynamics, behaviours and dependencies. As long as we approach this task with humility, care and openness, I believe we will stay on a path of positive impact. We will learn as we advance, correct our mistakes, and get better on the way. I start the course with the above passage from Harraway's *When Species*

Meet as I feel it sets the tone for an exploration that is, on the one hand, modest and grounded in everyday actions and interactions, and on the other, radically optimistic– the chance of getting on together with some grace. Imagine a world where we did!

Questions

- To what extent do the methods described here influence the development of students' approach to design in the long run, and how could this be measured?
- How could we demonstrate the practical application of Multispecies Design and the methods experienced in the course in the context of commercial design?
- What other pedagogical approaches to more-than-human design exist or could be developed, and what qualities could they highlight?

Note

1 In the original context the term *such meetings* refers to a meeting between a pack of wild wolves and a wolf *raised by scientists* carrying a technological communications box on her back – although it would appear to speak more broadly to encounters between what might be considered *natural* and *artificial*.

References

Brook, I. (1998). Goethean science as a way to read landscape. *Landscape Research*, *23*(1), 51–69. https://doi.org/10.1080/01426399808706525

Bunnell, F. L., Houde, I., Johnston, B., & Wind, E. (2002). How dead trees sustain live organisms in western forests. In *Proceedings of the Symposium on the Ecology and Management of Dead Wood in Western Forests* (Vol. 4).

French, F., Mancini, C., & Sharp, H. (2020). More than human aesthetics: Interactive enrichment for elephants. In *DIS 2020 – Proceedings of the 2020 ACM Designing Interactive Systems Conference*. https://doi.org/10.1145/3357236.3395445

Irwin, T. (n.d.). *Goethean Science and Design: A Phenomenological Approach to Understanding Whole Forms in Nature & the Built/Designed World*. Academia.Edu. Retrieved 3 January 2024, from www.academia.edu/18348890/Goethean_Science_and_Design_A_%20Phenomenological_Approach_to_Understanding_Whole_Forms_in_%20Nature_and_the_Built_Designed_World

Korenblat, J. (2021). Inspiration from Goethe's Tender Empiricism. In *Right Research*. https://doi.org/10.11647/obp.0213.08

Kouprie, M., & Visser, F. S. (2009). A framework for empathy in design: Stepping into and out of the user's life. *Journal of Engineering Design*, *20*(5), 437–448. https://doi.org/10.1080/09544820902875033

Mancini, C., Metcalfe, D., & Hirsch-Matsioulas, O. (2023). *Justice by Design: The Case for Equitable and Inclusive Smart Cities for Animal Dwellers*.

Marris, E. (2011). *Rambunctious Garden: Saving Nature in a Post-Wild World*, 1st ed. Bloomsbury Publishing.

Metcalfe, D. (2015). *Multispecies Design* [University of the Arts London and Falmouth University]. https://ualresearchonline.arts.ac.uk/id/eprint/13351/

Metcalfe, D. (2023). Empathic Multispecies Design – using empathy to design with animal perspectives in mind. In G. Goldschmidt & E. Tarazi (Eds), *Expanding the Frontiers of Design*. Routledge. https://doi.org/https://doi.org/10.1201/b22630

Pyle, R. M. (2002). Eden in a vacant lot: Special places, species, and kids in the neighborhood of life. In *Children and Nature* (pp. 305–327). The MIT Press. https://doi.org/10.7551/MITPRESS/1807.003.0013

Root-Bernstein, M., & Ladle, R. J. (2010). Conservation by design. *Conservation Biology*, *24*(5), 1205–1211. https://doi.org/10.1111/j.1523-1739.2010.01501.x

Rosenzweig, M. L. (2003). *Win-win Ecology: How the Earth's Species Can Survive in the Midst of Human Enterprise*. Oxford University Press.

Roy, K. (2018). Goethe and delicate empiricism. In *The Power of Philosophy*. https://doi.org/10.1007/978-3-319-96911-4_4

Schuttler, S. G., Sorensen, A. E., Jordan, R. C., Cooper, C., & Shwartz, A. (2018). Bridging the nature gap: Can citizen science reverse the extinction of experience? *Frontiers in Ecology and the Environment*, *16*(7), 405–411. https://doi.org/10.1002/fee.1826

Soga, M., & Gaston, K. J. (2016). Extinction of experience: The loss of human-nature interactions. *Frontiers in Ecology and the Environment*, *14*(2), 94–101. Ecological Society of America. https://doi.org/10.1002/fee.1225

The Life Science Centre (n.d.). *Goethean Scientific Process*. Retrieved 10 January 2024, from https://lifesciencecentre.co/goethean-scientific-process

von Goethe, J. W. (2009). *The Metamorphosis of Plants*. MIT Press.

von Uexküll, J. (1992). A stroll through the worlds of animals and men: A picture book of invisible worlds. *Semiotica*, *89*(4). https://doi.org/10.1515/semi.1992.89.4.319

Wahl, D. C. (2005). "Zarte Empirie": Goethean science as a way of knowing. *Janus Head*, *8*(1), 58–76. https://doi.org/10.5840/JH20058134

EPILOGUE

Towards Creaturely Ways of Designing

Ann Light

Reading through these chapters has me thinking, feeling and hoping that we are at the start of something new. That so many researchers, designers and educators can appreciate the need for remaking design practices (and the infrastructures these have led to) speaks to a moment of reappraisal in the best ways. We might treat a search for 'the new' with suspicion, linked, as it often is, with novelty for the sake of it and purely commercial imperatives. It is therefore exciting to be in the company of authors reaching for new worlds with an appreciation of the damage done by the manufacturing of need and cultures of accumulation. Designers, more than most, are aware of two faces to design: one serving business uncritically and ignoring the future; another dreaming of change, recognising the uncertainty of acting, yet feeling the force of responsibility to make a difference. The book has brought into conversation many perspectives on what that difference might be – from welcoming the Other to proposing new global systems, from energetic overhauling of world politics to patient listening to quiet voices. But all agree that business-as-usual can only result in further destruction; we have heavy machinery to shift, both literally and metaphorically.

From biblical texts offering humans dominion over other life to the legacy of Roman Europe's colonialism and property law, there is much cultural legacy to move on from. Design at its best – futures-orientated, inclusive and reaching towards other ways of doing and being – is necessary to make these transitions just and effective. But there is a paradox in this: "We need to be designerly enough to rethink and remake our world, but an element most in need of change is the nature of designing itself" (Bardzell et al., p. 1). An embrace of *anthropocentric functionalism* lies at the heart of what has been wrong with design and a turn to the more-than-human is needed "cohesively at the levels of theory, ideology, and practice" (Bardzell et al., p. 1). While *anthropocentricity* has been tackled throughout this book, perhaps *functionalism*[1] is not so obviously in need of replacement. Yet our cultures prioritise efficiency at the expense of care, sanctioning the appropriation of the world's resources without regard to vitality, finitude or joy. Productivity has no place in a design rewritten for custodianship. Ignoring the impact of extractive behaviour or labelling it 'externalities' is part of the same toxic failing. When we look, instead, with more-than-human sensibilities, we reveal "the trade-offs that are being made in prioritizing traditional human-centred perspectives" (Akama et al., 2020, p. 2). Doing so, allows us to make room for "being and interconnection [to be] respected and honoured in all humans and more-than-humans". This broad sweep of care extends design's capacities: designing *otherwise* honours "commitments to support people in their struggles" but offers "another, complementary, view of relations with which to develop practice" (Akama et al., 2020, p. 2).

Nonetheless, anthropocentric functionalism, with its "human-centered appropriation of the world's resources" (Bardzell et al., p. 1), is being celebrated even as climate crisis makes it unsustainable. The world's big technology players are proudly revealing 'human-centred AI' as an update on the opaque, perhaps intrinsically illegible, challenge to human-technology-world relations emerging with machine learning. Maybe the use of the term 'human-centred' in these contexts is there to reassure us. But I do not find it reassuring that we face a further wave of automation removing us from what is earthy, visceral and living.

Humans will be changed by AI. If we doubt this, we need only look at how humans were changed by mobile phones, social media, the bicycle and the first writing systems. We will be changed, in this case, by an intersection of technology, innovation and profit-seeking. The fact that I can assert this with confidence and yet no governance system is weighty enough to contend with these global changes is grounds enough to argue for new worlds. In the world in which we find ourselves, companies that hold centralised data sufficient to manage all life are led only by market concerns. This mode of designing, letting loose initiatives in competition for market capture, is deliberately unprincipled and thus prone to the defects of capitalism. Nowhere is it stipulated that these systems should value environmental factors, such as water, air and life, instead/as well as profit[2]. While ethics bodies scramble post-hoc to discuss possibilities for regulation, there is little thought to the long-term impacts of creating self-adapting models in a market where only pursuit of shareholder returns is optimised. And yet we know – for certain – that such developments have consequences.

For these reasons, my work has long concerned the *human* in the more-than-human. This is a response to years of looking at the impact of digital technology. While I care about other life forms as individuals and species – and I share an abhorrence of losing types of animal, fungi and plant, virgin forests (above and below sea) and ancient ecosystems – I regard the need to focus on human hearts and minds as a prerequisite for offering care more broadly. Work to mobilise the groundswell of human feeling is part of halting global patterns of exploitation and extraction. We need solidarity and the sense that action belongs to all of us, whatever we most wish to address. This might be the only force that can divert industry from profit- to planet-centred. Making spaces for that transition is an important role for designers. We are people who understand agency and how to push past the feelings of frustration that come with working for change. We might provide alternative leadership in dealing with the too-little-too-late response of the political classes. As I read yet another article showing how fossil fuel lobbies have manipulated governments into slowing progress towards other sources of energy (e.g. Niranjan, 2024), I am reminded that only *everyone together* can do this and that that also means countering polarisation and narratives that divide and rule. It means striving for social as well as ecological justice. Design has a significant role.

With this in mind, I have helped run an *eco-social* project that looked at how cultural change can be inspired by the processes of socially-engaged art called *CreaTures* (Creative Practices for Transformative Futures: www.creatures-eu.org and www.creaturesframework.org. We used the term *eco-social* to carry the weight of our sense that the social and ecological are entangled.

> In defining *eco-social*, we explicitly acknowledge that the world we have is not the one that we wish to sustain, but a planet damaged by extractivist and exploitative values and actions. Eco-social change means systemic change to produce futures of greater fairness and respect.

> Solidarity and social justice will be key to addressing the related crises that are in motion, fuelled by an unstable climate. And stabilizing the climate and global ecologies will be key to protecting the lives and opportunity for fulfilment of all beings.
>
> (www.creaturesframework.org/eco-social-change-and-eco-social-futures.html)

Fulfilment is key here, hoping for something richer than we have now, even as we contemplate designing from the ruins. We are on a search for new meanings but this search reveals the challenges still to be faced in embedding the more-than-human across design practices. First, more-than-human proposals are hard to speak of and diverse in their manifestations, as evidenced in these pages. Second, they invoke concepts and ways of being that run contrary to the mainstream, making this a struggle against norms and infrastructures as well as methods. Third, the goal is slippery. The *more-than-human* is not a one-time fix for an ailing planet, but a vibrant and fluid *being-with* and *becoming-with*. As intrinsically relational, it has no core or parameters, dancing as negative space between moving objects. It is always a journey.

CreaTures, as a project, was also a space for things *to become*. One output is a framework to evaluate arts-based transformation techniques centred on meaning, connection and power (see Vervoort et al., 2024). It is intended to support fluid responses to changing contexts and, at the same time, to look at evaluation as creative and constitutive and disrupt blinkered practices, such as that focused solely on 'net zero' or 'carbon saved'. But we sought too to wrestle with the practical challenges of being *more-than-human* in our designing. This may be best caught in the neologism *creaturely*, which speaks to the values in these pages. To offer a cut-down version of how colleagues and I presented it:

> Given the struggle for language that talks of changing/changed ways of being, we offer *creaturely* as a placeholder for gentle, connected, pluralist and life-focused ways of doing, being and making. The term is an adverb, not a noun – it points to a way, not a state. *Creaturely* ways attempt to capture the paradoxical connectedness and separateness of being alive, to enable us to become the change we want to see.
>
> Creaturely ways of being and doing involve a gentler, more connected, approach where the artificial lines between humans and nature are no longer drawn. To be 'creaturely' is to be alive to possibility. Being *creaturely* incorporates our understanding of eco-social change: it is post-capitalist, playful, more-than-human and fundamentally relational. It embodies change, recognizing the importance of our bodies, our emotions and 'sentipensar' (a term coined by Colombian sociologist Orlando Fals Borda to express a holistic, inclusive way of engaging through feeling-thinking).
>
> *CreaTures* research has been undertaken in this creaturely spirit. Multiple intellectual strands have been woven tightly together through an interdependence between ideas, processes and members. It has involved an ambitious plan for the co-creation of research between academic and practitioner partners, as well as recognising the whole person as researcher and learner (and with no expectation for emotional objectivity to accompany professional analytic objectivity). It is based upon a rhetoric of care, a resistance to abstraction for the sake of it and to extractive practices that give nothing back, to produce analyses that are situated and contextualized in the richness of our worlds.
>
> (www.creaturesframework.org/creaturely.html)

In other words, this was research that attempted *in the doing of it* to show the values it committed to. We conceived of the design of the project as a chance to challenge research norms, basing our work in collaborative action research and hoping that the artefact of the project itself (or, perhaps, more the aggregation of the people and other creatures that constituted it) would become significant beyond its outputs as an example of what something different might be (Light et al., 2024). In designing the project, in conducting and speaking about it, we sought to resist metaphors of scale and growth and adopt those of entanglement and interdependence (Light et al., 2024). In focusing on moments of transformation, we looked at what was evolving and how individuals found meaning together, recognising that these moments could not be reproduced or scaled. Instead, we drew attention to the conditions that enabled such participative intimacy to flourish and discussed how to stimulate it and connect it back into everyday life (Light, 2021). In one case, this was interactions in deciding the future of a park with people in role as animals and plants to learn about the importance of different non-human park dwellers. In another, a forest walk is led by a dog. A third involved paying back to the oceans that support us.

The project gave us room to co-create intimacies among researchers and partners and learn about the range of intentions for transformation that creative practices might support. While it might have been judicious to limit our work to focus on our countries' stated climate mitigation and net zero ambitions, this ran against our goal of motivating people to hope, dream and draw on new possible realities – a prospect that needs many starting and ending points. Instead of mainstream sustainability, we produced something unclassifiable, probably still more legible to artists than scientists in its lack of precise impacts, trying to interpret the energy of positive, collective re-thinking and re-feeling for professions that have evaluation and efficiency as their mantra – and I said to colleagues during our study that perhaps the measure of our success should be that *evaluation* and *success* would no longer dominate global thinking (Light et al., 2024). It seems a long stretch.

Our project was conducted within the constraints of the European Union's funding and evaluation mechanisms, which favour discretely packaged work streams, mappable impacts and the rhetoric of scaling. Remaining at odds with this was a major trial and we learnt how much effort must go into resisting norms to enact, even partially, what we conceived as a consistency between theory and practice. Doing more-than-human project management is tricky. Even the concept of 'project' is alien to a world beyond anthropocentric functionalism.

Since this exhausting but exhilarating experience, I have been developing my understanding through related work, such as more-than-human codesign of regenerative ocean cultures (www.bauhaus-seas.eu/co-design). I have been asking: Might philosophies of flow and acts of subversion (Light, 2022) come together in the *more-than-human* to produce an agenda inspiring people to find better ways to dwell together with all life? For that inspiration, I believe we need not just a vision of new systems, but a taste of new relations. I have become preoccupied by how we might experience relations that do not exist in our everyday: a sweet hint of what we might aspire to if cultures and relations relied less on the happenstance of new technologies, but were instead understood as something we notice, choose and commit to.

My latest work explores *affective prefiguration*, which is design of contexts in which different capacities for feeling as well as different feelings can be born ahead of more general changes in relations. "Affective prefiguration means giving experience in the present of what else might be emotionally and relationally possible" (Light, 2023) and examples of dialoguing with the more-than-human can enact this. I suggest that arts practice makes the space for this opportunity, inviting people into encounters that can surprise, move, or uplift them and leaving people feeling something special has happened between them, briefly or more enduringly. Compounded by the chance to speak together about this experience and learn what others have felt, these practices might be the stuff of transformation, existing alongside other processes to help articulate and find the new worlds we seek.

Writing about redesigning design (Light, 2020), I argued that:

> When we engage in democratic participatory design practices, we are co-designing ways of being, as well as staging encounters to learn about ways of being. It is a form of research through design where the product for iteration is a process. We literally make our futures by the practice of performing them and to design for this opportunity offers a profoundly important and meaningful future for design, despite its dangers".
>
> (n.p.)

I stand by that. And, as so often, I am reminded of Jane Bennett's discussion of enchantment (2001): "One must be enamored with existence and occasionally even enchanted in the face of it in order to be capable of donating some of one's scarce mortal resources to the service of others", she said. We all need to feel that enchantment and act on it. We will all interpret it in different ways, pushing for different futures, yet all these ways will treat the fabric of life with more dignity and care. And our very lack of consistency in responding to that enchantment will be part of our strength.

Acknowledgements

With thanks to my many co-travellers in CreaTures and beyond. CreaTures received funding from the European Union's Horizon 2020 research and innovation programme, grant agreement no. 870759. The European Commission has no liability in respect of the content of this work.

Notes

1 We used it to refer to 'the extractive nature of contemporary production' and accompanying productivity culture (Bardzell et al., 2021), whereas it can be used in psychology and sociology with other meanings.

2 The 'public limited company' is structured to pursue profit exclusively in the face of other concerns, such as social or ecological wellbeing. Shareholders may legally dismiss a company leader who attempts to challenge this single-minded pursuit.

References

Akama, Y., Light, A. and Kamihira, T. (2020) Expanding participation to design with more-than-human concerns. *Proceedings of PDC '20*. https://doi.org/10.1145/3385010.3385016

Bardzell, J., Bardzell, S. and Light, A. (2021) Wanting to live here: Design after anthropocentric functionalism. *Proceedings of the 2021 CHI Conference on Human Factors in Computing Systems*. Association for Computing Machinery, New York, Article 293, 1–24. https://doi.org/10.1145/3411764.3445167

Bennett, J. (2001). *The Enchantment of Modern Life*. Princeton University Press

Light, A. (2019, December 11–12). Redesigning design for culture change: Theory in the anthropocene. In *Design Research for Change*, Design Museum, London.

Light, A. (2020). Redesigning design for culture change: Theory in the anthropocene. *Digital Culture and Education*, 12(2). https://static1.squarespace.com/static/5cf15af7a259990001706378/t/5f02e48181fcf002f4f0e64a/1594025092372/Light-merged.pdf

Light, A. (2021, June 27). Participative intimacy. The Light Stuff blog: http://lightstuff.co.uk/participative-intimacy

Light, A. (2022). Ecologies of subversion: Troubling interaction design for climate care. *Interactions* 29(1), 34–38. https://doi.org/10.1145/3501301

Light, A. (2023). In dialogue with the more-than-human?: Affective prefiguration in encountering otherwise. *Interactions*, 78. doi:10.1145/3599956

Light, A., Hee-jeong Choi, J., Houston, L. and Botero A. (2024). Enacting entanglement: CreaTurely socio-technical collaboration and designing a transformative ethos. *Proceedings of 22nd European Conference on Computer-Supported Cooperative Work*, Rimini, Italy (ECSCW '24).

Niranjan, A. (2024, March 8). Oil industry has sought to block state backing for green tech since 1960s. *The Guardian*. www.theguardian.com/environment/2024/mar/08/oil-industry-has-sought-to-block-state-backing-for-green-tech-since-1960s

Vervoort, J. M., Smeenk, T., Zamuruieva, I. et al. (2024). 9 Dimensions for evaluating how art and creative practice stimulate societal transformations. *Journal of Ecology and Society*, *29*(1).

INDEX

For Product Safety Concerns and Information please contact our EU representative GPSR@taylorandfrancis.com Taylor & Francis Verlag GmbH, Kaufingerstraße 24, 80331 München, Germany

Batch number: 10733637

Printed by Printforce, the Netherlands